GILBERT WHITE'S

Gilbert White's House: The Wakes, from the garden front. The "Great Parlour" is on the left. (From the quarto edition of the *Nat. Hist.* 1813)

GILBERT WHITE'S JOURNALS

Edited by
WALTER JOHNSON

THE M. I. T. PRESS
CAMBRIDGE, MASSACHUSETTS

ISBN 0 252 60003 X

This is a reprint of a book first published in 1931 by
George Routledge & Sons (London). It contains extracts from
the *Journals* written by Gilbert White during the years from
1768 until his death in 1793. These *Journals* remained in
manuscript for nearly 140 years until Walter Johnson, FGS,
made and edited this selection. The original manuscript is
now in the British Museum.

Little has altered in Selborne since Walter Johnson
wrote his introduction to the 1931 edition, but two
changes are worth noting. The Well-head, though still
flowing, no longer serves as the village water supply, and
the oak seat at the top of the Hanger, worn out by the
ravages of time and weather, has been replaced by an
iron bench provided by the National Trust.

Library of Congress catalog card number: 70–133320

Printed in Great Britain in 1971 by
Latimer Trend & Company Limited Whitstable

May 20, 1992

Dear Dad,

I was browsing through one of Berkeley's many secondhand bookstores the other day (one of my favorite occupations). I saw this book and thought of you at once. My only anxiety is that you might possibly already have it. Anyway, happy birthday (belated). I think of you often and wish I could visit.

All my love,

Barbara

" This was thy daily task, to learn that man
   Is small, and not forget that man is great."
                    (A. C. BENSON : *Gilbert White.*)

" From flinty fields till curlews call the night
   And woodlarks welcome the immortal White."
               (ARMINE T. KENT : *The Eighteenth Century.*)

Year Selborne. Place Soil.	Therm.ʳ	Barom.ʳ	Wind	Inches of Rain, or Sn. Size of Hail fn.	Weather	Plants first in leaf—Fungi first appear vegetate.	Trees first in leaf. Flower first ap- pear	Birds and Insects first in Insects first appear, or in regard to fish, and disappear.	Miscellaneous Observations and Memorandums and other animals
Sunday. 8 12 4 8 Jan. 10.	43	29 8-10	S.		frost. deep fog. sun.			A ripe wood straw-berry on a bank, & several. Hepaticas. Moist & mild. Grass grows on the walks.	
Monday 8 12 4 8 11.		29 7-10	S.W.		fog. bel. int.			The white spotted Bantam hen lays.	
Tuesday 8 12 4 8 12.	49	29 5-10	S.W.		rain. driving rain.			Snow-drops blow. We have in the windows of the stair-case a flower-pot with several sorts of flowers, very sweet & fragrant.	
Wednsf 8 12 4 8 13.	49 49	29 3-10	S.W.		rain. rain. rain.			driving rain all day.	
Thurs 8 12 4 8 14.	49 49	29 5-10½	E. P.	83.	dark. mood.ᵏ mild. rain.			A large speckled diver or loon was just to be seen...	
Friday 8 12 4 8 15.	44	29 6-10	E. S.W.	33.	deep fog. sun. bright.			a large lake, or pond. These birds are seldom seen so far in mild winters.	
Saturday 8 12 4 8 16.	42	29 8-10½	W.	3.	sun & bright air.			Turnip-greens come in.	

FACSIMILE OF A PAGE OF WHITE'S JOURNALS (JAN. 1790)

PREFATORY NOTE

Of late years there has prevailed a growing wish that Gilbert White's *Naturalist's Journal,* which has remained in manuscript since 1793, should be made public. It has even been suggested that the *Journal* should be printed in full, but to anyone familiar with the manuscript this seems neither practicable nor desirable. Few persons can wish to read nearly ten thousand undigested daily records of temperature, wind, and weather, to say nothing of innumerable entries concerning the first appearances of birds and insects, and the flowering of plants. Nor is it of much value to trace every step in White's cultivation of vines or cucumbers ; the general notes on such subjects for a single year will doubtless be deemed sufficient.

Something more important forms the editor's task : to select all those passages, not already used by White in his *Natural History,* which have scientific worth, or which shed light on the social life of the old naturalist. This latter reason explains why details of personal visits and the number of White's nephews and nieces are constantly reproduced here. For equally good reasons, the recurring entries about sainfoin, the stone-curlew, and Timothy the tortoise, have led to a silent breaking of the rule. A selection of White's " records of first occurrences " has also been admitted.

The map, kindly re-drawn by my friend Mr Sydney Harrowing, has been carefully prepared to illustrate the localities mentioned in the *Journal,* so far as identifications could be made by personal inquiry and the inspection of old maps. To Mr Harrowing thanks are also due for most of the illustrations in the book. For the permission to use photographs, thanks are here accorded to Mr H. G. Osborne, of Alresford.

Square brackets occurring in the text indicate editorial notes ; when marginal, they enclose localities inserted in White's date column.

T.P. refers to White's Letters to Thomas Pennant (see *Natural History*).

D.B. refers to White's Letters to Daines Barrington

Antiq. refers to *Antiquities* appended to the *Natural History*.

R.H.W. refers to *Life and Letters of Gilbert White,* by Rashleigh Holt-White, 2 vols., 1901.

W.J. refers to *Gilbert White : Pioneer, Poet and Stylist.* by Walter Johnson, 1928.

LIST OF
CHIEF AUTHORITIES

Allen, Grant. Edition of " Selborne," 1900. Good illustrations, and an excellent introduction.

Bell, Thomas. Edition of " Selborne," 2 vols., 1877. Contains information derived, in part, from persons who remembered White.

Harting, J. E. Edition of "Selborne," 1875, 1877, &c.

Holt-White, Rashleigh. " Life and Letters of Gilbert White of Selborne," 2 vols., 1901. The standard work of reference, written by a collateral descendant.

Holt-White, Rashleigh. " Letters to Gilbert White of Selborne from John Mulso " [1907].

Jesse, Edward. " Gleanings in Natural History," 2nd series, 1834.

Johnson, Walter. " Gilbert White, Pioneer, Poet, and Stylist," 1928.

Martin, Edward A. " A Bibliography of Gilbert White " [1897].

Miall, L. C and Fowler, W. Warde. Edition of " Selborne," 1901. A splendid and trustworthy annotation.

Newton, Alfred. Art. " White, Gilbert," in *Dict. Nat. Biog.*

Selborne Magazine (Selborne Society), originally *Nature Notes, passim.*

Sharpe, R. Bowdler. Edition of "Selborne," 2 vols.,
    1900. Very complete, containing matters
    omitted in many editions. The *Garden Kalendar*
    is printed in full.

Shelley, H. C. "Gilbert White and Selborne," 1909.

Victoria History of Hampshire. 1908, vol. iii,
    pp. 3-16.

(*Mr. E. M. Nicholson's* fine edition of "Selborne,"
    1929, appeared too late for purposes of com-
    parison.)

LIST OF
ILLUSTRATIONS

CONTENTS

I

GILBERT WHITE AND HIS VILLAGE

" Remote from towns, he ran his godly race,
Nor e'er had changed, nor wished to change, his place."
(The Deserted Village.)

THE sequestered village of Selborne, in Hampshire, lying hidden almost midway between the Portsmouth and Southampton roads respectively, was a fit birthplace for the most gentle and unobtrusive of British naturalists. The village rests cosily at the foot of the steep, beech-clad escarpment of the Hampshire Downs, where the Chalk abruptly drops two hundred feet to the Malmstone and Gault below, whence, by gentle undulations, the landscape reaches out eastward to the hungry Greensand of Wolmer Forest. The dense beech-wood covering the Chalk scarp is known as the Hanger, although the term is now often applied both to the wood and the chalk hill itself. The parish of Selborne is wide and scattered, but the active part is the village street which runs parallel to the Hanger. The houses unfortunately are so near the base as to be screened from much of the afternoon sun.

Even to-day this beautifully quaint village is not seriously spoilt, and contains less than 700 people.* In White's heyday the population was about 600, about half of that number dwelling in the village street and the remainder in outlying farms and hamlets, of which Oakhanger was the chief. The houses, as we gather from White's book, were built of malmrock,—a sandy limestone,—or of brick, or

---

* The total population in 1921 was 2,004, but this included 1,393 in the modern ecclesiastical parish of Blackmoor, which is growing fast.

they were half-timbered, with thatched or tiled roofs.
All the houses were glazed. "Mud buildings we
have none," wrote White. There were many poor
folk among the inhabitants, but their pastor records
that they were sober and industrious. The majority
found occupation on the farms, but some were
employed in brickmaking, in spinning wool, and
preparing rushlights. During the hop season, all
spare hands were needed to secure the crop, which
was as important then as it is to-day. Life ran in even
channels ; a parish tea, a visit to a neighbouring fair,
or the quartering of twenty-six Highlanders in the
village for a whole winter, afforded exciting diver-
sion for the natives. A casual poacher, capturing
game from the moor or fish from the stream, a so-
called "leper," who for years lived on in tolerable
misery, a village idiot, who fell down a deep well
twice in one day, or a mad dog from Newton careering
down the street and biting the local dogs prevented
a dead monotony.

Taken all together, eighteenth-century Selborne
was a peaceful place, and a friendly. At least, no
festering religious strife embittered the little community. As late as 1783, in filling up a report preparatory
to an Episcopal Visitation, White was able to state
that "for more than a century past there does not
appear to have been one Papist or any Protestant
dissenter of any denomination." To this we may
add that White's life and character were such as not
to cause a break in this pleasant tradition.

The word "sequestered" has been used of set
purpose, and justly, to describe the situation of
Selborne. There were, indeed four ways of approaching the village, but three of these consisted of hollow
lanes, steep-sided and very narrow, scooped out of
the freestone by the traffic and weathering of ages.
The village street itself was but a hollow lane, though
wider than its fellows. Along these lanes vehicles
could pass each other only at certain definite

points, and timid horsemen went in some peril
during wet and frosty seasons, when the naked
rock was slippery for "unroughed" horses.   White's
visitors used to complain, with a mixture of
seriousness and jocularity, of the intricacy of these
routes.   His old college friend, John Mulso, averred
that Selborne was "more difficult to find than
ye Bower of Woodstock." Yet we know that, bad
roads notwithstanding, White was no recluse.   Until
he became elderly, he was a man of many journeys,
travelling both on horseback and in post-chaises.
For shorter distances, as when he was serving the
churches of Faringdon or Newton Valence, or for
those longer expeditions to West Deane or Fyfield,
he rode on a favourite pony.   For visits around
his native village, he relied upon what he ruefully
called his "short stumps."   Given the best inten-
tions to be active, however, the position of Selborne
was an invitation to parochiality.   One instance will
illustrate how it lay remote from the busy world
of events.   John Wesley records in his diary that
he once stayed at Lyss for a short time on his way
to Portsmouth.   Now Lyss is about six miles from
Selborne, yet these two parsons, Wesley and White,
each unique in church history of the eighteenth
century, and each destined, in vastly differing ways,
to gather posthumous fame, never met; nay, it is
doubtful whether Wesley had ever heard of White,
though the reverse is probable.*

In this quiet spot, then, at the Vicarage, where his
grandfather, another Gilbert White, lived as vicar,
the naturalist was born.   The year was 1720, that of
the South Sea Bubble, but none of the perturbations
seem to have reached Selborne.   John White, the
father of the new-comer, was a retired barrister,
who chanced to be living temporarily with his father
at the Vicarage.   Of John White there is little
known.   He was an affectionate, amiable man, but

* Wesley left Oxford in 1735; White went to Oriel in 1740.

he passes in and out of the picture erratically until his death in 1758, nineteen years after that of his wife. Anticipating matters, it may be said that Gilbert White was the first-born of a large family, mainly boys, of whom four,—John, Thomas, Benjamin, and Henry,—had scientific tastes, and their names are constantly appearing in the *Journal*. One of White's sisters, Rebecca, married a Sussex gentleman, Henry Woods, of Chilgrove, while his younger sister, Anne, married Thomas Barker, of Lyndon Hall, Rutland. These husbands are frequently heard of, for Woods was a naturalist in a mild way, and Barker was keenly interested in astronomy, physics, and meteorology.

How long John White remained in his father's house is uncertain, but we know that he dwelt with his wife and child for a few years at Compton, near Guildford. Thence he removed to East Harting, in Sussex, but somewhere about 1730 the family finally returned to Selborne, and settled in a neat, unassuming house, opposite the Plestor, or village green. This house, known as "The Wakes," from one Wake, who was either the original occupier or the owner of the land, was built of local freestone, edged with red bricks, and was already about sixty years old. Since White's day it has been enlarged and modified to such an extent as can only be realized by examining old prints.

From 1730 onwards, "The Wakes" was Gilbert White's home. During his vacations, and between the intervals of serving distant curacies, he was assiduously making alterations in the grounds and garden, and was plainly indicating that, whatever attractions the outside world might offer, here was his permanent habitation. In 1763 the ideal was attained, and White became not only the occupier, but the possessor, of "The Wakes." The annual rental seems to have been assessed at the modest sum of five guineas.

It is conjectured from incidental evidence that
Gilbert White was, for a short time, a pupil at Farnham
Grammar School. Solid fact is reached when he was
dispatched to Basingstoke Grammar School, about
the year 1735. This school had already achieved
fame. The headmaster was the Rev. Thomas Warton,
the father of two sons who afterwards became
eminent in the literary world, and who are mentioned
several times in White's correspondence. Joseph
the elder, was appointed headmaster of Winchester
College, and Thomas, the younger, became Poet-
Laureate and Professor of Poetry at Oxford.
Although White's comradeship with the Wartons was,
on account of their respective ages, of short duration,
the acquaintance was not forgotten, and we know
that the Wartons highly appraised White's writings.
There is extant a list of the books which the lad
carried with him to Basingstoke,—Greek and Latin
authors, dictionaries, works on theology, philosophy and poetry. That he early allowed his fancy
to roam among the poets is clear from his abundant
quotations from a miscellaneous group of first,
second, and even third rate men. That he bore his
part in boyish escapades, seems evident from the
26th Letter in *The Antiquities of Selborne,* in which
he describes how a party of lads undermined a portion
of the fine old ruin of the Holy Ghost Chapel at
Basingstoke. He admits pawkily that he " was eye-
witness, perhaps a party concerned."

In 1740 we find White in residence at Oriel College,
Oxford. There he entered upon a wide scheme of
reading, which afterwards bore fruit in his writings.
We note, too, that he attended " Dr Bradley's first
course of Mathematical Lectures," a rather curious
choice for one who was primarily a naturalist, with
leanings towards literature. Many forms of recrea-
tion were open to White ; he seems to have pre-
ferred riding and shooting. He kept a dog and carried
a gun, being especially fond of shooting partridges

and hares. His journeys from Oxford to Selborne were performed on horseback, for he loved horses. Yet his judgment in the matter of horse flesh provoked the laughter of his friends. A broken-winded, spavined nag well met his needs, provided that the creature was an old playmate. As White rode along, his eyes were wide open. He noted the landscape, the crops, the style of farming, the commoner plants and birds, the churches, ruins, and antiquities.

At Oxford White formed the habit of keeping an account book, in which his receipts and disbursements were carefully entered. The accounts are models of neatness. Purchases of books, gunflints, powder-horns, spurs, wine, and tea, payment for boat hire, subscriptions to a "music club," are among the entries which speak clearly of a happy, unruffled College life. He made several firm friends, chief among whom was John Mulso, who later became a country parson and kept up a continuous correspondence, supplemented by visits to Selborne, until his death in 1791, two years before that of White. Another of White's companions, or at least acquaintances, was William Collins, the poet, whose ill-starred career and early death were greatly lamented by the naturalist. When White took his bachelor's degree in 1743, no less a person than Alexander Pope presented him with his translation of the *Iliad* in six small volumes. What was the bond existing between the poet and the young student seems beyond inquiry, but the appropriate gift has, for us, a derivative interest of some importance. On fly-leaves of volumes of the *Iliad* are two pen-and-ink sketches, one certainly, and the other probably, intended to represent Gilbert White. The importance of these poor sketches lies in the fact that White is believed never to have sat for a portrait.

The M.A. degree followed in 1746, and in this year he spent six months in the Isle of Ely to settle an estate and dispose of live stock. In the succeeding

year White took deacon's orders, and became curate
to his uncle Charles, who was rector of Swarraton and Bradley, near Alresford.   He appears to have
ridden on his mare to his weekly duty, the stipend
for which was £25 yearly.   He continued his work
at Swarraton after he had quitted Oxford in 1748,
but the curacy does not seem to have lasted long.
In 1751, White was established permanently at
Selborne, and for five months he acted as curate-in-
charge for the vicar, Dr Bristowe, who was in ill-
health.   This undertaking gave him residence at
the Vicarage.   In 1752 White became Proctor to
the University, an office which had fallen to Oriel
for that term.   This was followed by his appointment as Dean of Oriel, but when the Provostship
fell vacant, he failed to secure election.

At this period of his life, in particular, White
could not be called an unemployed man.   He made
frequent journeys to Oxford in connection with his
office.   He undertook ill-paid curacies, on the principle that a clergyman should not be idle.   In 1753
he became curate at Durley, Hampshire, and retained
the office a year and a half, though the expenses
exceeded the receipts by nearly £20.   For a year he
served curacies at West Deane, Wiltshire, and Newton
Valence, a parish adjoining Selborne.   From 1756
to 1759 he was again curate at Selborne, and from
1761 to 1784 curate at the neighbouring village of
Faringdon, taking the place of a non-resident rector.
As soon as the Faringdon appointment terminated,
he again took up that of Selborne, and, the vicar being
non-resident, might himself have been considered
vicar, except in name and emoluments, until his
death in 1793.   It is a common error, however,
to style him " Vicar of Selborne " ; the living was
in the gift of Magdalen College, while White, as
already stated, belonged to Oriel.

White repeatedly refused promotion to other
livings lest he should have to leave Selborne.   Once,

indeed, he actually applied for a benefice, that of Bradley, where, as before mentioned, he had once been curate. White's uncle had died in 1763, and an application made to the private patron to become his uncle's successor was refused. Since White had just become owner of "The Wakes," it seems that he did not intend, in any event, to live at Bradley. One outside preferment, and that, one is unfortunately bound to admit, a sinecure, White did accept, namely, the College living of Moreton Pinkney, Northampton-shire. The net income from this living was pro-bably about £30, but all that White had to do for this sum was to ride to the village once or twice a year to collect his dues.

But let us not mistake. The average receipts from one of White's curacies might be from £20 to £30 ; in one instance, as we have seen, he was out of pocket. It is necessary to insist upon this, in order to remove any idea that White was a wealthy man, a rich pluralist, or a kind of territorial "squar-son." His total annual income, derived from his Oriel Fellowship, curacies, shares in farm property, and private investments, seems to have been only a few hundreds, perhaps, at most three. Being a bachelor, this steady income enabled him to live in comfort, even in some degree of luxury. He was rich only in the sense that his wants were really few. He was fond of good things, but he shunned extrava-gance and excess, knowing that by practising true economy he could afford to be generous to others. In his Account Books gifts are frequently mentioned, small gifts of money to old wives and sick persons, presents to men and boys who brought him some item of news in natural history, subscriptions to those who were in distress, through fire, epidemics, or unemployment. "Gave little Xtian, 1s."; "A warm wastecoat for old Lee, 2s.", are entries which need no gloss.

The picture we must form of White, then, is that

of a man who, while not disdaining comfort, was
acutely alive to his neighbours' misfortunes, but
who was no simpleton to be preyed upon in business
matters. In winding up an estate for a relative,
he knew that he must bide his markets to sell sheep
and oxen to the greatest advantage, that it was best
to dispose of plate in London by weight, that docu-
ments should be carefully preserved, that secrecy
in money matters was essential. Both in buying and
selling a field or a crop his judgment was sound.
A shifty tenant, or an exorbitant driver of a post-
chaise, or an impostor with a plausible tale of woe,
did not escape without feeling that his design had been
detected.

When young and vigorous, White had travelled
widely over most of the Midland and Southern
counties of England. Sometimes business, sometimes family ties, had caused him to seek out and examine many remote spots. His most north-westerly
point was Shrewsbury. Lancashire he never reached,
although his brother John was vicar of Blackburn,
and sent frequent pressing invitations. First, because
White could get no satisfactory deputy to take his
duties, and afterwards because he was becoming
elderly, and always because coach-sickness was a
genuine terror, the visit was postponed time after
time, with the result usual in such cases. But
Gilbert White gazed at Eldon Hole, in Derbyshire.
inspected the churches of South Lincolnshire and
Cambridgeshire, spent a long holiday in Rutland,
and periodically visited his parishioners in Northants.
Excursions to Essex, Huntingdon, Oxford, and Berkshire were accounted small matters. He knew
Gloucestershire, especially Bristol, where, at three-
and-twenty, he underwent treatment at the Hot
Well, to relieve some kind of fever. He was familiar
with Salisbury, Wilton and Stonehenge, and once
stayed for a short time in South Devon. When we
consider his many rides over the South Downs,

" that chain of majestic mountains," to Ringmer, or across country to Fyfield, near Andover, or his expeditions to London itself, we need not wonder that Mulso jocularly nicknamed him the " Huzzar Parson." Nor are we surprised that White's travels bore fruit in the form of unexpected and apposite comparisons concerning the " oeconomy " of Nature. No naturalist needed less to be preserved from mere book-knowledge ; had there been any such danger, these pilgrimages would have been a sufficient safe-guard.

A double attraction brought Gilbert White to town with fair frequency. Two of his brothers lived there,—Thomas White, four years Gilbert's junior, a merchant who had retired to a comfortable house in South Lambeth, and Benjamin, a year younger still, a bookseller in Fleet Street, who also later built himself a handsome dwelling in the same rural suburb. Thomas White had an interest in natural history, especially in British trees, while Benjamin, besides having a similar, but more general range, published books on birds and animal life. From his firm emanated the well-known works of such men as Colonel Montagu and Thomas Pennant.

It was through visits to the Fleet Street bookshop that White became acquainted (1767) with Thomas Pennant, at first it seems, by mutual messages, then in person. Pennant (1726-98) was a country gentle-man of Holywell, in Flintshire, and was publishing, through Benjamin White, an ambitious work entitled *British Zoology.* Presently White got into corres-pondence with Pennant, but it was not long before he had made another bookshop friend, the Hon. Daines Barrington (1727-1800) a Welsh barrister who later became a judge. Barrington was a philo-sophical naturalist, with a strong leaning to such disputed questions as. the hibernation of swallows and the extent to which the cuckoo is parasitic. He, too, became a correspondent of White's, and he

has the honour of having been the first to suggest
that White should write a history of his native parish.
The idea was not at the outset greeted warmly,
but eventually it began to take shape.   However, all
the preparations were leisurely and unforced. System-
atic notes were made year by year in the *Journal*
and some of these were afterwards extracted and
woven into letters to form the *Natural History*. The
majority of the "letters," as we see them in that
famous work, though considerably altered, are
genuine epistles ; a fair number, though addressed
to Pennant or Barrington, are letters only in form,
and never went through the post. Before the
Natural History appeared, in 1789, two important
instalments, in the form of papers on the Hirundines,
had already been published in the *Philosophical
Transactions* of the Royal Society, in 1774 and 1775
respectively.

During the quiet fruitful years in which the great
work was being prepared and matured, White's
journeys from home became fewer, and, though no
hermit, he began to cultivate the joys of his fireside.

From a study of his letters and his Journals, we
can easily reconstruct his life and its general setting.
We walk up the village street towards " The Wakes."
A few of the houses are actually older than White's
day, others, though less ancient, meet our gaze as
they met his.   One bears the date 1710, and another
1697, with a heart and the initials of the builder.
Some of them are built of that blue vein of malmstone which he describes so concisely, and which
was eagerly sought after by local masons.   Two or
three houses still show " garneting," that is, mortar-
joints lined with chips of red ferruginous grit (car-
stone), a practice which called forth the stranger's
joke about fastening the walls with tenpenny nails.
Some of the modern cottages, though substantial,
are not very pleasing.   The Plestor remains inviolate,
although the sycamore is later than White's day,

and the maypole has disappeared altogether. We
see the churchyard yew, Gracious Street, and the
Punfle, the Short Lith, the Long Lith, the Priory,
and the Temple, though these two buildings are not
original. The Zigzag and the Boſtal ſtill exiſt to
assiſt the wayfarer who would climb the Hanger, and
the Well-Head, now enclosed in masonry, ſtill feeds
the Selborne Stream, and supplies the villagers with
water. Of the four lindens which White planted
in 1756 to shut out the view of the slaughter-
house, three survive. And this liſt of relics of the
Whitean era might be further extended.

Natural conditions have conspired to make Selborne a Paradise for such a man as White. The
heavy rainfall which is so annoying to a visitor,
especially if he has chanced upon one of Selborne's
wet periods, together with soft winds and a mild
atmosphere, has encouraged artificial plantations
and has produced deep verdure and a rich natural flora.
The numerous ſtreams, the uneven country, the wide
diversity of soils,—bare chalk, loam, malmſtone,
"black malm," "white malm," ſtiff clay, peat, and
pure sand,—ſtill further aid in enriching both
the animal and the plant life. The diſtrict forms,
in fact, a microcosm of geographical, geological,
and botanical wealth. When White knew Selborne,
some of the downs were as yet unenclosed, and
much wild life was thriving in its coombes and
coppices. The parish had the usual proportion of
meadow, arable, and waſte, each with its own natural
products, and finally, there was the grim, yet attractive wilderness of Wolmer Foreſt. These then, were
the scenes which daily met White's eye as he walked
or rode to Newton or Faringdon, where he faithfully
read the service, and gave his flock a plain, if rather
dry, eighteenth century sermon. The prayers were
read in a clear resonant voice. All was done rever-
ently and in order. We infer, from his own writings,
that no formal or slovenly makeshifts would have

met with White's approval. The day may be Communion Sunday, when the services will be longer, but his pony is stabled ready to carry him home along the hollow lanes. Arrived at " The Wakes," White neatly endorses and dates the sermon ; it will be used again and again, at Fyfield, Chawton, or Tidworth, on occasion, but at Selborne and Faringdon several times, with intervals of two or three years between each reading. By day there was the usual routine work of a village parson to be performed. About a score of baptisms, less than a score of funerals, some half dozen weddings annually,—not a heavy task, the captious critic might object. But White was often abroad, preaching in other villages, and he visited his own flock faithfully. A sense of duty pervaded his work. We know that his poorer parishioners looked to him for help and advice, doubtless, too, for simple medical attention. Children's tea parties, calls at the dame schools in the village, and visits to the sick and dying, all helped to use up his time, but undoubtedly still left him a good deal of leisure for gardening and natural history.

White's garden, with its appurtenances, was one of his greatest joys. He planted timber trees and fruit trees. The names of some of the latter, as given in the *Journal*, have quaint, old-world memories, and are pleasant to the ear. He constructed melon-beds and cucumber-beds, laid out walks, arranged vistas, built an arbour and a ha-ha, erected a sun-dial, and made new walls for fruit trees. Along with his brother Thomas he carried out experiments in cropping and fertilizing his fields, both arable and meadow. Either on his own account, or in trust for others, he had the letting of two or three farms in Sussex. He bought, with money advanced by Thomas, John Wells's " farm,"—some fields behind his house, " that *angulus iste*, which the family have so long desired." He built a new parlour and a new " hermitage." This hermitage was a place for tea-drinking with friends, for dancing, and dressing in costume, in short, it belonged to the lighter side of White's social life. At three-and-twenty he had helped his brother John to construct the Zigzag path up the Hanger; at sixty he caused to be cut a Bostal, or steep, sloping oblique path which was warranted to be easier to climb, through far shorter in measurement. At the foot of the Hanger he levelled an "area," and transported "bridestones," or sarsens, from a lane three or four miles away, and set them up on the spot. Another sarsen, now known as the "Wishstone," was placed at the top of the Hanger.

White was constantly making little gardening experiments. He brought in foxglove and mullein from the Hanger, and planted them in his garden. Packets of seed were sent to far away friends, and exchanges were received for testing at "The Wakes." Various relatives were importuned to send a few seeds of any new species for trial. Gathering beech mast, he either sowed or caused his old factotum, Thomas Hoar, to sow his collected stores in the hedges and on the bare downs. There is little doubt that Thomas Hoar, and Goody Hammond, or Hampton, White's "weeding woman," were kept well employed. Yet the phraseology used in the entries indicates that White himself was a helper, and not a mere onlooker.

Within doors, too, there were signs of industry. Now he is brewing beer, again he is making raisin wine. He is devising a new kind of plaster for walls, or he is shearing his mongrel dog, Rover. The famous tortoise, Timothy, bequeathed to him by his aunt, Mrs. Snooke, of Ringmer, has to undergo several experiments, and is even immersed in water, with such resultant discomfort as to prove that it does not naturally belong to that element.

One matter appears to have escaped general notice. Although White was not actively interested in national affairs, four newspapers entered "The Wakes"

White's "obelisk" at the top of the Hanger (" Wishing stone" is a modern name)

[*face p.* **xxix**

every week.   Daytime occupations always included
the study of natural history.   Sometimes this work
was of a practical kind, such as searching the streams
for fish to send to Pennant, digging out mole crickets,
or beating eaves and bushes in the vain attempt to
discover hibernating swallows or martins.   When
night closed in White had his correspondence to consider.   Letters must be written to friends, to his
many relatives, especially the Barkers, to Pennant,
Barrington, Montagu, Chandler, Churton, or Sir
Joseph Banks.   These men were all associated,
to a greater or less degree, with White's hobbies.
Indeed, White might have a parcel of bird skins or
insects to examine and name, preparatory to packing
them up carefully for the next mail.   Then the
Journal had to be posted, and occasionally, one must
assume, a fresh sermon composed.   But all was
done without haste; the handwriting is tidy, the
style is clear, witty, concise, and vastly attractive.
The letters, full of the breath of life, are tinged with
old-fashioned courtesy.

We are not to suppose, however, that the bachelor
household was drowsy with the monotony of routine.
For twelve years before his death White gave a home
to the widow of his brother John.   Visitors were
frequent,—local families, the Ettys or the Yaldens;
relatives, particularly his brothers Thomas and John;
brother-students, like Skinner, the botanist, or Chandler, the antiquary;   old friends, like John Mulso, his
most regular, nay life-long correspondent.

We peep over White's shoulder as he is entering
records in his *Journal*.   He writes of his bantam sow,
of the wasps in his peaches, of the dogs which stole
his apricots, of tree-fellers and brick-makers, of
floods, frosts, and fires, of sowing fields and draining
meres, of Timothy the tortoise, of the coming into
the world of his nephews and nieces.   The stealthy
overlooker may be assured that he will espy no entries
of personal quarrels or upbraidings, no Pepysian

discontents, none of Cobbett's flaming outbursts, nor any accounts of gargantuan meals, like those of Parson Woodfoorde. White, it is true, delighted in good fare, the best of the farm and garden,—ham, cheese, fruit, beer,—but he was no sensualist, only a plain man who accepted with gratitude Nature's choicest gifts. Trials and calamities are often recorded ; those of his neighbours with sympathy, his own with a wise resignation. While White is writing, young folk are chatting, or playing music. In these days when speed, noise, and turmoil are popularly worshipped, it is a relief for the mind to transfer itself to the quietude and repose of that Selborne parlour. Presently, the old man rises, re-adjusts his horn-rimmed spectacles, snuffs the candle, and reads with his nephew John an epistle or two of Horace, all in an even-tempered way, for master and pupil differ only about the quantity of a Latin syllable.

The evening round of duties is ended. Conversation becomes general, and mirth prevails. To quote White's own words : " When the children are buzzing down at their spinnet, and we grave folks sit round the chimney, I am put in mind of the following couplet, which you [Samuel Barker] will remember :

". . .   all the distant din that world can keep
　　Rolls o'er my grotto, and improves my sleep."

A life, not of indolence, but of comfortable regulated leisure, with definite duties to be conscientiously performed,—that would adequately summarize White's career. It was certainly not a lazy life, and certainly not an idle one. On the very morrow he may have to ride into Sussex on farming business, or to an election at Oxford, or to London, to confer with his brothers at South Lambeth. Even in London he is not idle. He attends meetings of the Royal Society and the Society of Antiquaries. He

Studies the accessible museums, and inspects the markets and fishmongers' shops to become acquainted with new kinds of game or fish. These methods of observation he was constantly impressing on his brother John, when stationed at Gibraltar. Although Gilbert White's teaching failed finally in the publication of the proposed Natural History of Gibraltar, it was successful in bringing him abundance of foreign specimens to examine.

Thus life flowed peacefully by, unruffled, except by ordinary human vicissitudes. Halcyon days are mainly recorded in the annals, but there were also days of sadness, illness, or personal loss. Yet scarcely a whisper of Fortune's rebuffs is found in his letters. White's noble nature would not diffuse its pain. A hint of impaired eyesight or deafness is, as a rule, all we get. When the shades were actually closing in, White's illness was mercifully short. On June 10th, 1793, he read the funeral service over "Mary Burbey, aged 16," and his handwriting in the Register maintains its customary neatness, a fact difficult to explain, but often noticeable in a man's last days. On June 15th he penned his last letter to that congenial correspondent of his old age, Robert Marsham, of Stratton Strawless, in Norfolk. The letter shows no falling off in keenness for knowledge, whether in asking or imparting, but White explains, as a reason for delay in writing, that he had " been annoyed this spring with a bad nervous cough, and a wandering gout, that have pulled me down very much, and rendered me very languid, and indolent." The letter ends with noticeable, even ominous words : " The season with us is unhealthy."

White, by this time, was evidently feeling depressed in body and spirit. In the outside world, about which he is sometimes erroneously thought to have concerned himself but little, the skies were angry. War had been declared against France, and both the excesses in that country and their reactions in our

own, disturbed the ailing man, and made him appre-
hensive of the future. He was a strict constitutionalist,
and feared outbreaks of violence, yet he trusted in
the sturdy good sense of his countrymen. Two days
after White had written the letter to Marsham, the Alton
doctor had to be sent for. The exact nature of the
illness is not known, but White underwent much
suffering. He bore his pain with patience and
resignation, and on June 26th, in the family parlour on
the first floor of " The Wakes," in full view of his
beloved garden and the beech-clad Hanger, the gentle
old naturalist passed away.

The spiritual influence of White and his works
cannot be fully assessed : it would be folly to make the
attempt. His unseen memorials lie in the " Selborne tradition," that of a calm, yet vivacious, even-
tempered, Christian gentleman, whose life was well-
poised and his goal steadily pursued. We have, too,
the record of a parson who made Christmas presents,
paid the school fees of poor children, and caused the
humbler members of his flock to feel that he was their
friend. The late Edward Thomas once wrote,
" In this present year, 1906, at least, it is hard to find
a flaw in the life he led." If for 1906 we choose to
substitute any later date the statement will still hold
good ; the biographer has no fear of what are popularly called " bombshells " which may unexpectedly
destroy the reputation of his hero.

Among the material memorials of White we may
conventionally mention the *Natural History*, *The
Garden Kalendar*, and the *Naturalist's Journal*, though
it seems sacrilege to class these accomplishments as
material. John White's *Fauna Calpensis*, which
owed its inspiration to Gilbert, never passed out of
manuscript, and was ultimately lost. The remaining
lime-trees in front of the house which was formerly
a butcher's shop were planted by White. A section
of the cobbled path in the same vicinity was laid
down at his expense. As we have seen, he helped to

construct the Zigzag, and was responsible for the cutting of the Bostal. The improvements at "The Wakes," such as the great parlour and the brick pathway, need not be mentioned in detail. Towards the end of his life White bore his part in defeating an inquitous enclosure scheme, which might have hindered free access to the Hanger. Finally, and there can be little doubt of this proposition, though direct proof is absent, White's posthumous influence maintained the fabric of Selborne church in a clean and orderly condition, while the offices were performed in the reverent spirit which was reasonably demanded by the early founders. Of memorials set up in honour of White we must notice the village school, the water supply from the Well-Head, and the oak seat at the top of the Hanger, the last being an appreciation from the "Gilbert White Fellowship."

While it is comparatively easy, from White's own writings, to visualize his daily life, it is all but impossible, in the absence of an authentic portrait, to obtain from tradition any just idea of his personal appearance. In a case of this kind folk-memory often fails entirely. We know, from his own statements, that White was a short man. From his nephew, the Rev. Francis White, Bell gathered that he was of spare form and of remarkably erect carriage. Whether he bore any traces of the small-pox which attacked him in his twenty-eighth year is uncertain. Tradition, if possessing any merit at all, contradicts even the "spare" frame. Fifty years ago, the village sexton, whose grandfather had known White well, told the late Mr T. E. Kebbel what he had "overheard," namely, that White was "a square-built man, of what you'd call medium statue" (sic). Against such hearsay we must set Francis White's direct knowledge. Miss Phillips, a very old native of Selborne, who was still alive in 1926, told me over thirty years ago that her grandfather worked as a bricklayer for Gilbert White. White's account books corroborated this, yet not a tradition of any value had been handed down. Nor was this strange: where Professor Bell and Dr Bowdler Sharpe were unable to discover much ore, later inquirers can expect to win little from the spoil heaps.

Coming into possession of " The Wakes " in 1844, Bell could draw upon the memories of a few folk who still remembered White. Except one or two anecdotes, the flimsy shreds of information thus collected are so thin and outworn that it is ridiculous to use them again. They tell us virtually nothing. I must, however, recall what the Rev. Thomas Mozley records in his " Reminiscences." When he went to Moreton Pinkney, in 1832, there was still living a farmer and one-time churchwarden, John Stockley, who had a distinct recollection of Gilbert White. White used to come annually to receive his rent, and, as Stockley said, was " a pleasant, quiet man, easy to get on with."

To prove that White was gifted with a sly humour and a witty, epigrammatic mode of expression, one would only have to refer to his letters, public and private. These abound in whimsical phrases and happy jests. It is needless to quote specimens, but two anecdotes, showing White's sense of humour, and proportion, are not so well known as they should be. They are told by Bell, who received them directly from White's nephew.

One evening, when White was carrying out his proctorial duties at Oxford, he found an undergraduate, sound asleep, and intoxicated. He was partly undressed, but his outer garments were neatly folded up by his side. White awakened him, and the quickly sobered culprit was ordered to present himself next morning. " I was shocked and distressed, young man," said the proctor, " to see you in such a disgraceful position as you were in last evening, a situation of which I hope you are thoroughly ashamed. You deserve an exemplary punishment ; but I observe one circumstance which shows that you are not utterly degraded. Your clothes were neatly folded up by your side, indicating habits of care and neatness which appear incompatible with habitual degradation." A few more words, and the homily ended, the offender being let off with a warning.

The second story concerns Thomas Hoar, White's valued manservant, Hoar was perhaps a little spoilt by indulgence, and was some years older than his master, who showed his esteem by leaving him a legacy in his will. Hoar survived White by nearly four years, having reached the age of eighty-three. The old fellow came to his master one morning and said, " Please, Sir, I've been and broke a glass." " Broke a glass, Thomas ! How did you do that ? " " I'll show you, Sir," Thereupon he went and brought another wine-glass, which he threw on the floor, saying, " That's how I broke it, Sir." " There, go along, Thomas ! you are a great fool," said White, and then muttered, " And I was as great a one for asking such a foolish question." We must hasten to add that only on one or two other occasions is White known to have shown even such an approach to anger.

This slight sketch would be incomplete without a passing reference to White's solid contributions to natural history and zoology. He added the harvest mouse and the noctule bat to the list of British mammals, and was the first to distinguish clearly the three species of " willow wrens." He had grasped the interdependence of animals and plants, and insisted on the importance of love and hunger in what is now called the struggle " for existence." He understood the importance of sexual characters, and pointed out at least one instance of protective coloration,— that of the stone curlew. He rightly traced the lineage of the domestic pigeon to the blue rock dove. His chapter on earthworms possibly inspired Darwin's classical work on those creatures.

The catalogue might be much extended, but I have already done this elsewhere. Against White's chief blunder, his belief in the hibernation of some, or the majority of the individuals of the swallow family, may be set his discovery that the ring-ouzel is a "bird of passage" in Southern England. The reader of the Journals has the pleasure in witnessing how White was led, step by step, to form some of his sound conclusions. Voluminous as White's notes are, we might have enjoyed even greater delights had the naturalist not been in the phrase which he applied to his relative Henry Woods, "of an *unwriting* constitution," and sometimes a victim of the "Daemon of Procrastination." Those failings are still common among men, but we must honestly admit that White's actual accomplishments, not less than his ingenuous modesty and frank confession, lift him, at least, quite beyond the reach of reasonable censure.

II

THE "NATURALIST'S JOURNAL":

Its Genesis and History

"The air is full of sounds, the sky of tokens, the ground of memoranda and signatures; and every object is covered over with hints, which speak to the intelligent."—Hugh Miller.

In his very first letter to Thomas Pennant, dated 4 August, 1767, White declared that he had been attached to "natural knowledge" from his childhood. Curious and unexpected confirmation of this statement has been discovered by Mr Rashleigh Holt-White, the biographer of the naturalist, and the present head of the White family. The evidence occurs in a manuscript diary kept at Lyndon, in Rutland, by that Thomas Barker who afterwards became White's brother-in-law. In this diary, such events as the arrival of the first cuckoo, and a flock of wild geese going north, are noted, and initialled "G.W.", manifestly referring to Gilbert White, who is supposed to have been staying at the time with an uncle in a neighbouring village.

Further testimony of youthful ardour is found in White's first letter to Robert Marsham, in which he states, "In an humble way I have been an early planter myself." He proceeds to enumerate several trees which he planted in his father's grounds at Selborne, beginning with an oak and an ash in 1731, when he was only eleven years of age.

We may fairly suppose that White's inborn love of Nature was quickened and strengthened by his intimacy with Dr Stephen Hales (1677-1761), the clerical philosopher and social reformer, who, during White's early manhood, was rector of Faringdon, near to Selborne. Although Hales made his permanent

abode at Teddington on the Thames, he was
accustomed to spend about two months every year
at his Faringdon rectory. The two men, with a
difference of more than forty years between them,
became good friends. Hales, besides being a social
reformer, was pre-eminent as a meteorologist and a
student of physics, and White's constant references
to his friend's published works sufficiently show the
influence of the older man upon the younger. It
is quite possible that White's practice of keeping a
journal originated with Hales. If so, the position is
very striking, because White's correspondent, Marsham, asserted that it was "by the advice of his
most estimable friend the late Dr Hales," that he, too,
had kept "a poor imperfect journal" for more than
fifty years.

Whatever may have been the prime incentive, other
than a natural desire to observe and a passion for
methodical note-taking, White commenced in 1751 a
kind of diary, which he called a *Garden Kalendar*.
Four years previously he had bought a copy of Philip
Miller's *Gardener's Dictionary*, and this was the beginning of a series of purchases of books of reference,
of which Hudson's *Flora Anglica*, acquired many
years later, was perhaps the most important. White's
Garden Kalendar consisted of sheets of quarto letter
paper, stitched together, and in this cheaply made
book he entered the principal events in his horti-
cultural life : the sowing of seeds, the planting of
shrubs, the flowering of plants, the critical features of
certain genera, the introduction of new fodder crops,
and other kindred matters. There were also miscellaneous notes on birds, insects, and the weather,
with some items of private import. The *Kalendar*
was carefully posted every day, and, when absent
from home, he appears to have carried it with him,
the encumbrance being but slight. Occasionally,
however, during what we may assume were sudden
calls of duty or pleasure, entries were made by his

brothers Thomas or John, or even by his faithful servant, Thomas Hoar. The *Kalendar* went steadily on, changing its spelling to *Calendar* in 1765, until the year 1768, when a transformation took place. Before describing this, it should be stated that a small part of the *Garden Kalendar* was reprinted in Bell's edition of the *Natural History* in 1877, and that the whole was reproduced in Bowdler Sharpe's excellent edition of 1900.

In 1768 the *Garden Kalendar* was superseded* by the *Naturalist's Journal,* and since it is from this *Journal* that the present book has been extracted, something further must be said about the change. White's inscription on the title-page of the first year's journal runs, " The gift of the Honourable Mr Barrington, the Inventer." Barrington, as we have already seen, had heard of White on his visits to the bookshop in Fleet Street, most likely through an intermediary, Thomas Pennant, the zoologist and traveller, to whom White had been introduced in the preceding year (1767). On 22 January, 1768, White writes to Pennant, " Your friend Mr Barrington (to whom I am an entire stranger) has been so obliging as to make me a present of one of his Naturalist's Journals, which I hope to fill in the course of the year." Two months later, on March 30th, White asks Pennant " to pay my compliments and thanks to Mr Barrington for the agreeable present of his Journal, which I am filling up day by day." The friendship with Barrington afterwards became personal, and on 30 June, 1769 White addressed his first letter to the " Inventer " of the *Journal.* Here it may be remarked, that while Barrington's influence on White tended to encourage a somewhat speculative attitude to certain questions, yet his friendship was probably more disinterested than that of Pennant, who, at

---

* The *Kalendar* was continued in a very attenuated form until the end of 1771, but merely as a record of wine-making, beer-brewing, and transactions in manure for the garden.

this interval of time, impresses us chiefly as an author who was anxious to collect White's facts and ideas for his own purposes.

The *Journal* was oblong in shape, about 12 inches by 10 inches ; the paper was rather coarse and rough-edged. The imprint ran : " London : Printed for W. Sandby, in Fleet St. MDCCLXVII," and this continued until 1771, when the inscription was changed to " Printed by Benjamin White, at Horace's Head, in Fleet Street." (N.D.) The paper was now of better quality, and in 1775 a stock of the *Journals* seems to have been bought. Benjamin White was, as may be guessed, Gilbert's younger brother, and the address " Horace's Head," was the source of at least one curious blunder. Until the year 1768 the title-page of the *Journal* was printed ; from that year onwards, a somewhat elaborately engraved page took its place, the words *Naturalist's Journal* being executed with many flourishes.

On each title page White has made a written note : " The Insects are named according to Linnaeus : the plants according to the sexual system : the birds according to Ray." It may be added, in passing, that, in writing to Pennant, an adherent of the Linnaean nomenclature, White employed that system instead of following Ray ; his courtesy stood before his convenience.

At the foot of the title-page, for the years 1768 and 1769, White wrote this motto :

> " I solitary court
> The inspiring breeze : & meditate the book
> Of Nature ever open . . ." Thomson's Seasons.

The year 1770 brings us two mottoes, which White continued to copy for some years. The first runs :

> " Igneus est illis vigor, & caelestis origo
> Seminibus . . . Virg."

The second quotation is much longer, and was evidently a great favourite, because it also appears on the title-page of *Selborne* :

"Omnia bene describere, quae in hoc mundo a deo facta, aut naturae creatae viribus elaborata fuerunt, opus est non unius hominis, nec unius aevi. Hinc Faunae & Florae utilissimae ; hinc Monographi praestantissimi."
Joan : Anton : Scopoli annus 2^{dus}.

In 1773 first appears the quotation from Horace, "ego Apis Matinae" which is printed in a shortened form on the original title-page of the *Natural History*.

These maxims were inserted in no spirit of pedantry ; on the contrary, we know that they were only variations of White's own watchwords. Direct, personal observation he never failed to insist upon ; he was equally impressed by the belief that every province should have its own monographer. The field of Nature being beyond the power of any one man to cultivate in full, let everyone begin with his own parish, and till that area intensively.

We open the *Journal* and find that each week is represented by a page. This is ruled into columns of unequal width, each having a printed head-line. Most of the columns are narrow, being intended for figures and brief notes ; others leave room for short phrases. The right hand column is liberally spaced, for it has to accommodate "Miscellaneous Observations and Memorandums." On the top of the page, on the left hand, the diarist was to fill in the year, the day, the place of observation, and the nature of the soil. There are narrow columns for readings of the thermometer, barometer, and wind. These were supposed to be taken at 8, 12, 4, and 8 o'clock respectively. Then follow "Inches of Rain," "Weather," "Trees first in leaf," and so on,— an arrangement which strikes a naturalist as mechanical, and likely to bind him in fetters of iron. From 1774 onwards, the *Journal* was provided with four extra pages, lettered for an index, but White made no indexes.

The poor diarist was obviously expected to cramp his remarks into these restricted spaces. Gilbert White was a neat, painstaking penman, with the precise, tidy habits that accompany a well-ordered mind, but, as his biographer sagely observes, forms are " good servants but bad masters." At first, he was very strict in keeping well within the ruled limits. When recording the readings of the thermometer this was quite easy, though he did not give four readings per day, usually only one. With the barometer it was his custom to give the daily variations : thus, $29\frac{8\text{-}10\frac{1}{2}}{}$ or simply, $29\frac{3}{10}$. It was also convenient to insert " rises," " falls," or " sinks," instead of the actual figures. The amount of rainfall does not appear in the *Journal* until August, 1779, although the naturalist had been presented by his brother Thomas with a " rain-measurer " in the preceding April. When he came to record dates of phenomena, he could contrive to write " Blackcap sings," or *Fragaria sterilis* in flower," without unduly insulting his own intelligence. Soon, however, the naturalist broke through his bonds, and, while continuing to fill the smaller spaces in the prescribed way, he began to disregard the other divisions and write across them. The gain to posterity has been great. The *Journal* might have degenerated into a mere phenological record, whose statistics might never have been collated. Actually, it became, in a fairly acceptable sense, a diary of happenings, scientific and personal. The large space on the right of the page encouraged the habit, and White would let his entries extend to blank leaves, or even the backs of leaves. Barrington's preface ingenuously notes that, for lengthy observations, this course is very easy, and then White goes further, and sometimes inserts an extra leaf.

Though " invented " by the ingenuous Barrington, the *Journal* seems to have been first suggested by

Benjamin Stillingfleet's *Tracts* on natural history and husbandry; indeed, this is tacitly admitted in the printed preface. Barrington goes on to say, "It is hoped that the use [of the *Journal*] will not only be experienced by the Naturalist and Farmer, but by the Gardiner." Then he gives instructions for filling up the forms. The hours for observing the thermometer may be varied from those given in the margin, but the reading should be in the shade, and the degrees those of Fahrenheit, unless specially mentioned. The compiler should record such particulars as celestial phenomena, auroras, tides, animals, plants, fruits, blights, and the age of trees. A page is partly filled up as a specimen, and under the heading "*Observations*" we read:

> "Salmon come up the Thames."
> "An ox of ———— weight killed at ————."
> "Mem. I find by the having mowed ———— pasture as a meadow, for two years together, I have destroyed the thistles."
> "William was cured of an ague by the use of . . . plant."

Plainly, details of this description may become trivial and scarcely important enough for a naturalist's diary. An exceptionally heavy ox might be usefully noted in a farmer's account book, and the dates when salmon come up the Thames are important only when averaged over a long period of years. It is well for us that Gilbert White did not follow the suggested models slavishly. We may as well face the truth. Even as it is, many of the entries are of no permanent value, and there was a danger that the whole diary might be occupied with ephemeral details. White continued, as in the *Garden Kalendar*, to insert such matter as the dates of the arrival of migratory birds and the blooming of plants. These facts, though of much private interest to a diarist, have little scientific value until accurately digested and compared with other phenomena over long periods.

At great labour, Dr John Aikin (1747-1822) collected, somewhat carelessly, from the Journals records of first occurrences of natural phenomena throughout the year, and published in 1795 a *Naturalist's Calendar*. Side by side with White's dates are placed those of William Markwick, of Catsfield, in Sussex.

This is not the place to discuss the usefulness of such Calendars. Some modern naturalists decry them as barren and valueless. The present writer holds a middle opinion : when, after many years, the observer will industriously marshal and classify his dates, as Robert Marsham and his successors did, the future student will find such summaries both a present help and a stimulus to wider inquiries.

What counts, in the *Naturalist's Journal*, is the vast mass of miscellaneous comment. The space devoted to marginalia tends to increase as the *Garden Kalendar* is left behind. The diary habit became engrained in White, and while the trite diurnal details are never dropped, all kinds of interesting items began to appear. These might be classed under two heads, first, those commonplace and private records which normally have individual interest only. Now and then notes occur which rest only on the authority of correspondents, or even of the public press. Had White been an ordinary man, such notes might as well be waste paper ; but because he was Gilbert White, and for this reason alone, we eagerly seize and treasure them as aids in understanding the man. The second class of comment comprises many exceedingly acute judgments, hints and solutions, as well as bits of scientific foresight. Many of these entries were entirely incorporated in the " Natural History," and have not been re-copied here, with a present but inevitable loss of pith and substance. The use of the Journals as a quarry for *Selborne* and for some of his private letters was continued by White until the year 1787 or 1788. Consequently, it has sometimes been difficult to make a clean line of separation between published and unpublished matter. This is really no drawback; it is better to have a narrative repeated in another guise, than, by attempting to shear portions away, to lose some central fact.

The information given in the manuscript Journals from 1788 to 1793, after the elaboration of the "Natural History," was perforce left untouched by White. After his death the Journals were lent to Aikin, who as already stated, loosely compiled from them a *Naturalist's Calendar*, and this has appeared in most, though not all, of the subsequent editions of *Selborne*. Aikin also selected from the manuscript many important passages which, under the head of "Observations," have been reproduced by most of the later editors. In spite of this, and also because Aikin took considerable liberties with the text, it has been thought advisable to re-print these notes in their original form and setting, with their full context.

In 1802, the manuscript Journals were in the hands of Gilbert White's nephew, John, who in that year published the second edition of *Selborne*. It must have been this John White (1765-1855) who had lent the Journals to Aikin, and who afterwards possibly lent them to Edward Jesse. Jesse (1780-1868), surveyor of the Royal Parks and Palaces, was a lover of natural history, and in 1834 published the first edition of his *Gleanings*. He had manifestly read the Journals, and had " picked and culled " a number of extracts, which he printed, not too faithfully, as throwing light upon White's private life and habits. Perhaps it was Jesse who inserted sundry notes in red pencil, occasionally but not always pertinent. The manuscript may, by this time, have come into the hands of Gilbert White's niece, Mary, who then occupied " The Wakes," and continued to do so until her death in 1839. But of this supposition there is no direct evidence.

The subsequent history of the manuscript is mostly a matter for surmise. One suspects that the Journals passed out of the family at the sale of "The Wakes" estate in 1840. Then, at some unknown date, the volumes were beautifully bound, perhaps by George Soaper, whose bookplate is found pasted inside the cover of each volume. From scanty bits of evidence, one may conjecture that this was the George Soaper, of "Stoke, by Guildford," who, in 1839, reprinted, for private presentation, "The Institucion of a Gentleman, 1568." Little more can be added, save that, in 1881, the British Museum purchased the manuscripts from the Rev. George Taylor, a curate at Pulborough, and formerly a curate at Guildford. Mr Taylor in a letter to the Museum authorities, stated that the manuscript had been in his wife's possession many years. In the Manuscript Department the Journals have remained, little having been done in the meanwhile to make their contents public, except by Mr Rashleigh Holt-White, and one or two other writers who have made partial drafts for special purposes.

The *Journal* could be easily carried in an overcoat pocket or travelling bag, and undoubtedly it was sometimes thus carried on White's travels. At other times it was left at home and entries were made on his return. Even, however, when he took the *Journal* with him, some of the insertions represented details gathered from letters received from home or from distant friends. Back entries and foot-notes often indicate the arrival of later information. Now and then, items of expenditure are noted, but in general, White's business transactions, like his charitable disbursements, were sedulously recorded in separate account books.

Those readers who have never seen the manuscript may be glad to know something about the minutiae of White's work. The handwriting, to take that important matter first, is painstakingly and

uniformly neat, reminding us of the diarist's admoni-
tion to his niece Molly : " Your handwriting is very
fair, and handsome ; pray keep it up ; & don't
scribble it away." When White's quill was freshly
trimmed the delicate uniformity was very noticeable.
John White once posted the *Journal*, in bare outline,
for ten days, and his writing much resembles that
of Gilbert. The few entries made by Thomas White
are easily detected by the smaller and more delicate
characters. Thomas Hoar's observations, made
during his master's absence, were re-copied by White ;
the old servant was not himself allowed to make
entries.

The *Journal* is beautifully free from blots, and the
insertions and erasures are managed skilfully. The
diarist allows himself no flourishes, except an occasional fan-like curve at the end of a paragraph. The
spelling, alike in its oddity and its inconsistency,
is that of the eighteenth century : *lightening, lightning ;
wallnut, walnut* ; *mackrel, mackarel, mackerel* ; *color,
colour,* and so on. White's own spelling varied as
the years passed ; *turnep* becomes *turnip*, and the
obstinate *cuckow* changes at last to *cuckoo.* White's
use of italics, with respect to both scientific and
popular names, is hopelessly without rule, but his
curious punctuation is fairly consistent. Indeed,
he averred that punctuation was " a thing much
neglected."* We find a great fondness for colons
and semi-colons, and his punctuation of a series of
descriptive words is comical. He writes *it's* for *its*,
but does not employ the apostrophe for the posses-
sive plural in s, agreeing in this with the rule of some
grammarians of the period. Invariably we get &
for *and*, and nearly always *'til* or *'till, tho'*, and a few
similar abbreviations. White likes to note that
work is *finish'd*, and ocasionally amuses himself with
*y*ᵉ and *y*ⁱʳ for *the* and *their.* Diagrams are almost
wanting, but there is an approach to these in White's

* (Letter to his brother, John White, 29 April 1774.)

sign for veering winds, and in his account of the artificial "fairy ring."

As with the Parish Registers, very little falling off in White's handwriting can be detected, even on the last page. His interests, too, despite his declining health, were still potent. He was cutting cucumbers on June 12th, 13th and 14th, and admiring the beauty of his dame's violets, Provence roses, and ten weeks' stocks. On the 14th, " Mr John Mulso [the son of his old friend] came." The next day White observes that " Men wash their sheep," and that " Mr J. Mulso left us." There is a note on the blank page opposite which reads, " The ground is as hard as iron : we can sow nothing, nor plant out," and it is a little doubtful whether this is not the last entry, rather than that just quoted. Whichever way we take it, the matter is trivial. There is an abrupt check, and as already stated, on June 26th 1793, the gentle old diarist passed beyond mortal gaze.

CHAPTER I

"*Nulla dies sine linea.*" (Latin proverb: adopted by Wesley.)

1768

Jan. 3. Horses are still falling with their general disorder. It freezes under peoples beds.

Jan. 4. The birds must suffer greatly as there are no Haws. Meat frozen so hard it can't be spitted. Several of the thrush kind are frozen to death.

Jan. 6. Coughs and colds are general. Provisions freeze within.

Jan. 7. Laurels begin to suffer. Laurustines suffer.

Jan. 8. My provisions are kept in the Cellar. Birds pull the moss from ye trees.

Jan. 9. Lambs begin to fall. Nothing frozen in my cellar. Titmice pull straws from the eaves. [A severe frost had set in on Jan. 1st; on the 7th the thermometer stood at $17\frac{1}{2}°$ F; snow fell and a thaw set in on the 9th.]

Jan. 12-16. A cock-pheasant appeared on the dunghill at the end of my stable; tamed by hunger. Laurustines appear as if scorched in the fire. Portugal laurel, red American Juniper untouched.[1] Nuthatch, *sitta*, chatters. Garden-plants were well preserved under the snow. Turneps in general little damaged. Wheat, being secured by the snow, looks finely.

Jan. 21. Ananas[2] budding for bloom. First crop of kidney-beans gather'd in the Hothouse at Hartley.[3]

Jan. 28. Sowed more cucumber seeds. *Fringillado*, great titmouse, begins some of his spring notes. Pricked out the cucumbers.

Feb. 1. Jack-daw, *monedula*, chatters on churches. Went to London.

Feb. 11. Went to Oxford from London. [Returned to Selborne on Feb. 13.]

Feb. 15, 16, 19. Cucumber plants show two rough leaves. Forward turneps rot. Evergreens appear more damaged than at first was imagined, especially those in sunny aspects. Bees gather on ye winter aconite. Arbutus but little damaged by the frost. Ilex much hurt. Hollies, pinched by the frost, cast their leaves. Laurustinus killed to the ground.

Feb. 23. [Selborne] Great rain. Prodigious floods in Yorkshire, which have swept away all the bridges.

Feb. 24. Cucumb^r plants thrive, and show their claspers.

Feb. 25. The missel-thrush, *turdus viscivorus major* (called in Hants & Sussex the stormcock) sings.

Feb. 28. Wet continues still : has lasted three weeks this day. Pinched off the tops of the cucumber plants, which have several joints.

March 5. Cucum^rs shew side-shoots. Female yew tree shows rudiments of fruit.

March 7-8. The Ground and paths drie very fast. Wheat is fed down by sheep. Beans are planted in ye fields. Pease sown. Cut down the new planted nectarines.

March 10-11. Made the four-light Cucum^r bed with 8 cart-loads of dung. Cucum^r blows in male bloom.

March 13-14. Cucumber shows rudiments of fruit. Turned out pots of cucum^rs into the great bed.

March 24. Blue mist ; & the smell (as the Country people say) of London smoke.

March 25. Apricot is covered with boards. Lucern^4 is 6 inches & ¾ high ; burnet 5 inches & ½.

March 26. Ground is all dust. Sowed various sorts of seeds from the physic-garden at Oxford.

March 29. Cock pheasant crows. Blue stinking mist.

March 30. Raw fog. Canes femininæ catuliunt.

March 31. Black weather. Cucumber fruit swells. Rooks sit. This day the dry weather has lasted a month.

April 2. Bombylius medius. Musca bombylii-
formis densè pilosa nigra abdomine obtuso, ad latera
rufo, longissimum spiculum quoddam ceu linguam
ex ore protendit.[5] Ray's Hist : Insect : p. 273.
April 5. Luscinia !
April 9. The titlark,[6] *Alauda pratorum,* first sings.
It is a delicate songster; flying from tree to tree, &
spreading out it's wings it chants in it's descent. It
also sings on trees, & on the ground walking in pasture fields.
April 13. Hirundo[7] domestica ! ! !
April 17. Rooks have young. Young ravens
fledged. Forked-tailed kite lays three eggs. Redstart sings for the first time.
April 18. Nuthatch, *sitta,* makes its jarring,
clattering noise in the trees.
April 22. Cut a brace of large cucumbers.
April 29. Grass-hopper lark[8] chirps at eight o'
the Clock in the evening.
April 30. The grass-hopper lark chirps concealed
at the bottoms of Hedges.
May 3-14. [London. On May 4th Herrings,
mackerel; May 5th green gooseberries; May 8th
Sturgeon.]
May 14. Returned to Selborne. Melon-fruit in
bloom. A brace of sand-pipers (*tringa minor*)[9] at
James Knight's ponds.[10]
May 18. Young wood-larks come forth. My
apple trees are but poorly blown.
May 19. Rudiments of wasps' nests are found.
No chaffers, or tree beetles, appear yet. The wasp's
nest contained eleven eggs in eleven cells.
May 21. *Lanius minor ruffus,* red-backed butcher-
bird, shot near the village. It's gizzard was full of
the legs & parts of beetles.
May 29. The female viper has a string of eggs
within her as large as those of a blackbird : but no
rudiments of the young are yet formed within the
egg. The viper is ἔσω μὲν ᾠοτόκα, ἔξω δὲ ζῳοτόκα.[11]

June 3-4. The reed-sparrow,[12] passer torquatus in aurundinetis nidificans, sings at Liss, [E. Hants] near M[rs] Cole's ponds. It sings night and day while breeding & has a fine variety of notes. [The following side-note was evidently added later.] As it appears since, this was the *passer arundinaceus minor* of Ray : a thin-billed bird, and probably a bird of summer passage.

June 10. The nightingale, having young, leaves off singing, & makes a plaintive & a jarring noise.

June 12. Glow-worms abound. Phallus[13] stinks in the hedges.

June 17. Wheat that was lodged [beaten down] rises again.

June 28. Showers about. Dryed & cocked my S[t] foin.

June 29. Ricked my S[t] foin in good order. The ears of wheat in general are very long. Wheat blows still.

July 1. Great storm of thunder and lightening. Tiled the succades.

July 3-16. [Headed, " Fyfield [Hants] Hasel loam on chalk ".]

July 4. First young swallows. Cut the first succade-melon. Grasshopper lark sings day & night.

July 6-7. No sun for several days. Bad time for corn. No cucumbers under hand-glasses will set.

July 9. The capsule of the twayblade[14] bursts at a touch, & scatters the dust-like seed on all sides. [From July 10 to nearly the end of the year, exact barometric readings are not given, but the words " sinks," " falls," " rises," are inserted.]

July 11. Cut my great meadow.

July 13. Truffles began to be taken for y[e] first time in my Brother Henry White's[15] grove ; & will continue to be found in great abundance every fortnight till about Lady day.

July 14. Thomas[16] brings down Succade-melons

from Selborne; & says he has cut four brace. They
are very fine.

July 16. Grasshopper-lark sings at Bradley [Hants].

July 17. [Selborne] Succade-melons come in heaps.

July 18. The country is drenched with wet, and
quantities of hay were spoiled.

July 19. Young swallows are able to take flies for
themselves.

July 20. Vast aurora borealis. The white owl
has young. It brings a mouse to its nest about every
five minutes beginning at sunset. Hay in tollerable
order. Cut my little mead.

July 23. Martins begin to congregate on the may-
pole[17]. Ricked my little mead, & finish'd my Hay-
making.

July 25. Cut the first cantelupe-melon.

July 26. Threat'ning clouds at a distance, but
most delicate ripening weather.

July 28. Gathered frenchbeans.

July 29. Vast storms about. Crickets are silent.

Aug. 1. Rock-like clouds. Oats & pease are
cutting.

Aug. 3. The whame, or barrel-fly[18] of Derham,
lays nits or eggs on the legs and sides of Horses at
Grass. See physicotheology.

Aug. 4. Thunder at a distance. The thermometer, which stood at 63 in the dining room, sank
only to 62 in the wine-vault.

Aug. 7. Cold dew. Mulberry begins to cast
some leaves. Tops of beeches in the hanger begin
to look pale.

Aug. 10. Young pheasants are flyers. White
butter-flies gather in flocks on the mud of puddles.

Aug. 11. Wheat harvest is pretty general. The
male & female flying ants, leaving their nests, fill the
air. See Gould on ants.

Aug. 13. Sweet harvest weather. Helleborus
viridis [green hellebore] begins to wither. Brisk
gale of wind.

Aug. 15. Young broods of goldfinches appear.

Aug. 18. Martins continue to hatch new broods. Flies begin to abound in the windows.

Aug. 19. White wheat begins to grow. Plums ripe.

Aug. 22-3. Young gold-finches come forth. Wheat in very bad condition.

Aug. 24. Much wheat bound up in the afternoon. Goldfinch sings. Oats are cutting.

Aug. 25. Cucumber plants begin to decline. Tyed up endive. Large showers about.

Aug. 26. White dew. Peaches ripen. Barley begins to be cut. Much wheat housed.

Aug. 27. Much wheat housed. Blue mist. Yellow-hammers have young still, which they feed with tipulae [daddly-long legs].

Aug. 30. The goatsucker still appears.

Aug. 31. Grapes begin to turn colour. Nectarines ripe. Stoparola[19] brings out it's young.

Sept. 1. Transplanted some plants of the Helleborus viridis from the Honey-lane near Norton [farm] to the shrubbery in the orchard. Smallest Regulus non cristatus [chiffchaff] chirps. Owls have young still in the nest.

Sept. 3. Much wheat still abroad. Hop-picking becomes general : there is a vast crop.

Sept. 7. First blanched endive. Some wheat standing still. A few wasps. Inyx [wryneck] still appears.

Sept. 10. Hedge-hogs bore holes in the grass-walks to come at the plantain roots, which they eat upwards.[20]

Sept. 12. Sheep die frequently on the common, tho' so wholsome a spot. Ravens flock on the hanger [Hanging beech-wood on the down].

Sept. 13. Nectarines rot on ye trees. Ravens are continually playing by pairs in the air.

Sept. 15. Black warty water-efts [*Triton cristatus*] with fin tails & yellow bellies are drawn up in the well-bucket.

Sept. 17. Wheat still abroad. The fields are drenched with rains, and almost all the spring corn is abroad. Sheep die.

Sept. 19. First blanched Celeri. Wheat still abroad : oats & barley much grown.

Sept. 20. A few wasps which spoil ye grapes.

Sept. 21. Nectarines all water. Great rain at night.

Sept. 23. The whame, or barrel-fly, Oestrus bovis, still lays it's nits on the horses sides.

Sept. 24. Much wheat still out, & spoiled. Much barley and oats spoiled. Young martins still in their nest.

Sept. 25. A few of these rare birds (rock-ouzels)[21] appeared, just this time twelve months, in orchards about yew-trees. I have not been able yet to procure a cock.

Sept. 26. I saw a small Ichneumon-fly laying it's eggs on, or in the aurelia [pupa] of a papilio.

Sept. 27. People are now housing corn after 27 days interruption.

Sept. 28. These ring-ouzels are seen again in the spring in their return to the north.

Sept. 29. Swallows cluster on the bushes in the barnet.[22] Redstart.

Sept. 30. Stares [Starlings] flock at Chilgrove [Sussex]. Oedicnemus [stone-curlew] does not flock yet.

Oct. 1. Harvest pretty well finished this evening. Some wheat out at Harting [Sussex]. Roads are much dryed.

Oct. 2. Swallows still. Glow-worms shine.

Oct. 4. Grapes are good. The ash and mulberry cast their leaves.

Oct. 5. Rooks carry off the nuts from ye wall-nut trees.

Oct. 12. Lapwings begin to congregate in the uplands. Fields of barley abroad.

Oct. 13. Swallows and martins at Streatley [Berks.]

Oct. 14-22. [At Oxford.]

Oct. 14. Meadows flooded.

Oct. 19. Herrings.

Oct. 21. Swallow.

Oct. 23. Fallows in a sad, wet, weedy condition : scarce any wheat sown.

Oct. 28. Some wheat is now sowing.

Oct. 29. Grapes are very good, but decay apace.

Oct. 30. Fine grey day. Fallows glutted with water, and full of weeds. Wells rise very fast.

Nov. 1. Bucks grunt.

Nov. 3. Bat appears. Hedge-hogs cease to dig the walks.

Nov. 5. Glass rises violently. Planted a plot of cabbages to stand the winter. Wheat is sown.

Nov. 9. Fallows begin to work well. Woodcocks in the high wood.

Nov. 13. The Ground dries much. Wheat continues to be sown. Elms are still in full leaf.

Nov. 19. Many sorts of flies still appear, the musca carnaria, meridiana, tenax, &c.

Nov. 22. The barometer unusually low [$28\frac{1}{10}$] considering there is little wind. This astonishing fall of the glass was remarked all the kingdom over : we had no wind, & not much rain ; only vast swagging rock-like clouds appear'd at a distance.

Nov. 23. The ground in a sad drowned condition. The low fallows can never be sowed.

Nov. 26. Ice, fine day. Soft afternoon. Many gnats appear. A martin seen : it was very brisk, & lively.

Nov. 30. Crysanthemums still in bloom. Crocuss, Jonquils, winter aconite, snow-drops peep out of Ground.

Dec. 1. Vast floods. Vast rain & stormy wind all night.

Dec. 2. Thunder and hail. Incredible quantities of rain have fallen this week.

Dec. 7. Lavants rise very fast at Farindon.[23] and Chawton [near Alton].

Dec. 8. Stock-doves, or wood-pigeons [*Columba oenas*] appear in great flocks : they are winter-birds of passage, never breeding in these parts.

Dec. 9. Wells run over at the bottom of the village.

Dec. 10. Paths get firm and dry. Rooks frequent their nest-trees. People sow wheat again briskly.

Dec. 11. The first great frost: the ground carries horse and man. Wheat comes up well.

Dec. 12. Moles work. The ground was very white in the morning. Ice bears.

Dec. 13. Wood-pigeons appear in flocks. Ground very hard.

Dec. 14. Milk freezes within. Some snow all day.

Dec. 15. Still, but very sharp air. Immundi meminere sues jactare maniplos.[24] The thermometer which was at 27 in the dining room, rose to 44½ in the wine vaults.

Dec. 19. Wren sings briskly. Smoke beats down.

Dec. 20. Rain & wind all night. Toad appears crawling.

Dec. 21. Rooks feed earnestly in ye stubbles. Red-breast sings.

Dec. 22. French-beans are planted in the hot-house at Hartley. Pines are still cutting.

Dec. 24. Gnats appear much, & some flies. A dry, mild season.

Dec. 25. Wheat comes up well. Lavants seem to abate.

Dec. 27. Weather more like April than ye end of December. Hedge-sparrow sings.

Dec. 31. A wet season began about the 9th of June, which lasted thro' haymaking, harvest, & seed-time, & did infinite mischief to the country. It appears from my Brother[25] Barker's instrument, with which he measures the quantity of rain, that more water fell in the county of Rutland in the year 1768 from Jan : the 1 to Decr 31 than in any other Calendar year for 30 years paſt; viz. 30$\frac{9}{10}$ in. Yet he has known more rain fall with 12 months: from Feb: 1: 1763 to Jan: 31, 1764 there fell 32$\frac{1}{8}$ in. A mean year's rain in Rutland is about 20$\frac{3}{4}$ in.

CHAPTER II

"To every thing there is a season, and a time to every purpose under the heaven." ECCLES. iii., 1.

1769

Jan. 1. Nuthatch chatters. It chatters as it flies.

Jan. 6. Hen chaffinches flock.

Jan. 7. The ground is much dryed : people plow comfortably. Wheat comes up well.

[*Jan.* 9-27. Fyfield.]

Jan. 9. The bunting,[1] emberiza alba, appears in great flocks about Bradley [Hants]. Linnets congregate in vast flocks, & make a kind of singing as they sit on trees. Rooks resort to their nest-trees. Hepaticas, winter-aconite, wall-flowers, daiseys, polyanths, black hellebores blow. Wheat looks well on ye downs.

Jan. 15. Foul, stormy day. Mezereon & groundsel blow.

Jan. 17. Wood-lark whistles. Hogs carry straw.

Jan. 18. The sheep on the downs are very ragged & their coats much torn : the shepherds say they tear their fleeces with their own mouths, & horns : & that they are always in that way in mild wet winters, being teized & tickled with a kind of lice.

Jan. 22. Ice in the roads bears horse & man. Vast halo round the moon. The landsprings in part of N. Tidworth [Hants] street not fordable : they run like a vast river.

Jan. 25. Soft day. Bunting sings. A snipe appears on the high downs among the wheat. Royston crow.[2] Skylark sings.

Jan. 31. Sowed the meadows with ashes.

Feb. 4. Fog, rain, sun, grey. Hedge sparrows sing vehemently.

Feb. 7.   Helleborus viridis, planted in my orchard from the stony-lane, begins to spring.   It rises from the earth with it's flower-buds formed : & differs from Helleborus foetidus, that it dies down to the ground

SETTERWORT, or STINKING HELLEBORE
(Helleborus foetidus.)
(after Step).

in the autumn, while that maintains a large handsome plant all the winter.

Feb. 12.   Snow, fog, sleet.   Icicles.   Snow on the hills.

Feb. 16.   Daws to the churches.

Feb. 17. Pease are sown in the fields. Land-springs abate.

Feb. 18. The missel-bird, turdus viscivorus (called by the country people the ſtorm-cock) sings. Some flies appear in the windows. Gnats abound.

Feb. 20. Bees gather on the Crocuss.

Feb. 26. Vaſt rain in the night. Vaſt aurora borealis.

Feb. 28. Raven sits.

March 1. Sheep rot in a moſt terrible manner in the low grounds.

March 2. Wheat on the clays looks sadly poor & thin. Stormy wind by fits.

March 4. Spring day. Young chickens. Crocuss makes a gallant shew.

March 6. The cock swan at two year's old. " Between his white wings mantling proudly, rows."[3]

March 7. Green woodpecker begins to laugh. Laſt night I heard that short quick note of birds flying in the dark : if this should be the voice of Oedicnemus [ſtone curlew], as is supposed : then that bird, which is not seen in the dead of winter, is returned. Blood-worms appear in the water : they are gnats in one ſtate.[4]

March 10. Oats are sown. Crows build : rooks build. Ewes & lambs are turned into the wheat to eat it down.

March 11. Made the bearing cucumber-bed for four lights with seven loads of dung. The bed was much wetted in making by the snow.

March 12. Golden-creſted wren, regulus criſtatus, sings. His voice is as minute as his body.

March 15. Made cucumber-bed over again ; & added many barrows of fresh dung : it was so drenched with snow & rain that it would not heat. Great hail-ſtorm.

March 18. Planted out the cucum^rs in the two-light frames : the plants are ſtout, but pretty long. Several fruit have bloom in the firſt bed.

March 20. Young cucumber swells. The great bed heats well.

March 21. Goose sits ; while the gander with vast assiduity keeps guard ; & takes the fiercest sow by the ear & leads her away crying.

March 23. Regulus non cristatus [chiffchaff] mini-mus. This bird appears the first of any of the summer-birds of passage, the jynx, or wryneck sometimes excepted. It has only two harsh shrill notes. Fine season for the husbandman.

March 25. Frogs croak : spawn abounds.

[*March* 28—*April* 7. Oxfordshire.]

March 28. Oedicnemus appears & whistles.

March 31. [Witney] Small flights of snow.

Apr. 5. [Oxford] Anemone pulsatilla budds. This plant, the pasque flower, which is just emerging and budding for bloom, abounds on the sheep-down just above Streatley in Berks.

Apr. 9. Atricapilla. The black-cap is usually the second bird of passage that appears. Some snow under the hedges.

Apr. 13. Regulus non cristatus medius [willow wren]. The second-sized willow-wren has a plain-tive, but pleasing note, widely different from that of the first [chiffchaff] which is harsh and sharp. Merula torquata. The ring-ouzels appear again on Noar-hill[5] in their return to the northward : they make but a few days stay in their spring visit ; but rest with us near a fortnight as they go to the Southward at Mich-aelmas.[6]

Apr. 18. Oedicnemus sings late at night.

Apr. 19. Regulus non cristatus voce stridulâ locu-stae [wood wren]. This is the largest of the three willow wrens : it haunts the tops of tall trees making a shivering noise & shaking its wings. It's colours are more vivid than those of the other two species.

Apr. 21. Hirundo apus ! ! ! [The swift]. Black-cap has a most sweet and mellow note. The red-starts frequent orchards & gardens : the white throats are scattered all over the fields far from neighbourhoods. Their notes are mean & much a like; short & without much variety. The white-throat is a most common bird. Young thrushes. The large species of bats appears. Nightingales abound.

Apr. 26-*May* 13. [London.]

Apr. 26. Herrings lately abound, & are the usual forerunners of mackrels.

Apr. 27. Dutch plaise abound. Turbots.

Apr. 28. Some mackrels.

Apr. 30. Fresh ling. Hollibut. Inyx.⁷

May 1. Received from Selborne 12 brace of fair cucumbers.

May 2. Cabbages begin to turn in. Prawns plenty.

May 3. Mackrels cryed in the streets. Asparagus falls to 4ˢ pr hundred. Apricots, small green.

May 4. Crayfish in high season. Smelts in season.

May 7. Caprimulgus !

May 8. Green goose-berries. Lapwing's eggs at the poulterers.

May 9. Green geese are driven along the streets in great droves.

May 14. One shower only for a full month.

May 15. The ground dryed-up in a very extraordinary manner. Much barley lying in the dust without vegetating. Apple-trees well blown. Grass very short.

May 21. Musca meridiana. This fly smells strongly of musk. White owls have young.

May 22. Flesh-flies buz about the room. Melon-fruit begins to blow.

May 23. Sultry. Thunder at a distance. Mole-cricket churs. Not one chaffer appears yet.

May 24. Thunder & rain in the night. Fat sheep are shorn. Young misslethrushes.

May 26. Fern-owl [goatsucker] chatters in yᵉ hanger.

May 27.   If the bough of a vine be cut late in the
spring just before the shoots push out, it will bleed
miserably ; but after the leaf is out any part may be
taken off without the least inconvenience.   So oaks
may be barked while the leaf is budding ; but as soon
as they are expanded the bark will no longer part from
the wood ; because the sap, that lubricates the bark
& makes it part, is evaporated off thro' the leaves.

June 3.   Saw the planet Venus enter the disk of
the sun.[8]   Just as the sun was setting the spot was
very visible to the naked eye.   Nightingale sings ;
wood-owl hoots ; fern-owl chatters.

June 4.   Bees swarm.   Turtle-dove cooes.

June 10.   Young   hedge-hogs.   Wood-straw-
berries.

June 11.   Great species of bat appears ; it flies
very high.   The fern-owl begins chattering just at
three quarters after 8 o'the clock at night.

June 15.   The bank-martin [sand martin] brings
out its young : they were so helpless that we took
one as it sate on a rail.   Young swallows appear.

June 16.   The less reed-sparrow, passer arundinaceus minor Raii,[9] sings sweetly, imitating the notes
of several birds : it haunts near waters, & sings all
night long.   Cold weather : nothing grows well.
S^t foin wants to be cut.   A distinct lunar rain-bow.

June 21.   Vast rain, cold wind.   Quite a winter's
day.

June 22.   Swallows begin to feed their young
flying.

June 23.   Thistles begin to blow.   Young wheat-
ears, birds so called.[10]

[*June* 25-*July* 10.   Newton, near Selborne.]

July 1.   Fine haymaking : hay-carting.   Young
hedge-hogs are frequently found, four or five in a
litter.   At five or six days old their spines (which
are then white) grow stiff enough to wound any
body's hands.   They, I see, are born blind, like
puppies ; have small external ears ; & can in part

draw their skins down over their faces : but are not
able to contract themselves into a ball, as they do for
defence when well-grown.

July 2.   Ricked my S^t foin in curious order : there
were five small loads without a drop of rain.

July 6.   Finished my hay-rick consisting of about
seven tons without one drop of rain.

July 11.   [Whitchurch, Hants.]   Butomus um-
bellatus [flowering rush].   The stint, cinclus, Aldro :
appears about the banks of the Thames.   At Oxford
it is called the summer snipe.

July 13.   [Oxford]   Vast flocks of young wag-
tails on the banks of the charwel.

[*July* 15-*July* 27.   Newton.]

July 16.   Great showers in sight to the E. & N.E.
The ground is very much burnt up, no rain having
fallen, very small showers excepted, since June 27.

July 18.   Moor-buzzard,[11] milvus aeruginosus, has
young.   It builds in low shrubs on wild heaths.
Five young.

July 27.   Some grapes are got pretty large.   Finished cutting the small hedges.

July 28.   The showers do not at all moisten the
ground, which remains as hard as iron.   No savoys,
endives, &c. can be planted-out.

Aug. 2.   Male-ants flie away & leave y^{ir} nests.

Aug. 7.   Showers to the S.   Wheat, rye, oats,
barley cutting round the forest.

Aug. 9.   Wheat begins to be cut at Selborne.
Swifts appear to be gone.   Swallows congregate in
trees with their young & whistle much.   Young
martins begin to congregate on ye wallnut trees.
Nuthatch chirps much.   One swift appears.   Caprimulgus [Goatsucker] chatters.

Aug. 10.   Mr Sheffield[12] of Worcester Coll : went
into Wolmer-forest & procured me a green sand-
piper, Tringa Aldrov ; Tringa ochropus Lin : They
were in pairs & had been seen about by many people
on the streams, & banks of the ponds.

Aug. 18.   Martins congregate on the roofs of houses.

Aug. 20.   Bulls begin to make their shrill autumnal note.

Aug. 21.   Vast showers about.   People here housed all day.   Vine-leaves begin to turn purple.

Aug. 25.   Great showers about.   Male & female ants migrate at a great rate filling the ground and air.

Aug. 28.   Much wheat abroad in this parish. Plums and pears crack with the rain.

Sept. 2.   Wheat harvest finished.   A comet,[13] having a tail about six degrees in length, appears nightly in the constellation of Aries, between the 24, 29, & 51 stars of that constellation in the English catalogue.

Sept. 3.   Winged ants migrate from their nests. Tame buzzard eats the winged ants.   Swallows congregate in vast flocks.

Sept. 4.   Hop-picking begins.   A very slender crop.

Sept. 6.   White sweet-water grapes begin to get clear ; & the wasps to eat them.

Sept. 10.   Land rail.

[*Sept.* 13-*Sept.* 29.   Ringmer, Sussex.]

Sept. 14.   Papilio Machaon [swallow-tail butterfly] is found here in May.

Sept. 17.   Gryllus gryllotalpa [mole-cricket] works. Rooks frequent their nest-trees & repair their nests.

Sept. 18.   Bustards on the downs.

Sept. 26.   Sweet day.   The sheep about Lewes are all without Horns : & have black faces & legs. Sheep have horns & white faces again west of Bramber.

Sept. 29.   Swallows and martins all the way on the downs.

Sept. 30.   [Selborne]   The ring-ouzels, merulae torquatae, are most punctual in their migration, & appear again in a considerable flock.

Oct. 11.   Grapes begin to be very good.   Ground white & dirt a little crisped.

Oct. 13. The dry fit has lasted a fortnight, sprinklings excepted.

Oct. 15. Hedge sparrow whistles. Sprinkling rain. Three martins appear, & settle under the eves of the stable.

Oct. 17. One martin appears.

Oct. 19. Large flock of goldfinches. The sun is very hot. The air is full of spider's webs.

Oct. 20. Linnets, chaffinches, yellow-hammers congregate. Skylark sings sweetly. Glowworms appear.

Oct. 21. Merulae torquatae still about: they abound more, & stay longer than in former autumns. Oedicnemus clamours very loudly. Leaves fall apace. Barometer falls apace.

Oct. 25. A vivid aurora borealis, which like a broad belt stretched across the welkin from East to West. This extraordinary phenomenon was seen the same evening at Gibraltar.

Oct. 27. The weather has been dry just a month this day, one wet day excepted. The fields are so dry that farmers decline sowing. The lapwing, vanellus, congregates in great flocks on the downs, & uplands.

Oct. 28. M^rs J: W: sailed.[14]

Oct. 29. North lights every evening. Six martins appeared flying under ye hanger. Thunder and lightning with vast rain.

Oct. 31. Swan's egg-pears in high perfection. This was the last time the Oedicnemus was heard for this year.

Nov. 2. Golden-crowned wren on the tops of trees.

Nov. 3. Five or six swallows appear.[15]

Nov. 4. Vast storm that broke some boughs, & tore thatch.

Nov. 5. Grass grows. Ricks much torn at Faringdon.

Nov. 8. Goldfinch & Red-breast sing.

Nov. 10.   A few stock-doves, Oenas, sive vinago, or wood-pigeons appear.

Nov. 12.   Glass sinks very fast.   Sheep feed in the night.

Nov. 13.   The hedge-sparrow makes its winter-piping.

Nov. 18.   The ground as hard as a stone.

Nov. 19.   Bearing ice.

Nov. 27.   Green-finches in a vast flock : they seem to feed on the seeds of echium vulgare [viper's bugloss].

Dec. 4.   White water-wagtail appears.[16]

Dec. 11.   Damp moist air.   Large broods of long-tailed titmice.

Dec. 12.   Shell-less snails crawl forth.   Red-breast, hedge-sparrow, wren sing.   Worms come out on the turf by night.   Great rain.

Dec. 14.   Night phalaenae.[17]   Flood.

Dec. 18.   Roses bud in hothouses ; french beans thrive : Ananas carry some late fruit.

Dec. 19.   Wood-pigeons abound in the fields. Some lambs fall.

Dec. 22.   Thunder, lightening & hail before day-break.   Hen chaffinches congregate.

Dec. 27.   Here & there a lamb.

Dec. 30.   Papilio Io [peacock butterfly] appears within doors, & is very brisk.

CHAPTER III

"He is richly endowed who is cheaply amused." (Tuscan Proverb.)

1770

[On flyleaf] A proper antiseptic substance for the preservation of birds, &c.

Black pepper, & ginger, each one Ounce :
Camphire, cloves, each half an Oz :
Allom, one drachm :
Nitre, common salt, each half a drachm :

The intestines of the animal are to be carefully drawn out ; & the abdomen is to be filled with this antiseptic mixture ; or a layer of it, & a thin layer of tow alternately : then let the incision of the abdomen be sewed-up carefully. A little of the seasoning may also be thrust down the throat with a quill or skewer.

A motto for my brother John's Naturalist's Journal kept at Gibraltar.[1]

"Certè si aliquis Naturae consultus *in maxime australi Hispaniâ* in aves observaret, quando accedant aut recedant austrum & septentrionem versus, notatis scilicet diebus mensis & speciebus ; res haec adeo obscura brevi maximè illustraretur."

'Amoeni : Academ. [Linn] vol. iv.'

My brother John's birds are preserved with salt, allom, & pepper.

Jan. 1-5. [Bradley.]

Jan. 2. Grey, small rain. Storm-cock, turdus viscivorus, sings.

Jan. 8. [Selborne] Frost begins to come in a door. The thermometer abroad sunk to 25 ; & in the wine-vault rose to 44.

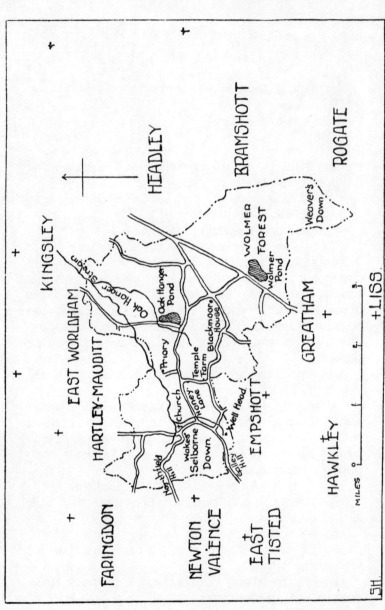

MAP OF SELBORNE PARISH, with the position of Surrounding Parishes referred to in the *Journal*.

Jan. 9. Cocks crow much. The sky promises for fall.

Jan. 18. Vast aurora: a red fiery broad belt from E. to W.

Jan. 21. Peziza acetabulum.[2]

Jan. 23. Scarabaeus stercorarius.[3]
Saw a bird which I suspected to be an Aberdevine,[4] or siskin : it was the passer torquatus, or reed sparrow. Woodlark, great titmouse, chaffinch sing. Blackbird whistles. Woodlark sings in the air before daybreak. Thrush sings. Missel-bird sings.

Jan. 25. Chaffinches in vast flocks : mostly hens : some bramblings among them.

Jan. 27. Papilio urticae ,[5] Partial fog.

Jan. 28. White wagtails sing a sort of song. Paths are steady.

Jan. 29. Corylus avellana. masc : fem. [hazel].[6]

Feb. 5. Daphne mezereon. Turn-pikes are dusty.

Feb. 6-12. [Fyfield.]   *Feb.* 6. Crocus vernus. Hedge-sparrow, curruca, sings. Vast halo round the moon.

Feb. 7. Most vehement wind, with snow ! ! ! Wind blows off tiles & thatch.

Feb. 8. Bunting, emberiza alba [corn bunting] in small flocks.

Feb. 11. Linnets whistle inwardly as they sit in flocks.

Feb. 12. Yellow-hammer, emberiza flava, sings. Bee gathers on the snow-drops. Buntings sing.

Feb. 13-14. [Charlton, in Wilts.]

Feb. 13. Saw bustards on Salisbury plain : they resemble fallow-deer at a distance.[7] Partridges pair. Wild-geese in the winter do damage to the green wheat on Salisbury plain.

Feb. 15-16. [Fyfield.]

Feb. 16. Grey crows[8] are not seen until we come to about Andover from the eastward. As you go thence westward into Wilts they abound. Buntings abound in this part of Wilts.

Feb. 17. [Selborne]   Storm, rain, sleet.

Feb. 23.   Blue mist.   Vulg. called London smoke.
Quae : Does this meteorous appearance shew itself
on the N :E. side of London when the wind is N:E ?
If that is the case then that mist cannot proceed from
the smoke of the metropolis.   This mist has a strong
smell, & is supposed to occasion blights.   When
such mists appear they are usually followed by dry
weather.   They have somewhat the smell of coal-
smoke & therefore are supposed to come from London
as they always come to us with a N:E: wind.

Mar. 1.   Calculus aegagropila was found in the
stomach of a fat ox.   It was black, shining, & round,
& about the size of a large Sevil-orange.   See Syst:
Nat: vol: 4: p: 176 n: 5⁹ Pheasant crows.

Mar. 4.   Chrysoplenium oppositifolium [Oppos-
ite-leaved golden Saxifrage].   Rooks seem to have
finished new nests.   Crocuss make a gay appearance.

Mar. 6.   Taxus baccata [yew] Marsh titmouse,
parus palustris, chirps.   This species is not so common as the great ox-eye or the blue nun.   It frequents hedges & bushes.

Mar. 12.   The golden-crowned wren, regulus
cristatus, sings.   His voice is as minute as his body.

Mar. 16.   Tussilago farfara [Coltsfoot].   Thick
ice.   Ground as hard as a stone.

Mar. 18.   Milk frozen in the pantry.   Vast rock-
like clouds in the horizon.

Mar. 19-21.   [Bradley.]

Mar. 19.   Viper, Coluber Berus, appears.

Mar. 20.   Swan-goose, anser cygneus guineensis,
sits.   The peacock, pavo, asserts his gallantry when
the hens appear :

. . .   whose gay train
Adorns him color'd with the florid hue
Of rainbows, & starry eyes,¹⁰—Milton.

Mar. 22.   [Selborne]   Ice very thick : ground
growing dusty.   Blossom-buds of the pear-trees
seem to be injured by the frost.

Mar. 23. Thermometer abroad sunk to 29. Plows are frozen out. Great Northern aurora.

Mar. 26. Sowed carrots, parsneps, onions, coss-lettuce, leeks, lark-spurs.

Mar. 27. Planted potatoes, five rows. Flights of snow. Red-wings congregate on trees & whistle inwardly. In their breeding-country they are good songsters : See Fauna Suecica.

Mar. 29. Dirt bears horse and man. Boys slide on the ice.

Mar. 30. Papilio rhamni [brimstone butterfly] sucks the bloom of ye primrose. Polyanths coddled with ye frost.

Mar. 31. Turkey & duck lay. Goose sits.

Apr. 3. Snipe pipes in the moors. Bat appears.

Apr. 5. Mercuralis perennis [dog's mercury] Oxalis acetosella [wood sorrel]. Sour, cold day. Great storms about.

Apr. 7. Cut the first cucumber : full old. Snow covers the ground.

Apr. 9. No birds sing, & no insects appear during this wintry, sharp season.

Apr. 11. Kite sits. Raven has young.[11] Swallow amidst frost & snow.

Apr. 12. Rooks have young. Snow melts away very fast. Peaches & nectarines are in full bloom. Apricot bloom seems to be cut off. Barley begins to be sown.

Apr. 14. Butterflies abound. Flies swarm. Grass lamb. Cut a brace more of large cucumbers. Ants appear. Primula veris [Cowslip]. Draba verna [whitlow grass].

Apr. 16. Green wood-pecker laughs at all the world. Storm-cock sings.

Apr. 17-18. [Whitchurch.]

Apr. 17. Aberdavines in Oxfordshire. These were passeres torquati, or reed-sparrows.[12]

Apr. 19-21. [Oxford.]

Apr. 21. Vast storm that did much damage.
[Foot-note, evidently added after return to Selborne.]
In my absence the ring-ouzels made their regular
spring visit en passant, but they seemed to be few
that passed this way.

Apr. 22. [Witney] Snow, stormy.

Apr. 29. Two swifts.

Apr. 30. Titlark sings : frogs migrate.

May 2. [Oxford] Cuculus [cuckoo] Swallows
abound. Great snowstorms.

May 4. Nightingales abound.

May 5. [Selborne] Clouds, great rain.

May 7. Grass-hopper lark.

May 9. Nightingales in my outlet [grounds,
prospect]. A brace of green sandpipers at James
Knight's ponds. Tringa Aldrov : tringa ochropus
Lin :

May 13. Fly-catcher, Stoparola, of Brit: zool:
appears.   Sedge-bird, Passer arundinaceus minor,
Sedge-bird of Brit: zool: sings.

May 15. Chafers begin to abound. Grass-hopper
lark chirps.

May 16. Mole-cricket churs.

May 17. No redstarts whistle yet about the village.

May 19. Black-cap sings sweetly, but rather inwardly : it is a songster of the first rate. It's notes
are deep & sweet. Called in Norfolk the mock
nightingale.

May 20. Rooks have carry'd off their young from
the nest-trees.

May 21. No flesh-flies yet. Cartway runs.

May 26. Caprimulgus [goatsucker, nightjar] su-
surrat. Chafers have not prevailed for some years
as now—they seldom abound oftener than once in
three or four years. When they swarm so, they
deface the trees & hedges.

May 27. Cold & black. Harsh, hazy day.

May 31. Backward apples begin to blow. The
chafers seem much incommoded by the cold weather.

June 1. St. foin is large, & thick, & lodged by the rain.

June 2. Many sorts of dragon-flies appear for the firśt time. Swifts devour the small dragon-flies as they firśt take their flight from out their aurelias [pupae], which are lodged on the weeds of ponds. Chafers are eaten by the turkey, the rook, & the house-sparrow.

June 3. Chafers much suppressed by the cold & rain.

June 4. Fleas [pulex irritans] abound on the śteep sand-banks where the bank-martins build.[13]

June 6. Chafers abound. Sanicula europea [Wood sanicle] in flower.

June 7. Polygala vulg. [common milkwort] in flower. Mole-cricket churs.

June 11-23. [London] Only one entry.

June 11. Hinds on Bagshot-heath.

[Selborne.]

June 23. Wheat is very backward : hardly any ears appear. It is worthy of notice that on my clayey soil horses prefer the grass that grows on a sand-walk, tho' shaded & dripped by a tall hedge, to that which springs from the natural ground in a sunny & open situation.

June 28. Trufles begin to be found. Chafers śtill appear.

June 29. A pound of trufles were found by a trufle-hunter in my Brother's grove.

June 30. Farmers do not care to persiśt in cutting their St. foin. The thermometer fluctuates between 29 & 29 & ½. The Rooks pursue & catch the chafers as they flie, whole woods of oaks are śtripped bare by the chafers.

[*July* 1 to 7. Fyfield.]

July 1. Cuckow sings. Quail calls. Wheat begins to blow.

July 3. Red pinks begin to blow. Blackcap sings sweetly. Titlark sings, & black bird.

July 4. Sultry. Thunder-like clouds rising on all sides. Heavy rain. Roses blow but poorly. Large titmouse makes his spring note.

July 5. Sultry. Showers at a distance. The therm.ʳ 73 abroad in the shade.

July 6. Phallus impudicus olet.¹⁴ Young daws come forth. Cut my St. foin: a vast crop. Vast showers about.

[Selborne.]

July 7. Orchis conopsea [fragrant orchis] abounds in the dry banks of the corn-fields. Wornils¹⁵ are grown very large in the backs of cows. If they could be watched, so as to be taken when going into the pupa state, perhaps it might be discovered from what insect they are derived. See Derham Physico-theol : . . . wornils.

July 10. Very few apples or pears. Cherries hardly begin to turn. Wood straw berries turn.

July 11. Vast showers about but no rain. Turn'd the St. foin twice, & cocked it in small cock.

July 12. Ricked five loads & ½ of St. foin in good order. Fern owl chatters.

July 13. Cut my great mead, a good crop. Young bank-martins are flyers : this species every year is the first that brings forth it's young. Quer : Do they feed their young flying, or not ?

July 15. Heavy showers. Young frogs migrate from their ponds. Young partridges.

July 17. First young swallows appear. Young Goldfinches. Turned the grass-cocks about the last week in June. Vine begins to blow very late ! in good summers.

July 20. Spread the hay. Stopped & tacked yᵉ vines. Cut the tall hedges.

July 21. Cut first melon. Apis longicornis¹⁶ carries wax on it's thighs into it's hole in the walks : in this wax it deposits it's eggs. Cocked the hay in large cocks. Martins tread in their nest, & flie out one on the back of the other.

July 23. Housed three loads of hay perfectly dry but discolored.

July 24. Swallows begin to feed yir young ones flying.

July 26. Turneps begin to be hoed. Redbreast's note begins to be distinguishable, other birds being more silent.

July 30. Cut my little mead. Vines in bloom. Showers about.

Aug. 1. Hay makes in the afternoon. Cocked ye hay. Martins (young) peep out of their nests. Bulfinches devour all the rasps. Ricked last load of hay in fine order.

Aug. 3. Sweet day. Vast dew. Somewhat of an autumnal temperament seems to take place. Young martins come out. Young swifts seem to be out.

Aug. 5. Hops promise well, & throw out branches at every joint.

Aug. 6. Levant weather: a brisk gale all day that dies away at sunset.

Aug. 7. [Ante-dated July 8] Those maggots that make worm-holes in tables, chairs, bed-posts, &c., & destroy wooden-furniture, especially where there is any sap, are the larvae of the ptinus pectinicornis. This insect, it is probable, deposits its eggs on the surface, & the worms eat their way in. In their holes they turn into their pupa state, & so come forth winged in July: eating their way thro' the valences or curtains of a bed, or any other furniture that happens to obstruct their passage. They seem to be most inclined to breed in beech; hence beech will not make lasting utensils, or furniture. If their eggs are deposited on the surface, frequent rubbings will preserve wooden furniture.

Aug. 8. Housed the grass of shrubbery. Oestrus σκολιουρος, sive curvicauda.[17] This fly which lays it's nits on the hairs of horses flanks & legs, is very busy this sultry weather. See Derham's Physicotheol: 250. This species of Oestrus seems to be

unknown to Linnaeus, & Geoffroy. Oedicnemi
abound these moonshine nights on Selborne down :
they come out most probably for worms. Quer :
Are not hops, which are dioecious plants, subject to
blites, & more frequent failures in their crops from
the great over care that is taken to root out every
male plant ? Male plants will not bear good hops :
but may not some of their farina be necessary towards
rendering the female productions more perfect ?

Aug. 12. Lapwings flie in parties to the downs
as it grows dusk.

Aug. 13. Swifts to be partly gone. Martins congregate.

Aug. 14. Pease begin to be hacked. Saw two
swifts.

Aug. 16. Nuthatch chirps much.

Aug. 19. Ponds begin to fail. Hops are perfectly free from lice.

Aug. 21. Sowed spinnage, & lettuces to stand ye
winter.

Aug. 25. Wheat begins to be housed. Trenched
out celeri.

Aug. 26. Young swallows & martins congregate
in prodigious swarms.

Aug. 27. Sweet harvest-weather. Wheat in
general is light. Hops grow very fast : a vast crop.

Aug. 28. Delicate harvest weather. Many loads
of wheat housed. Great bat appears ; flies strongly
and vigorously & very high. I call this rare species
vespertilio altivolans.[18]

Aug. 31. Hop-picking begins. Plants in the garden suffer from want of moisture. Great N. Aurora
considering the bright moon.

Sept. 1. Not one wasp appears notwithstanding
the long dry season. Cuckows skim over the ponds at
Oakhanger, & catch libellulae [dragon-flies] on the
weeds, & as they flie in the air. I can give no credit
to the motion that they are birds of prey. They have
a weak bill & no talons.

Sept. 4. The ring-ousel appears again in it's autumnal visit; but about twenty days earlier than usual.

[*Sept.* 10 to 14. Farnham Castle.]

Sept. 10. The hop-picking at Farnham is just beginning. About 8,000 people besides natives are employed. A vast crop. Much wall fruit at Farnham castle; but void of flower.

Sept. 13. Fly-catcher & white throat appear.

Sept. 14. [Selborne] Several fields of wheat unhoused.

Sept. 15. Tyed-up a large plot of well-grown endive. Young swallows come out. Peaches, but not well-flavoured.

Sept. 18-22. [Fyfield.]

Sept. 18. Heavy showers after 'tis dark.

Sept. 19. Stormy all night. Aequinoctial weather. Wheat begins to be sown.

Sept. 23. [Selborne.]

Sept. 25. Barley grows in the swarth. Thunder, lightening, & hail.

Sept. 26. Annuals are spoiled in the gardens.

Sept. 27. Gardens are torn to pieces, & great boughs off trees.

Oct. 2-20. [Chilgrove and Ringmer.]

Oct. 2. Ring-ouzel on Harting Hills.

Oct. 3. Ring-ouzels again on the downs eastward.

Oct. 4. Ring-ouzels near Ringmer. Swallows abound.

Oct. 5. [Ringmer] Crossbills, loxiae curvirostrae among Mrs Snooke's [19] Scotch pines.

Oct. 6. Harvest not finished. Not one wasp or hornet.

Oct. 9. Fog on the hills.

Oct. 10. Several very young nestling swallows with square tails. Oestrus curvicauda still appears. Apples gathering. Grapes begin to be eatable.

Oct. 18. Cornix cinerea. Swallows. Some Martins at Findon. Vast floods on the Sussex

rivers : the meadows all under water. Vast flood at
Houghton.[20] Martins, crossbeaks. The Sussex-
rivers are very liable to floods, which occasion great
loss & inconvenience to the Farmers. The cattle
from this time must be taken into the yards to live
on straw, because the meads, which would have
maintained them many weeks longer, are all under
water. The standing grass is often flooded in summer.
They call their meads by the rivers-sides, brooks.

Oct. 20. [Selborne] Turdus iliacus [redwing].
Rain all night.

Oct. 25. A young swallow appears.

Oct. 27. Ice. Cobwebs float in the air & cover
the ground.

Oct. 29. Trees carry their leaves well for the
season.

Oct. 30. Rooks & jays carry away the acorns
from the oaks.

Oct. 31. Flights of skie-larks go westward.

Nov. 2. Wallnut, & ash leaves fall at a vast rate.

Nov. 3. Misling rain all day.

Nov. 8. Heavy rain for 24 hours. Vast flood at
Gracious street, & dorton.[21]

Nov. 9. Lime-trees leaves fall all at once. Floods :
torrents & cataracts in the lanes.

Nov. 14. Bee on the asters.

Nov. 15. Vast rain at night. The ground so
wet that no sowing goes forward. Much ground
unsown.

Nov. 20. Hard frost. Severe wind. Plows are
frozen out.

Nov. 24. The wild wood-pigeon, or stock-dove
begins to appear. They leave us all to a bird in the
spring, & do not breed in these parts ; perhaps not
in this island. If they are birds of passage, they are
the last winter bird of passage that appears. The
numbers that come to these parts are strangely diminished within these twenty years. For about that
distance of time such multitudes used to be observed,

as they went to & from roost, that they filled the air
for a mile together : but now seldom more than
40 or 50 are to be anywhere seen. They feed on
acorns, beech mast, & turneps. They are much
smaller than the ring-dove, which stays with us all
the year.[22]

Nov. 25. Linnets flock in prodigious numbers.

Nov. 28. The planet Mercury appears above the
sun.

Dec. 1. Some oaks have yet some green leaves.
Those oaks that were eaten bare by the chafers
leafed about midsum[r] & continued unusually green
late into Novem[r].

Dec. 2. The earth in a sad wet condition. Wells
strangely risen : one well runs over : our wells are
about ten fathoms deep.

Dec. 4. Most owls seem to hoot exactly in B flat
according to several pitch-pipes used in tuning of
harpsichords, & sold as strictly at concert pitch.[23]

Dec. 8. Wild fowl abounds in the ponds on Woolmere forest : they lie in the great waters by day &
feed in the streams & plashes by night.

Dec. 9. Hail in the night. Frost almost constantly succeeds hail.

Dec. 17. Young lambs begin to fall on the sands
round the forest.

Dec. 19. Tempestuous wind all night. This
storm did great mischief at sea, especially among the
colliers.

Dec. 21. Musca tenax does not die as winter
comes on, but lays itself up.

Dec. 22. Trenched up the quarters of the garden
for the winter.

Dec. 23. Linnets flock & haunt the oat stubbles &
pease-fields.

Dec. 28. The lavants, or land-springs[24] run very
strong between Faringdon & Chawton.

Dec. 29. Wrens whistle all the winter except in
severe frost.[25] Wrens whistle much more than any

English bird in a wild ſtate. The redbreaſt sings great part of the year ; but at intervals is silent. This year concludes with a very wet season, which has laſted from the middle of OⱯr laſt, & has occasioned vaſt floods, & desolation both at home & abroad. Much wheat-land in wet countries remains unsown.

CHAPTER IV

"And well the tiny things of earth
  Repay the watching eye."
      (E. ELLIOTT : *The Wonders of the Lane.*)

1771

Jan. 3. Wood-lice, onisci aselli, appear all the winter in mild weather : spiders appear all the winter in moist weather; lepismae[1] appear all the winter round hearths & in warm places. Some kinds of gnats appear all the winter in mild weather, as do earth-worms, after it is dark, when there is no frost.

Jan. 9. Frost comes within doors. Thermometer within 28, in the wine vault 43½, abroad 24.

Jan. 11. Small snow on the ground. Water bottles freeze in chambers.

Jan. 17. Paths very dusty. Blue mist still.

Jan. 18. Barometer sinks apace. Dark sun, stars dark.

Jan. 19. Small snow. More snow. By a letter from town it appears that in London in an hard frost Martin's thermometer is just at the same pitch abroad in the area, that it is with me in the dining-room without a fire.

Jan. 24. Sky strangely streaked with blue and red. Wind & rain in the night. Larks rise & essay to sing. Daws begin to come to churches.

Jan. 28. Turneps are very small this year, and are on the decay.

Jan. 30. Hedge-sparrow essays to sing.

Feb. 3. Hens sit. Soft, spring like weather. Rooks resort to their nest-trees.

Feb. 5 to 24. [Fyfield.]

Feb. 5. Warm fog. Grey crows. Creeping mist over the meadows.

Feb. 6. Frost, sun, fog, rain, snow. Bunting twitters.

Feb. 9. A Decanter of water froze in my chamber. Thermom^r abroad 28. Eggs in the ovary of a turkey-pullet about the size of mustard-seeds. Mem : to enquire when the pullets of the same brood will begin to lay.

[Selborne.]

Feb. 28. Blackbird whistles. Helleborus viridis [green hellebore] emerges, & shows it's flower budds.

Footnote to February. Bro^r Henry's field [at Fy-field] opposite his house, was fallowed for barley before the two frosts, all save the headlands : mem : to enquire if the earlier fallowing in that part proved of any advantage.

Mar. 2. Glass sinks steadily tho' ye weather looks like dry. Turneps are all rotten, & the wheat-fields look quite bare, & destitute of all verdure. Farmer parsons sows wheat in his fallow behind Beacher's shop, which was drowned in the winter. Mem : to observe what crop he gets from this spring-sowing. The spring sowing round this village proved the finest wheat & best crop.[2]

Mar. 3. Rain, dark & raw. Considerable rain in the night. Harsh wind.

Mar. 4. Great distress among the flocks : the turneps are all rotten. The ewes have little milk, & the lambs die.

Mar. 8. Wind N.W. Large flocks of wild-geese go over the forest to the Eastward.

Mar. 10. Hard frost, grey, severe wind. The ground thawed much in the middle of the day. Rooks build notwithstanding the severe weather.

Mar. 11. Crocuss at this time used to be in full bloom. Only one or two roots blowed before this frost began. Made the bearing-cucumber bed with 8 cartloads of dung.

Mar. 13. Wild fowls on Woollmere pond. Some large white fowls also: qu: what? They had black heads. Snipes begin to pipe in the moors.

Mar. 13-16. [Bramshot.]

Mar. 16. [Selborne] Crocuss begins to blow & make a show. Upon examination it seems probable that the gulls which I saw were the pewit-gulls, or black caps, the larus ridibundus Linn: They haunt, it seems, inland pools, & sometimes breed on them.[3] See Brit. zööl: vol: 2nd.

Mar. 19. Cucumber-plants thrive & shew the rudiments of bloom & fruit. Farmer Turner sows wheat. Crocuss figure [takes full shape].

Mar. 23. Severe frost, sun, & flights of snow. Cutting wind. Dr. Johnson says "that in 1771 the season was so severe in the Island of Sky, that it is remembered by the name of the *black spring*. The snow, which seldom lies at all, covered the ground for eight weeks, many cattle dyed, & those that survived were so emaciated & dispirited that they did not require the male at the usual season." The case was just the same with us here in the South: never were so many barren cows known as in the spring following that dreadful period. Whole dairies missed being in calf together.

Mar. 26. Thermomr at sunrise down at 17 abroad: at 10 o'clock at night 25 : at sun rise 23½.

Mar. 28. Snow at night. A flock of lapwings haunt about the common.

Mar. 30. Ground hard, & thick ice. Crocuss in full bloom. Birds mute. Farmers feed yir sheep with bran & oates.

Mar. 31. The face of the earth naked to a surprising degree. Wheat hardly to be seen, & no signs of any grass : turneps all gone, & sheep in a starving way. All provisions rising in price. Farmers cannot sow for want of rain.

Apr. 1. Mr Woods,[4] of Chilgrove, had on this day 27 acres of spring-sown wheat not then sprouted

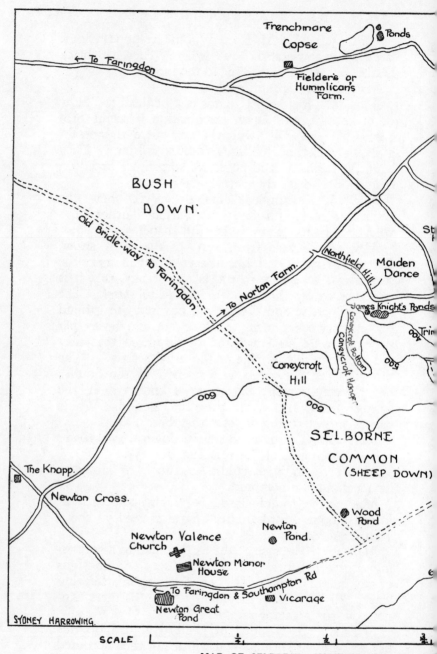

MAP OF SELBORNE, SHOWING THE PLACE-

SYDNEY HARROWING.

SCALE

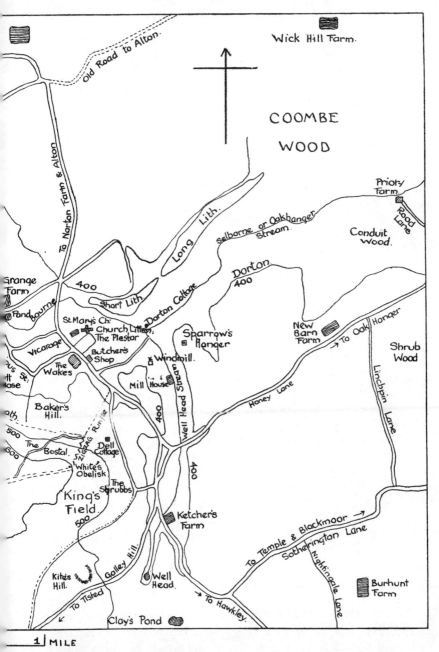

COOMBE WOOD

Wick Hill Farm.

Old Road to Alton.

Priory Farm

Rood Lane

To Norton Farm & Alton

Long Lith.

Selborne or Oakhanger Stream.

Conduit Wood.

Dorton 400

Grange Farm

400

Short Lith.

Dorton Cottage

St Mary's Ch:
Church Litten.
The Plestor

Vicarage

Butcher's Shop

Sparrow's Hanger

New Barn Farm → To Oak Hanger

Shrub Wood

Pond Bourne

The Wakes

Windmill.

Mill House

Well Head Stream

Honey Lane

Linchpin Lane

Baker's Hill.

400

500

The Bostal.

Zig-zag Runte

Dell Cottage

White's Obelisk

The Shrubbs

400

King's Field.

500

Ketcher's Farm

To Temple & Blackmoor →

Sotherington Lane

Nightingale Lane

Burhunt Farm

Kite's Hill.

Galley Hill.

To Tisted

Well Head.

→ To Hawkley.

Clay's Pond

1 MILE

out of the ground : & yet he had a good crop from those fields, no less than 4 quarters on an acre !

Apr. 2. Butterflies appear again. Some flies begin to appear. Spring-like day, sharp in the morning.

Apr. 3. Planted potatoes, & sowed carrots, parsneps, onions, coss-lettuce, leeks.

Apr. 4. Ring-ouzel. Pleasant day, but every thing quite dryed up. No lambs frolic & play as usual. . . . acrior illos Cura domat[5] . . . Virg.

Apr. 7. Began to be confined.[6]

Apr. 9. Wryneck pipes about in orchards. The first spring-bird of passage.

Apr. 11. Regulus non crist : minor [chiffchaff]. The second spring-bird of passage. No rain since the 16th of March : dirty lanes all dryed up.

Apr. 13. The dry weather has lasted just a month this day. Dry weather is always supposed to help the wheat in the clays : but the wheat in general is so poor this year, that it is hardly seen on the ground. It will be worth remarking at harvest how the crop will turn-out.

Apr. 14. Swallow appears as last year amidst frost & snow !

Apr. 15. Hail in the night. Flights of snow at times : harsh biting day. Thermomr abroad at 10 at night, down to 14 degrees ! ! !

Apr. 17. Snow on the ground. No oedicnemus (land curlew) has been heard yet.

Apr. 18. Luscinia [nightingale]. Cut the first cucumber ; not a very fair fruit. Swallow. Colds & coughs universal.

Apr. 19. Hay risen to four pounds per ton. High rumbling wind.

Apr. 20. The dry weather has lasted five weeks this day. Just rain enough to discolour the pavement. Myriads of minute frogs, encouraged by those few drops of rain, migrate from the ponds & pools where

they were hatched. Hence it appears that severe
frost doth not interrupt the hatching & growth of
young frogs.

Apr. 22.   Nightingale.   No more swallows
appear.   Great showers about.

Apr. 24.   Stone-curlew returns & clamours.

Apr. 26.   Wheat begins to mead.   Redstart
whistles.   Cuckow sings this year long before yᵉ
leaf appears.

Apr. 27 to *May* 1.   [Newton.]

Apr. 27.   Farmers feed their ewes with bran &
oats, or white pease !   Some few fat lambs are killed.
Wood-cock, & some field-fares still appear.

Apr. 29.   White throat.   Grass begins to grow.

May 1.   [Selborne] Plenty of rain in the night.
Trees as bare as at Xmass.

May 3.   The turtle-dove returns & cooes.   Sowed
white cucumbers under a hand-glass.

May 4.   Showers, spring weather.   Cucumbers
swell away, & set apace.   Black-caps appear &
begin to sing.   Sowed white dwarf kidney beans.

May 7.   Grass & corn grow very fast.

May 8.   Asparagus begins to sprout.

May 9.   Summer-like weather.   Some beeches
begin to leaf.

May 10.   Fly-catcher.   This bird is usually the
last of the summer birds of passage.

May 11.   Cherry-trees begin to blossom.   The
sedge bird, Passer arundinaceus minor, of the Brit:
zool: sings about waters: variety of notes; but
the manner is hurrying.

May 12.   Ground very moist.   Barley, & all
spring-corn comes up well.   Vast Aurora in the
N.E. and S.W. & all round.

May 13.   Swallows & martins collect dirt for
building. Regulus non crist: major cantat voce
stridulâ locustae [wood wren]. Usually a late bird
of passage.   The horizon looks dark & louring.

May 15. Vines begin to sprout, & shew leaves. Distant thunder, & showers about.

May 16. Hanger in leaf this day: yesterday but a few trees were green. Trees & hedges in general begin to leaf.

May 18. Began to cut grass for the horses. The side-fly[7] on horses. Mole-cricket churs in the moist meadows.

June 6. Ephemera vulgata [angler's may-fly] Meridie choreas aireas [aerial dances] instituit, sursum recte tendens, rediensque eâdem fere viâ: Scopoli. A mole-cricket's nest full of small eggs was discovered just under the turf in the garden near the pond. They were of a dirty yellow colour, & of an oval shape, surrounded with a tough skin, & too small to have any rudiments of young within them, being full of a viscous substance. There might be an hundred eggs in this one nest; they lay very shallow just under a little fresh-moved mould in an hollow formed for that purpose.[8]

June 10-12. [Wintoñ.]

June 10. Small rain in the night. Ephemera caudâ bisetâ. The angler's may-fly. Myriads of may-flies appear for the first time on the Alresford stream. The air was crowded with them, & the surface of the water covered. Large trouts sucked them in as they lay struggling on the surface of the stream, unable to rise till their wings were dryed. This appearance reconciled me in some measure to the wonderful account that Scopoli gives of the quantities emerging from the rivers of Carniola. See his Entomologia.

June 12-22. [Fyfield.]

June 13. Sphinx filipendula.[9] Emerges from it's aurelia state. Fixes it's cods [tips of abdomen] to the dry twigs in hedges; is called in Hants the S^t foin fly; & is in its crawling state said to be very pernicious to that plant.

June 15.   Bar: falls all day.   Wheat-ears peep. St foin begins to be cut.

June 16.   Tempestuous wind & vast rain for 28 hours.

June 17.   Tearing wind which damaged all the gardens.

June 19.   Swifts sits, & comes out of an evening to feed for a few minutes.

June 20.   Sheep are shorn.   St foin cut.

June 21.   St foin housed about Winton.

June 23.   [Selborne] Dark & cold, sun, clouds.

June 24-26.   [Bradley.]

June 24.   Cut my St foin.

June 25.   Rain-bow.   Rock-like clouds.   Sweet evening.   Moonshine.

June 26.   [Selborne]   Phallus impudicus olet. Showers about.

June 29.   Ricked in two summer-cocks five jobbs[10] of St foin in most curious order.   Young martins hatched.   Apples, & pears but few.   Titlark whistles still.

June 30.   Nothing grows in the garden.

July 1.   Cut part of the mead : a good crop.   Young goldfinches.

July 4.   Ricked 6 jobbs of meadow hay in curious[11] order, and added the St foin to it.

July 5.   Cut the slip [strip of pasture] and part of the mead.   Elder in full bloom.

July 6.   Young swallows appear.   Cocked the hay in large cocks.   No kindly, regular dews all the summer ; so that the walks & grass-plots were seldom well mowed.

July 7.   Myriads of frogs, a second brood, migrate from J : Knight's ponds.

July 8.   Ricked the two jobbs of hay, and finish'd my rick in delicate order.

July 10.   Vine begins to blow.   Very late again ! it blowed last year on July 17 : usually blows the last week in June.

July 12.   Vine-bloom smells sweetly.

July 14.   Young martins & swallows begin to congregate.   Young Swifts are fledge[d].

July 15.   Lovely weather for the blowing of wheat.

July 16.   Sultry, sunny day.   Good dew.   Gardens suffer from want of moisture.   Dark clouds round the horizon.

July 17.   Sun sultry, sweet even.   Good dew. Stopp'd the vines.   White cucumbers begin to bear : the green are still barren.   Clouds threaten.

July 19.   Tabanus bovinus [great horse fly]. Trenched out celeri.   Wind tears the hedges & flowers.

July 21.   Considerable rain in the night.   Frogs continue to migrate from the ponds.

July 26.   Turneps fail in many places, & are sown over again.

July 27.   Cucumbers begin to bear again.

July 30.   Sun chilly.   Cold white dew.   Rain.

July 31.   Considerable rain in the night.   Clap of thunder.   Showers about.

Aug. 5.   Young partridgers, strong flyers.   Soft showers.   Swifts.   Pease are hacking.[12]

Aug. 6.   Nuthatch chirps ; is very loquacious at this time of the year.   Large bat appears, vespertilio altivolans.

Aug. 7.   Rye-harvest begins.   Procured the above-mentioned specimen of the bat, a male.

Aug. 8.   Rain in the night, with wind.   Swifts. Sultry & moist :   Cucumbers bear abundantly. Showers about.   Procured a second large bat, a male.

Aug. 10.   Flying ants, male & female.

Aug. 11.   Heavy clouds round the horizon. Lambs play & frolick.

Aug. 16.   Rain, driving rain, dry.   Four swifts still.

Aug. 18.   No dew, rain, rain, rain.   Swans flounce & dive.   Chilly & dark.

Aug. 19.   Swifts abound.   Swallows & martins bring out their second broods which are perchers. Thunder :   wind.

Aug. 22. Bank-martins [sand-martins] bring out their second brood. Swifts. No swifts seen after this day.

Aug. 23. Young swallows & martins come out every day. Still weather. Wheat-harvest becomes pretty general.

Aug. 25. Wheat not ripe at Faringdon. Winter weather. Oats & barley ripe before wheat.

Aug. 26. Nuthatch chirps much. No swifts since 22nd.

Aug. 28. Dark, grey, & soft. People bind their wheat.

Aug. 29. Fog, sun, brisk wind. Sweet day. Wheat begins to be housed.

Aug. 30. Young Stoparolas abound. Swallows congregate in vast flocks. Wheat housed.

Sept. 2. Corn is housed. Swallows feed their young flying.

Sept. 3. Nuthatch chirps flying. Swallows feed their young perchers.

Sept. 4. Hop-picking begins. Hops small. Much wheat not ripe yet.

Sept. 5. Dark. Sun. Wheat pretty well cut-down. Soft & still.

Sept. 8. Blowing & winter-like.

Sept. 9. Missel-thrushes flock.

Sept. 10. Spring sown wheat is cut. Hirundines swarm under the hanger.[13]

Sept. 13. Grapes begin to turn colour. Mild.

Sept. 14. Great rain in the night. Spring sown wheat still standing. Regulus non cristatus minimus chirps.

Sept. 15-19. [Winton.]

Sept. 15. Muscae & papiliones abound on the asters.

Sept. 19. [Bradley] Lapwings congregate on the downs.

Sept. 20. [Selborne] Rain in the night. Spring sown wheat all housed.

Sept. 22. Swallows abound. Tops of the beeches are fringed with yellow. This morning the swallows rendezvoused in a neighbour's wallnut tree. At the dawn of the day they arose altogether in infinite numbers occasioning such a rushing with the strokes of their wings as might be heard to a considerable distance [Note inserted later]. Since that no flock has appeared, only some late broods, & stragglers.[14]

Sept. 23. Sprinkling rain & rumbling wind.

Sept. 24. Hardly any swallows have appeared since sunday.

Sept. 25. Hedge-sparrow begins it's winter note.

Sept. 26. Ring ouzels, merula torquata, begin to appear on their autumnal migration.

Sept. 27. Black cap. Few martins over oak-hanger ponds. Woodlark whistles.

Sept. 29. Woodcock, Scolopax, appears early. Glow worms shine.

Oct. 2. Woodlark whistles. Few swallows. One martin's nest with young in it. Some few martins about.

Oct. 3. Grapes turn black. Vetches, & seed-clover housed. Barom^r sinks very fast. Ring-ouzels. Ring-ouzels affect to perch on the top twigs of tall trees, like field-fares. When they flie off they chatter like black birds. Apples are gathering.

Oct. 5. White frost, grey, & clouds. Ashen leaves begin to fall.

Oct. 9. Several swallows & martins.

Oct. 13. Stormy winds, & gluts of rain. Floods.

Oct. 14. Some swallows. Grapes large & black, but not high-flavoured yet. Several martins.

Oct. 20. Mild, & sun. Sweet day, large halo round the moon.

Oct. 22-26. [Fyfield.]

Oct. 25. White frost, sun, tempest. Vast rain & wind.

Oct. 26-29. [Selborne.]

Oct. 29-*Nov.* 1. [Chilgrove] *Footnote :* Mr Woods saw many redwings about the 31st of October.

Oct. 30. White frost, cloudless. Curlews have cryed here within these few days. Haws fail here.

Nov. 1-14. [Ringmer.]

Nov. 1. An imperfect rainbow on the fog; a more vivid one on the dewy grass. Grey crows, Cornix cinerea frugilega, near South Wick. Mrs Snooke's tortoise[15] begins to scrape an hole in the ground in order for laying up.

Nov. 2. Mrs Snooke's tortoise begins to dig in order to hide himself for the winter. The vale of Bramber, & the river e[n]veloped in a vast fog : the downs were clear.

Nov. 4. Saw three house swallows flying briskly at Newhaven at the mouth of the Lewes river ! ! [River Ouse.]

Nov. 5. Phyteuma orbicularis [round-leaved rampion] in bloom on the downs S. of Lewes.

Nov. 6. Whitings in high season : herrings going out.

Nov. 8. Few petrifactions [fossils] about Ringmer & Lewes. Ringmer soil not clay at top but brick-loam : bears good apples, pears, & grapes. Clay under, which holds water like a dish. The trees are mostly elms.

Nov. 10. Tortoise comes out in the sun about noon, but soon returns to his work of digging a hole to retire into.

Nov. 13. Saw 16 forked-tail kites at once on the downs.[16]

Nov. 14. An epidemic disease among the dogs in Sussex, which proves fatal to many. They pine away, & die moping.

Nov. 15. [Chilgrove] Tortoise at Ringmer had not finish'd his hybernaculum, being interrupted by the sunny weather, which tempted him out.

Nov. 16. [Selborne.]

Nov. 17. A most astonishing, & destructive flood at Newcastle on Tyne.[17]

Nov. 18. Crocuss begins to spring.

Nov. 19. Hen chaffinches begin to congregate.

Nov. 23. Turdus pilaris [fieldfare]. Hardly any field-fares appear : there are no haws.

Nov. 26. September-like weather. Footpaths dry like march.

Nov. 27. A large flock of red-wings, Turdus iliacus, appear.

Nov. 28. Spitting fog, dark & cold. The reed-sparrow, passer torquatus, forsaking the reeds, & water side in the winter, roves about among the fields, & hedges. This bird which I sometimes saw, but never could procure 'til now, I mistook for the aberdavine.

Dec. 1. Hot sun. Cloudless and still. Dark clouds to the S.W. Bats about.

Dec. 2. Cole-mouse [cole titmouse] roosts in the eaves of a thatched house.

Dec. 11. Much rain towards morning. Redwings.

Dec. 14. Glass sinks all day.

Dec. 15. Song thrush sings. Daisey, wallflower, hepatica, mesereon, pot-marigold, spring flower blow. Thrush sings.

Dec. 21. Storm, rain & hail, thunder.

Dec. 24. Many sorts of flies are out & very brisk.

Dec. 26. Thrush and redbreast sing. Bunting, emberiza alba, at Faringdon. I never saw one in the parish of Selborne. They affect a champion [un-enclosed] country, & abound in the downy open parts. Ducks, teals, and wigeons have appeared on Wulmere-pond about three weeks : one pewit-gull, larus cinereus, appears.[18] A pike was taken lately in this pond measuring 3 feet & 3 inch : in length ; & 21 inch : in circumference : in it's belly were 3 considerable carps. When fit for the table it weighed 24 pounds.

Dec. 30. Lambs begin to fall. Harsh wind all day.

CHAPTER V

"The eye is for ever drawn onward, and knows no end."
(RICHARD JEFFERIES : *Wild Flowers*.)

1772

Jan. 2. Froſt, clouds, sprinkling, dark.

Jan. 5. Hedge-sparrow whiſtles : paths get dry. An extraordinary concussion in the air which shook peoples windows, & doors round the neighbourhood. *Footnote.* The concussion felt Jan : 6 [Qy. Jan. 5] was occasioned by the blowing-up of the powder-mills near Hounslow. Incredible damage was done in that neighbourhood.

Jan. 6-10. [Bradley.]

Jan. 9. Snow on the ground, & thick ice.

Jan. 10. [Selborne.]

Jan. 14. Vaſt white dew.

Jan. 16. Severe air. Icicles. Cutting wind & froſt.

Jan. 18. Snow covers the ground. Larks congregate in vaſt flocks.

Jan. 20. Thermom^r abroad at sun rise 11 : in the wine-vault 43. Snow dry & frozen : very deep.

Jan. 21. Snow does not melt. Snipes come up the ſtream.

Jan. 23. A gentle thaw all day : leaves drip all day.

Jan. 27. [Winton] Dark, fog, and thaw.

Jan. 29. [Newton] Thermom^r abroad before sun rise at 11. Bright sun.

Jan. 30. [Selborne] Much snow on the ground.

Feb. 2. Much old snow remaining, & the bare places now covered again. Tom-tit attempts it's spring note.

Feb. 3. In the evening of Feb: the 3rd the sheep were ravenous after their hay: & before bed-time came a great flight of snow with wind. Sheep are desirous of filling their bellies against bad weather: & are by their voraciousness prognostic of that bad weather. They also frolic & gambol about at such seasons.

Feb. 4. Considerable driving snow in the night, which powdered the trees & woods in a most beautiful, and romantic manner. Ground all covered.

Feb. 6. Hard frost, sunshine. Deep snow covers the ground. Beautiful winter-pieces.

Feb. 7. Cole-mouse [cole-tit] picks bones in the yard. The snow has lain on the ground this evening just 21 days; a long period for England!

Feb. 9. Red-breasts and hedge-sparrows whistle. Snow gone, save under hedges. Ravens seem paired.

Feb. 10. Made cucumber bed. Snow gone on the hills. Winter aconite blows.

Feb. 11. Large titmouse sings. Chaffinch sings. Hot sunshine. Snowdrops blow.

Feb. 13. Wood-pecker laughs. Spring-like weather. Skylark mounts & sings. Crocus begins to blow.

Feb. 22. No snow lies on the ground.

Feb. 25-*Mar.* 13. [London.][1] *Footnote* for February. An ash-coloured butcher-bird[2] was shot this winter in Rotherfield Park:[3] lanius seu collurio cinereus major: the only one I ever heard of in these parts.

Mar. 13. [Newton] Flights of snow, severe wind. Water-bottle freezes.

Mar. 14. [Selborne] Most severe flights of snow. Plows are frozen out, dirt carries (*sic*) horse, & man.

Mar. 17. Wild geese appear in a flock, flying to the Southward.

Mar. 22. Least uncrest: wren appears: First summer bird of passage.

Mar. 23. Tussilago farfara [coltsfoot]. Considerable mischief was done by this storm near & in London.

Mar. 26. Planted-out some stout cucum^r plants into the bearing beds : rudiments of fruit show.

Mar. 28. The ground too wet for ploughing.

Mar. 30. Merula torquata [ring-ouzel] on it's spring visit. Thunder. *Footnote.* One cock ringouzel appears on Nore-hill on it's spring visit, but earlier than common.

Apr. 4. Mackril sky, wheel round the sun. Clouds in horison.

Apr. 5. Uncrested wren chirps. Barometer falls apace. Ants appear.

Apr. 6. Wood lark sits. Hirundo domestica! Swallow comes early. Cock snipe pipes & hums in the air. Is the latter sound ventriloquous, or from the rapid motion of the wings? The bird always descends when that noise is made, & the wings are violently agitated.[4]

Apr. 7. Termes pulsatorium raps. Deathwatch vulg :[5]

Apr. 9. Titlark whistles.

Apr. 12. The cuckow is heard in the forest of Bere.[6] Grass grows apace. The great black & white Gull, larus maximus ex albo & nigro seu caeruleo nigricante varius Raii, was shot lately near Chawton [South of Alton] : Larus marinus Linn : The head & part of the neck of this bird is dotted with black small spots.

Apr. 13. Redstart appears and whistles. Swallow. Garden too wet for sowing.

Apr. 14. Began mowing the grass walks.

Apr. 15. Luscinia! Nightingale sings sweetly. Grass grows. Wheat looks well.

Apr. 16. The ground is in a sad wet condition ; & the farmers much behind on their spring sowing. No seeds sown yet in my garden. An high rumbling wind.

Apr. 17. Regulus non cristatus [willow wren] a pretty plaintive note. Chilly air. Ice. Martins appear.

Apr. 18. Snow covers the ground.

Apr. 19. Severe wind. Snow on the ground. Swallows abound.

Apr. 20. Thick ice. No swallows appear.

Apr. 21. The turtle-dove returns. Swallows again.

Apr. 22. The bloom of the fruit-trees on the wall does not seem to be destroyed. Sowed all sorts of garden seeds as carrots, parsneps, &c. Cucumbers swell.

Apr. 24. Martins appear but do not frequent houses. Black cap whistles. Showers about. Swift returns. Planted potatoes, four rows. Sowed box of polyanth-seed from London. Sowed annuals.

Apr. 25. Conops calcitrans [stinging fly]. Grass very forward in the fields. The ring dove cooes, & hangs about on the wing in the air in a toying manner.

Apr. 26. Barley-fields like to be very wet, & lumpy.

Apr. 27. Ground dries, & binds up very hard.

Apr. 28. Drying & cold. Black-caps abound.

Apr. 29. Grass crisp, with white frost. Cut first cucumber, a large fruit. Harsh wind. Sowed annuals.

Apr. 30. White-throat returns & whistles. Golden-crested wren whistles : his note is as minute as his person.

May 1. Some few beeches in the Hanger shew a small tinge of verdure.

May 2. Sand-martins abound at the sand-pit at short-heath.

May 3. Regulus non crist : major [wood wren] : Shaking its wings it makes at intervals a sibilous noise on the tops of the tallest beeches.

May 4. Ground very hard & cloddy ; & wants rain before it can be sown.

May 6. Dark, sun hot & harsh air. Rumbling wind.

May 7. No dews: so that the grass-walks get rough for want of mowing. Gardens suffer from want of rain.

May 8. Fields & gardens suffer by the severe harsh winds. Farmers in stiff ground can sow no barley: not one grain is yet sown on Newton great farm.

May 10. Drought has lasted three weeks this day.

May 11. Showers about.

May 12. The sedge-bird sings: variety of notes, but it's manner is hurrying.

May 13. Musca vomitoria [blue flesh fly]. Mason's morter frozen. Wheat looks yellow. Fruit-trees of all sorts blow much. Chill air.

May 15. The country dry as powder.

May 16. 20 horses with vast labour cannot on moderate ground sow more than three acres of barley in a day, instead of seven or eight. The ground wants endless rolling & dragging. The drought has lasted one month.

May 17. Very little barley above ground.

May 21. [Midhurst & Findon (Sussex)] Fogs, shower. Soaking shower.

May 22. [Brighton, Ringmer]⁷ Tortoise eats. Fly-catcher appears, and builds.

May 23 to *June* 3. [Ringmer] Wryneck pipes. The Ringmer-tortoise came forth from it's hybernaculum on the 6th of April, but did not appear to eat 'til May the 5th: it does not eat but on hot days. As far as I could find it has no perceptible pulse. The mole-cricket seems to chur all night.

May 29. Scarabaeus melolontha [cockchafer] Grasshopper-lark [grasshopper warbler] chirps.

May 30. Tortoise eats all day. In Mrs Snooke's ponds are vast spiders, which dive & conceal themselves on the underside of plants, lying on the water: perhaps aranea aquatica Linn: urinatoria. The swallow seems to be the only bird which washes itself as it flies, by dropping into the water.

June 4.   [Arundel] Rain, dark and windy, driving rain, stormy.

June 5.   [Chilgrove] Windy, showers.

June 6.   [Selborne] Showers, showers, clouds & wind.

June 7.   Field cricket makes its shrilling noise.

June 9.   Apis longicornis.   The long-horned bees bore their nests in the ground where it is trodden the hardest.

June 10.   Brisk wind all day which falls with the sun.

June 12.   St foin blows, & gets very tall.

June 14.   [Bradley] Bright, sweet afternoon.

June 15.   [Fyfield] Carduus nutans [musk thistle] Digitalis purpurea [foxglove].   Sheep shorn.

June 17.   [Charlton in Wilts] Bro^r John set-out on horse-back for Cadiz[8].   Polygonum bistorta [bistort or snake-weed].   Colchicum autumnale. [meadow saffron] in seed.   Ephemerae and phryganeae abound on the stream.

June 18.   [Fyfield] Thomas cut my St foin.

June 19.   [Winton] Vast fog, hot sun.   Thermom^r abroad in the shade 78.   [John White] arrived at Cadiz.

June 20.   [Winton : Selborne] Ephemerae innumerable on the Alresford stream.   When the swifts play very low over the water they are feeding on emphemerae and phryganeae.

June 21.   Brother John sailed from Cadiz for England.

June 22.   Sweet hay-making day.   Put all the St. foin up in a large cock in excellent order : four large jobs.

June 23.   A brood of swallows, flyers, appear for the first time.   Cut great part of the great meadow.

June 24.   Hay makes well.   Flisky[9] clouds, & some rock-like clouds.   Sambucus nigra [elder]. When the elder blows summer is established.

June 25. Hay in beautiful order. Gardens suffer much for want of rain.

June 26. Sun sultry, cloudless, severe heat. Hottest day. Ricked all the hay, save one job, in most excellent order. Ground much burnt.

June 27. Wheat begins to blow. Finished my hay, which is curious. Watered the garden : nothing grows. Cucumbers cease to bear. The drought has lasted three weeks this day.

June 28. Not rain enought to lay the dust.

June 29. Light showers, not enough to lay the dust, chilly. Stachys germanica [downy woundwort.]

June 30. Ground much chopped and burnt. Gave the garden many hoghs : of water : watered the rasps well with the engine.

July 1. Watered the pease. Some nectarines and peaches, two or three apricots, few apples & pears. Small walnuts fall off by thousands. Few nuts. Chilly.

July 3. Field-pease suffer. Watered the garden well.

July 4. Shattering,[10] soft showers all day. Dry weather has lasted just a month. Ground not wetted-in, half an inch.

July 5. Frogs migrate with the showers of yesterday. Dust flies. No appearance of rain left.

July 6. Young partridges are flyers. Vines continue to blow. Monotropa hypopithys [bird's nest orchis] emerges & blows.

July 7. Watered the ground for planting of annuals. Watered the garden plentifully. Planted out a double row of China-asters.

July 8. Planted out African & french marrigolds.

July 9. Meadow-hay begins to be cut. Some barley in ear : wheat uneven. Watered annuals. Finished cutting the tall hedges.

July 10. Woodstrawberries come. Rasps begin to ripen. Sprinkling shower. Showers at a distance.

July 11. Drought has continued five weeks this day. Watered the rasp and annuals well. *Footnote.* There is a sort of wild bee[11] frequenting the garden-campion for the sake of its tomentum, which probably it turns to some purpose in the business of nidification. It is very pleasant to see with what address it strips off the pubes, running from the top to the bottom of a branch, & shaving it bare with all the dexterity of a hoop-shaver. When it has got a vast bundle, almost as large as itself, it flies away, holding it secure between it's chin & it's fore legs.

July 12. Barley & pease suffer much. Frogs continue to migrate from the ponds.

July 13. Lime blows, & smells sweetly, & is much frequented by bees.

July 14. The grass-walks burnt to powder.

July 15. Scarabaeus solstitialis [fern, or summer-chafer]. The fern-owl preys on the fern-chafer.

July 16. Rasps & strawberries abound.

July 18. Frequent sprinklings, but not enough all day to lay the dust. The dry fit has lasted six weeks this day.

July 19. Some thunder & hail. Smart showers.

July 20. Vast showers about to the S.E. & N.W. Dust hardly laid in the roads.

July 21. Heavy clouds around. Roads are dusty.

July 22. Pease begin to he hacked.

July 23. Young martins begin to congregate on the tower.

July 25. Wheat turns yellowish. Mercury falls very fast.

July 26. Fine shower in the night. Distant thunder. Frogs migrate in myriads from the ponds.

July 27. [Bradley] Small shower at Selborne. Young swallows abound.

July 28. Veratrum rubrum [false hellebore]. Brother John arrived at Gravesend in 37 days from Cadiz. He went from Gibraltar to Cadiz by land

to get a ship.   Ponds fail.   Wheat turns.   Hardly
any rain at Selb :

July 29.   [Selborne] Dipsacus pilosus [shepherd's
rod].   Sky looks turbid.   Field-cricket chirps.

July 30.   Vast aurora borealis.

July 31.   The ground dryed to powder.

Aug. 1.   Clouds of dust attend the drags and
harrows.   Great rain.   No such rain at this place
since June 6th.

Aug. 2.   Ground well moistened.   The frogs from
James Knight's ponds travel in troops to the top of
the Hanger.

Aug. 3.   Red-breast sings.   Hops are perfectly
free from distemper, & promise a moderate crop.

Aug. 4.   Young black-caps abound, & eat the
rasps.   Trimmed the vines of their side-shoots.

Aug. 5.   Grass-walks look the better for the rain.

Aug. 6.   Wheat begins to be cut.   Not a breath of
air.   The nights are hot.

Aug. 8.   Fog, sun, & brisk wind, serene.   Ripening weather.   Young martins (the first brood) congregate & are very numerous ;   the old ones breed again.

Aug. 10.   An autumnal coolness begins to take
place, morning & evening.

Aug. 11.   Wheat-harvest becomes pretty general.
Barom : sinks & rises to it's former pitch.

Aug. 13.   Some few wasps begin to appear.

Aug. 14.   [Meonstoke (Hants)] Cloudless, sultry,
dark.

Aug. 15.   [Bp's Waltham (Hants), Selborne :] On
this day at 10 in the morning some sober & intelligent people felt at Noar hill what they thought to be
a slight shock of an earthquake.   A mother and her
son perceived the house to tremble at the same time
while one was above stairs & the other below ;   &
each called to the other to know what was the
matter.   A young man, in the field near, heard a
strange rumbling.   Notwithstanding the long severe
drought the little pond on the common[12] contains a

considerable share of water in spite of evaporation, &
the multitude of cattle that drink at it. Have ponds
on such high situations a power, unknown to us, of
recruiting from the air? Evaporation is probably
less on the tops of hills; but cattle use a vast proportion of the whole stock of water of a small pond.

Aug. 16. Several birds begin to resume their
spring notes, such as the wren, redbreast, smaller
Reg: non crist:

Aug. 18. The swifts seem for some days to have
taken their leave. Apricots. None seen after that
time.

Aug. 19. All the pastures are burnt up, & scarce
any butter made. Wheat in fine order, & heavy.

Aug. 20. Barometer falls very fast. Vast rock-
like clouds around. The drought lasted 10 weeks &
four days.

Aug. 21. Young swallows come forth. Orleans
plums begin to change color. Dark clouds in the
S.E.

Aug. 22. Planted-out endive, & trenched some
celeri. Ground strangely hard, & bound: will
require much rain to soften it. Invigorated by this
burning season such legions of Chrysomelae oleraceae
saltatoriae (vulg: called turnep-flies) swarm in the
fields that they destroy every turnep as fast as it
springs: they abound also in gardens, & devour not
only the tender plants, but the tough outer leaves
of cabbages. When disturbed on the cabbages they
leap in such multitudes as to make a pattering noise
on the leaves like a shower of rain. They seem to
relish the leaves of horse-radish.[13]

Aug. 23. Sun. Showers with wind. Vast
showers. Young stoparolas come forth.

Aug. 24. Trench more celeri. Sowed spinage.
Hops suffer from the wind. Planted small cabbages.

Aug. 25. Much wheat abroad. Strong gusts.
Much rain. The ground is well-moistened.

Aug. 26. Wheat begins to grow under hedges.

Aug. 29.   Hop-picking begins.   Sultry.   Wheat housed in cold condition.   Orleans-plums become ripe.

Aug. 30.   Mich. daisy begins to blow.

Sept. 2.   Hibiscus syriacus [*Althaea frutex*].   Grapes begin to change colour.   Some wheat ſtill out.   The weather bad for hop-picking.

Sept. 3.   Some wasps appear.   Museli plums become ripe.

Sept. 4.   Spring corn in a bad way where cut. Barley in general not ripe.   Hot & moiſt.   Grass grows.

Sept. 5.   Rain.   Oats grow as they lie.   Some wheat abroad.   Bad for hop-picking.   A ſtrange yellow tint in the sky at sunset.   Diſtant thunder & lightening in the evening.

Sept. 6.   The hops by the late winds are much injured & blown into flyers : at beſt they are very brown.   Crop large in some gardens.

Sept. 7.   Peaches begin to ripen.

Sept. 10.   Swallows & martins congregate in vaſt clouds.

Sept. 11.   Ring-ouzel appears on it's autumnal visit : several seen.   Stoparolas seem to be gone for three days paſt.

Sept. 14.   Oats rot as they lie : a very poor scanty crop.   Little barley cut ; but dead ripe.

Sept. 15.   Papilio Atalanta [red admiral] abounds.

Sept. 16.   Vaſt dews.   Chrysomeleae oleraceae ſtill abound on the cabbages.   Some corn housed.

Sept. 18.   Ivy begins to blow : & is the laſt flower which supports the hymenopterous, & dipterous, Inseɕs.   On sunny days, quite on to Novr they swarm on trees covered with this plant ; & when they disappear probably retire under the shelter of it's leaves, concealing themselves between it's fibres, & the tree that it entwines.

Sept. 21.   Few swallows about.

Sept. 22.   Began parlour-fires.   Martins abound under the hanger.   No swallows.

Sept. **23.** A miserable crop of barley round these parts. Grapes eatable.

Sept. **24.** Great rain, stormy. Some swallows & many martins under the hanger.

Sept. **25.** Vast tempest in the night that broke boughs from the trees, & blowed down much of the apples & pears. Gathered some apples.

Sept. **26.** Apples & pears large & fine. Chilly air. Swallows and martins. The tempest on thursday night [Sept. 24] did considerable damage in London, & at Oxford, & in many parts of the Kingdom.

Sept. **28.** Swallows & martins. Gathered first grapes : large & good. Some wasps damage them.

Oct. **1.** [Lassam (Hants)] Young martins in their nest at Lassam [Lasham.]

Oct. **2.** [Oxford.]

Oct. **19.** [Oxford : Lassam].

Oct. **20.** [Selborne] Woodcock returns. Papiliones and muscae abound on the asters. Redwings return.

Oct. **21.** Under the eaves of a neighbouring house is a martin's nest full of young ready to flie. The old ones hawk for flies with great alertness.

Oct. **22.** This morning the young martins forsook their nest & were flying round the village. Grapes delicate, & plenty.

Oct. **23.** The martins about. Glow-worms shine. *Footnote.* Information about ring-ouzels at Lewes and the Forest of Bere [Incorporated in D.B., xxxviii].

Oct. **26.** Swallow appears still. Vast rains.

Oct. **27.** Grapes decay with rain : are most highly ripened.

Oct. **29.** Vast quantities of rain has fallen lately.

Oct. **30.** Grass grows. Medlars shaken off the tree by the wind.

Nov. **2.** Fieldfare is seen.

Nov. **3.** 20 or perhaps 30 martins were playing all day along by the side of the hanger, & over my fields. Will these house-martins, some of which were

nestlings 12 days ago, shift their quarters at this late
season of the year to the other side of the northern
tropic! Or rather is it not more probable that the
next church, ruin, cliff, sand bank (a Northern naturalist would say) lake or pool will prove their

Helvella crispa.
White's *E. mitra* (actual height 2-3 inches,
cap, 1½-2½ inches across).

hybernaculum & afford them a ready, & obvious
retreat ?[14]

Nov. 4. Saw one martin.

Nov. 7. Warm air. Flesh-flies blow the meat in
the larder still.

Nov. 10. Vast quantities of rain.

Nov. 11. Nasturtiums & other Indian flowers[15] still in bloom : a sure token that there has been no frost.

Nov. 12. Oenas, sive vinago. The stock-dove, or wood-pigeon appears. Where they breed is uncertain. They leave us in spring, & do not return 'til about this time. Before the beechen woods were so much destroyed we had every winter prodigious flocks, reaching for a mile together as they went out from their roost of a morning. Hartley-wood used to abound with them. They are considerably less than the ring-dove, or queest, which breeds with us, & stays the whole year round.[16]

Nov. 15. Harsh air.

Nov. 16. Appears in my fields . . .Elvela pileo deflexo, adnato, lobato, difformi : Linn : flo : Suec : Elvela petiolata, lamina in formam capituli deorsum plicato-laciniata & crispa ; petiolo fistuloso, striato, & rimoso : Gleditsch methodus fungorum.[17]

Nov. 18. Nasturtiums blow yet ; some few leaves are decayed. Grapes delicate ; but many bunches decay. Paths dry.

Nov. 20. Snipes come up into the meadows.

Nov. 24. Nasturtiums nipped but still in bloom.

Nov. 25. [Bramshot place (Hants)] Rain with wind.

Nov. 26. At M[r] Pink's at Faringdon is a rook's nest with young in it.

Nov. 27. [Selborne] Vast flocks of wild fowls in the forest. They are probably migraters newly arrived.

Nov. 28. Vast rains in the night! Some few grapes left on the vines.

Dec. 2. Trimmed the vines. Their shools were by no means good, nor well-ripened, notwithstanding the hot summer.

Dec. 4. Dark and spitting. Nasturtiums blow yet. Indian flowers in Dec[r] ! Song thrush sings.

Dec. 6. A dead young rook, about half-grown was found in a nest on one of Mr Pink's trees near his house.

Dec. 7. Earthed asparagus beds. No ice yet.

Dec. 8. Brother & sister John [sic] arrived.

Dec. 11. Young lamb falls.

Dec. 12. Bats appear, & many phalaenae.

Dec. 13. Female chaffinches congregate.

Dec. 14. Nasturtiums blow still.

Dec. 16. Missel-thrush, or storm cock sings. Polyanth and annual stock blow.

Dec. 18. Thrush whistles.

Dec. 22. Vast fog. Nasturtiums still.

Dec. 22. First ice. Icicles. Ground very white. Nasturtiums cut all down, & rotten.

Dec. 24. Ground very white. Thermom^r abroad 27.

Dec. 26. Rime, dark, thaw.

CHAPTER VI

"Day unto day uttereth speech, and night unto night sheweth knowledge." (PSALM xix., v. 2.)

1773

Jan. 6. Hard frost, warm sun, dark. Great rime.[1]

Jan. 10. Nuthatch chirps. Hen chaffinches congregate in vast flocks.[2]

Jan. 17. Vast rain. Blackbird whistles.

Jan. 21. Thrush sings. Titmouse begins it's first spring notes.

Jan. 24. Parus major [great titmouse] sings. Hedge-sparrow essays to sing.

Jan. 27. Wood-lark sings.

Jan. 29. Vast halo round the moon.

Feb. 2. No snow lies.

Feb. 3. Ice within doors. Some flakes of snow.

Feb. 4. Larks congregate.

Feb. 5. Primula vulgaris [primrose]. Ground hard frozen.

Feb. 6. Snow all melted in the morning. Vast flocks of hen chaffinches.

Feb. 8. Severe frost. Ground hard all day.

Feb. 9. Made hot bed.

Feb. 10. Severe frost. Bottles of water freeze in chambers. Snow in the night. Cutting air.

Feb. 11. Reduced my barometer to the true standard of 28 inches, lowering it about two degrees.

Feb. 15. Helleborus viridis emerges & blows.

Feb. 16-25. [Fyfield] Grey crows near Andover.

Feb. 18. Lambs fall strong, & thrive very fast. Last year numbers perished.

Feb. 20. Trufles[3] continue to be found in my Bro: Henry's grove of beeches [at Fyfield] : tho' the season is near at an end. It is supposed that seven or eight pounds are taken annually at that little spot. My Bro: & the trufle-hunter divide them equally between them.

Feb. 26. [Winton] Stormy night, with vast rains, fierce wind all day. This storm did considerable damage in many places.

Feb. 27. [Selborne] Sun & clouds, showers & wind.

Mar. 2. Crocuss in high beauty.

Mar. 4. Pulmonaria officinalis [lungwort]. Papilio rhamni [brimstone butterfly].

Mar. 5. The rooks at Faringdon have built several nests since sunday.

Mar. 6. Mild, still, pleasant weather.

Mar. 8. Seed barley sells at 38ˢ pʳ quarter! a price never heard of before.

Mar. 11. Sun begins to look down the hanger.

Mar. 13. Bees abound on the crocuss. Humble-bees & some flies appear.

Mar. 14. Wh: frost, thick ice, cloudless sky. Skylarks rise & sing a little.

Mar. 15. Mild & grey, sun. Ants. Chryso-mela Gottingensis [one of the "golden apple" beetles]. This insect is very common with us.

Mar. 18. Many sorts of insects begin to come out. Water-insects begin to move. Milvus aeruginosus? [kite]. Hot in the sun.

Mar. 20. Lacerta [lizard] Sky thickens with flisky clouds.

Mar. 22. Gossamer floats about.

Mar. 23. Coluber natrix [common grass snake]. Summer weather with a brisk wind. Cock & hen wheatear.

Mar. 26. Grass begins to grow. A large flock of titlarks on the common,⁴ feeding & flitting on, probably going down to the forest to the moory moist places.

Mar. 28. Sharp air. Three swallows were seen I hear this day over the paper-mill pond at Bramshot.

Mar. 29. Turned out the cucumbers into their hills. Beds still too hot. The dry weather has lasted just a month. Roads all dryed up.

Mar. 30. Hard frost, ice, cloudless, sharp wind. No larks in the fields, & few birds to be heard or seen ; probably this harsh dry air renders their food scarce, & sends them to the lower moister grounds.

Apr. 3. Apricot blossoms seem mostly cut off : peaches & nectarines are well-blown, & look well. Sowed a box of polyanth-seed, & a bed of Celeri.

Apr. 6. I am informed that three swallows appeared over a mill-pond at Bramshot on Sunday, March 28. They were seen over the paper-mill pond by Mr Pym.

Apr. 8. Fritillaria imperialis meleagris [snake's head, or chequered daffodil].

Apr. 10. It appears from good information that sometimes the osprey falco haliaetus, Linn : is known to hawk the great pond at Frinsham. It darts down with great violence on a fish, so as to plunge itself quite under water. The man at the ale-house adjoining shot one as it was devouring its prey on the handle of a plough. [cf. T.P., xxxix]. This man shot also a sea-pie, haematopus ostralegus [oyster-catcher] on the banks of this pond.

Apr. 11. Goose-berry buds in leaf. Anemone nemorosa [wood anemone]. Cardamine pratensis [cuckoo flower].

Apr. 15. Titlark begins to whistle. Wind changing with every shower : soft growing weather.

Apr. 16. Redstart returns. Most soft growing weather. Thomas begins to mow the walks.

Apr. 17. Bank martin appears. House martin appears. Many swallows. Grass grows very fast. Ring-ouzels are first seen on their spring migration. They are very late this year.

Apr. 18. Ground very wet. Nightingale sings.

Apr. 19. Blackcap sings. The sedge-bird a delicate polyglott.[5]

Apr. 20. Regulus non cristatus medius [willow wren] sings : a pretty plaintive note : some call it a joyous note : it begins with an high note & runs down. The titlark, a sweet songster, not only sings flying in its descent, & on trees ; but also on the ground, as it walks about feeding in pastures.

Apr. 21. Field-crickets have opened their holes : they are full-grown, but have only the rudiments of wings, & are probably in their larva state ; yet they certainly eat, as appears by their dung. It seems likely that they die every winter, leaving eggs behind them. About Septem^r all the mouths of their holes are obliterated. They do not cry 'til about the middle of May. Their noise is shrill & loud. This is by no means a common insect. They probably cast another coat before their wings are perfect, & they capable of shrilling. [Cf. D.B., xlvi.] *Footnotes.* Whooping coughs have been general among children the winter thro' : and now putrid sore throats begin to be common among young people of the female sex. In the wings of the ring-ouzel eight of the secondary feathers have an unusual sharp point at the extremity : is not this irregularity peculiar to this species ? It seems to me that some peculiarity attends the ends of the secondaries of most birds. Might not this be used as a specific by nice observers.

Apr. 22. Grasshopper-lark chirps. Alauda minima locustae voce stridet.

Apr. 26. Went to London with Bro : & sister J. W. [No further entries until May 21.]

May 22. [Selborne] Stoparola builds. *Footnote.* May 12. First swifts were seen, many together. On May 19 at night was a vast rain with thunder & lightening : frequent showers before & since ; so that the ground is very moist ; & the corn & grass grow. The floods are much out at Staines. In the beginning of the month there were frosts, hail, & some snow. Apricots continue to fall off : peaches, & nectarines decent crop. Apples blow well : pears

seem hurt by the frosts.   Vine-shoots very backward :
they were pinched by the frost.

May 23.   Lathraea squammaria [toothwort] in seed.
Turtle-dove about.   Measles prevail in this neighbourhood.

May 24.   Scotch and spruce firs beautifully illuminated by the male & female blossoms !

May 28.   Apis longicornis bores it's nest in the
field-walks.

May 31.   Ashes & walnut-trees naked yet.   Fern-
owl chatters.   Thunder.

June 1.   Field cricket sings : sings all night.

June 2.   Lampyris noctiluca [glow-worm].   Thunder
& lightening & moderate rain half the night.   The
corn & grass & gardens look well after ye rain.

June 3.   A dozen pair of swifts appear at times.
Some heads of S^t. foin begin to blow.

June 4.   Began to tack the vines, which are backward.   Crataegus aria⁶ blows beautifully.

June 6.   Here & there a single chafer this year.

June 8.   Rye in ears.

June 9.   Swifts sit hard.

June 11.   Elder begins to blow.   When the elder
blows-out the summer is at it's height.

June 13.   Sanicula europaea [sanicle].   Lysim-
achia nemorum [yellow pimpernel].   Bees swarm.
Pease in the fields thrive wonderfully.   Thunder.

June 15.   Great rains in the night.   Planted-out a
bed of Savoys.   No apples or pears.

June 16.   Sheep are shorn.

June 18.   Some ears of wheat begin to appear.
Measles epidemic to a wonderful degree : whole
families down at a time.   Several children that
had been reduced by the whooping-cough dyed of
them.

June 20.   Young wild-ducks, or flappers are taken
at Oakhanger-pond ; & a small Anas olive, which
seemed to me to be a young teal : turned it into
James Knight's ponds.

June 21. First brood of young swallows comes forth more early than usual. They commonly appear about the first week in July.

June 22. The King [George iii] came down to Portsmouth to see the fleet. *Footnote.* June the 22 : 23 : 24. The firings at Spithead were so great that they shook this house. They were heard on those days at Ringmer two miles east of Lewes in Sussex; & at Epsom in Surry.

June 26. Great hail-storm at Alton St foin not yet turned.

June 28. St foin begins to be damaged.

July 1. Portugal-laurel blows in a most beautiful manner.

July 3. Ricked my St foin, five jobbs, into a large cock. It has suffered less than could be expected. Has lost it's smell. Is got full of coarse grass. This is the sixth crop.

July 4. Hops do not cover their poles well, checked perhaps by the cold, black weather: they are pretty much infested by aphides, that begin to abound.

July 5. Cold starving weather: nothing grows.

July 6. All vegetation in gardens seems to stand still.

July 7. Cut great part of my great mead.

July 9. Hay makes well. Flocks of lapwings on the common. After breeding they forsake the moory places, & take to the high grounds.

July 10. Wood strawberries begin to ripen. Hay makes well. Cock great part of the hay in very large cock. Many young bank-martins seem to be flown in the forest. The old ones carry dragon-flies into their nests to their young.

July 11. Partridges young, flyers.

July 12. Ricked all my hay. The st foin has lost all smell: the meadow-hay is most delicate. A large crop.

July 13. Finished stopping the vines: much bloom & much fruit set. Finished cutting the tall hedges.

July 14. Swallows & martins bring-out their first broods of young very fast. *Footnote.* Most of the sᵗ foin & clover this year damaged by rain : most of the meadow-hay ricked in a very curious condition. Good hay sold this year in the place for 25ˢ pᵣ load : last year for £2-7-0 pᵣ load in the place.

July 18. Loud thunder shower. Mʳˢ. Snooke of Ringmer near Lewes had a coach-horse killed by this tempest : the horse was at grass just before the house.

July 21. A bunting, emberiza alba, sitting about on the bushes in the North field. Probably has a nest there. This is a very rare bird in this parish : a very common one in the open champain country. I am not sure that I ever saw one before in Selborne.

July 22. Wheat is now at 17ˢ pds. per load, & very little left in the kingdom.

July 23. Turnips begin to be hoed. In general a good crop. The young clover among the corn is fine this year.

July 24. Wheat at Farnham £17-12-6 pᵣ load. Several fields of cone, or bearded wheat⁷ growing this year round the village : the bloom of this wheat is of a brimstone colour. The bloom of some beardless wheat is purple : qu : what sort ? The bloom of wheat in general is whitish.

July 25. Some hops much infested with aphides.

July 27. Some wheat seems to be blighted.

July 31. The lightening beat down a chimney on the Barnet : no person was hurt. Measles still about. *Footnote.* Thro' this month the Caprimulgi [nightjars, fern-owls] are busy every evening in catching the solstitial chafers which abound on chalky soils on the tops of hills. These birds certainly do, as I suspected last year, take these insects with their feet, & pick them to pieces as they flie along, & so pouch them for their young. Any person that has a quick eye may see them bend their heads downwards, & push out their short feet forwards as they pull their prey to pieces. The chafer may also be discerned in their claws. The serrated claw therefore on their longest toe is no doubt for the purpose of holding their prey. This is the only insectivorous bird that I know which takes it's prey flying with it's feet.[8]

Aug. 1. Turneps thrive at a vast rate : a fine crop. A prospect of much after-grass.

Aug. Apis manicata [carder bee]. This bee is never observed by me 'til the Stachys germanica blows, on which it feeds all day : tho' doubtless it had other plants to feed on before I introduced that Stachys.

Aug. 6. The male, & female ants of the little dusky sort come forth by myriads, & course about with great agility.

Aug. 7. The flight of the scarabaeus solstitiales seems to be over. Measles still in some families.

Aug. 8. Hops have been some time in bloom, & do not promise for much of a crop : they are lousy and do not run the poles well.

Aug. 10. [Meonstoke] Most sultry night.

Aug. 13. Great thunder, and lightening.

Aug. 14. [Selborne] Wheat-harvest pretty general. Dark heavy clouds to the N.W. Heat unusually severe all this week ! This storm [Aug. 13] did great damage in & about London.

Aug. 15. Hops visibly improved by the thunder.[9] If the swifts are gone, as they seem to be, they can never breed but once in a summer ; since the swallows & martins in general are now but laying their eggs for a second brood. As young swifts never perch or congregate on buildings I can never be sure exactly when they come forth. The retreat of swifts so early is a wonderful fact : & yet it is more strange still, that they withdraw full as soon in the summer at Gibraltar ! Swifts sat hard June 9th.

Aug. 16. Wind covers the walks with leaves, & blows down the annuals.

Aug. 17. Swifts seem to be gone ; very early. Vast clouds on the horizon. Wheat bound.

Aug. 18. Wheat lies in a bad way. Much cut, little bound, & scarce any housed.

Aug. 19. Terrible storm all night, which made sad havock among the hops, & broke off boughs from the trees.

Aug. 20. Wasps begin to appear. No swifts since last week.

Aug. 21. Sweet harvest day. Wheat housed all this afternoon. With respect to the singing of birds Aug: is much the most silent month: for many species begin to reassume their notes in Septemr. [Cf. T.P., XL.] The goldfinch sings now every day.

Aug. 24. Peaches & nectarines redden. China-asters begin to blow.

Aug. 25. Tho' there was a brisk air from the S. all the afternoon; yet the clouds in an upper region flew swiftly all the while from ye N. in great quantities.

Aug. 27. Vast quantities of nuts & filberts.

Aug. 28. Some few grapes begin to turn red. Peaches begin to ripen & are large & good. Nectarines look well: they are very ruddy & large.

Aug. 29. A little black curculio[10] damages the peaches by boring holes in them before they are quite ripe. I do not remember this insect on my wall-fruit before. They damage the leaves also.

Aug. 30. Tyed up endives. Some people have finished wheat-harvest.

Sept. 1. Orleans plums begin to ripen. Hops continue small & have not grown kindly since the storm.

Sept. 2. Some perennial asters begin to blow. Grapes rather small as yet. Several bunches begin to turn.

Sept. 3. Young wagtails come out. Delicate peaches & nectarines in plenty. *Footnote.* The beards of cone-wheat do not long preserve the grain from sparrows: for as the corn gets ripe those aristae [awns, or beards] are shaken off by the wind.

Sept. 7. [Winton] People begin to pick hops.

Sept. 9. Little barley housed.

Sept. 10. [Selborne] Sad harvest & hop-picking weather! Rain damages the wall-fruit.

Sept. 11. Wasps encrease, & injure peaches & nect: & begin on the grapes. Young martins come out.

Sept. 13. Young swallows in the nest. Hops very ordinary: very small.

Sept. 14. Young swallows come out. Barley & oats housed. Some wheat out still.

Sept. 16. Gathered first grapes: small but good. Last wheat housed.

Sept. 18. Linnets begin to congregate: they feed on the seeds of centaurea jacea [brown-rayed knap-weed].

Sept. 22. [Lasham] Stormy, with rain, sun, shower, windy.

Sept. 23. [Basingstoke, Lasham] Wh: frost, showers.

Sept. 25. [Selborne] Much barley abroad. Wet fit ever since the first of Sepr. Wall-fruit fine still.

Sept. 26. Barley not so much damaged as might be expected, being soon dryed by continual wind.

Sept. 27. Gathered the last nectarines: very good. Large aurora: very vivid in the S.W.

Sept. 28. Stoparola, flycatcher, still appears.

Sept. 29. Multitudes of martins, but I think not many swallows. Grapes are eatable, but not curious yet: are damaged by the wasps.

Sept. 30. Some barley abroad that has been cut a month. Earwigs cast their skins & come forth white. 10 or 12 ring-ouzels appear on their autumn migration round Noar hill. Martins are seldom seen at any distance from neighbourhoods. They feed over waters or under the shelter of an hanging wood. Swallows often hawk about on naked downs & fields, even in very windy seasons at a great distance from houses.

Oct. 2. Swallows do not resort to chimnies for some time before they retire. Titlarks abound on the common. Martins are the shortest-winged & least agile of all the swallow-tribe. They take their prey in a middle region, not so high as the swifts: nor do they usually sweep the ground so low as the swallows. Breed the latest of all the swallow genus: last year they had young nestlings on to the 21 of Octr. They usually stay later than their congeners. Last year 20 or 30 were playing all day long by the side of the hanger, & over my fields on Novr 3rd. After that they were seen no more.

Oct. 3. Grey, dark showers, dry & windy. Glass falls at a vast rate.

Oct. 4. Vetches and pease are mostly spoiled. Martins. Mr Yalden[11] has 10 acres of barley abroad.

Oct. 7. Wasps cease to appear. Swallows & Martins seem to be gone.

Oct. 8. Rooks frequent wallnut trees, & carry off the fruit.

Oct. 9. Many martins appear again. Mr Yalden's barley abroad: it was large corn & full of clover. *Footnote.* The breed of partridges was good this year: pheasants are very scarce; hardly any eyes[12] to be found. We abound usually in pheasants. In some counties pheasants are so scarce that the Gent: have agreed to refrain from killing any. Rains ever since the first of Sep^r.

Oct. 10. Storm that broke boughs from the hedges. Many swallows & martins. Much barley & vetches abroad. The housed & ricked barley in wet condition; it heats much.

Oct. 11. [Worting (near Basingstoke)] fine day, clouds & wind at night.

Oct. 12-21. [Oxford] Hops sold at Weyhill fair much cheaper than people expected: from £6-10-0 to £7-10-0.

Oct. 16. Mr Yalden finished his barley-harvest, some of which had been cut more than six weeks.

In general the grain is not spoiled, but by drying &
frequent turning in a floor will be tolerable.

Oct. 19. Venus is become an evening star. Vivid
Aurora bor:

Oct. 20. Aurora.

Oct. 21. [Worting] No swallows or martins
observed. *Footnote.* From Oct[r] 11th to Oct. 21st
inclusive the journal was kept by John White at
Selborne.[13]

Oct. 22. [Selborne] Fog, rain, fog, fog. *Footnote.* Oct[r] 21. Saw several martins at Dorchester
in Oxfordshire round the church. It is remarkable
that the swallow kind appear full as late in the midland
counties, as in the maritime : a circumstance this more
favourable to hiding than migration. As it proved
these martins were the last that I saw.

Oct. 24. Woodlark sings. Great titmouse re-
assumes it's spring note.

Oct. 25. Began levelling my grass-plot & walks
at the garden-door, & bringing them down to the
level of the floor of my house.

Oct. 27. Hares abound, but pheasants are very
scarce this year. One of the vines to the S.W. casts
its leaves & looks sickly.

Oct. 30. Grapes are very curious.

Nov. 1. Seed-clover cutting. A ring-ouzel was
shot in the high wood with a russet gorget, & russet
spots on its wings. Three or four more were seen.

Nov. 3. [Newton] Stock-dove, or wood-pigeon
appears. Redwing appears.

Nov. 4. Scolopax : Woodcock returns.

Nov. 5. Cornix cinerea. Flying over Faringdon
heath. The first grey crow that I ever saw in the
district of Selborne. They are common on the
downs about Andover : about Winton, & Bagshot.
Most of the earth to be removed in levelling the garden
is taken away ; in some places to the depth of 18
inches. The continued rains much interrupt the
work, & make it a nasty jobb. The best mould is

laid on the quarters of the garden, the clayey soil is wheeled into the meadow.

Nov. 8. Snipes leave the moors & marshes which are flooded, & get up into the uplands.

Nov. 9. Ground to be levelled is under water. Wood-cocks pretty common. The country all in a flood.

Nov. 10. Rains have lasted ten weeks. Saw a flock of seven or 8 stone-curlews. These birds generally retire before this time.

Nov. 11. All our levelling-work is under water. The barom has been unusually low for many days past; & yet with little wind: but the rains have been prodigious. Most of the rain has fallen by night.

Nov. 13. The turfing the level ground goes on briskly. No late martins have appeared this Nov: a flight sometimes is seen about the first week in this month.

Nov. 14. Green plovers now appear in small companies on the uplands. They flie high and make a whistling. They do not breed in these parts.

Nov. 15. Helleborus foetidus buds for bloom.

Nov. 16. Two grey crows flew over my garden to the hanger: a sight I never saw before.

Nov. 17. The turfing the walks advances apace.

Nov. 18. Stone Curlews appear still on Temple farm.

Nov. 19. Ring-ouzels still remain. Gathered in the last grapes: ye crop was very large, & the grapes delicate. And yet the vine-shoots were much pinched at their tops by frost the first week in May; & moreover Septem was a season of continual clouds & rain. Ring-ouzels, & stone-curlews stay late with us this year.

Nov. 21. Yellow water wagtail.[14]

Nov. 22. Beautiful rimes all day on the hanger.

Nov. 23. While my people move earth in the garden the redbreasts in pursuit of worms are very tame, & familiar, settling on the very wheel-barrows, while filling.

Nov. 24. Finished the levelling, & turfing of
garden. The alteration has a good effect. The
weather & rains considered, the turf lies pretty well.

Nov. 25. Considerable snow on the ground.

Nov. 26. [Alton] A profusion of turneps probably all the kingdom over : on which account lean
sheep are very dear. Hops at present lie on hand :
were carried to Weyhill, then to Andover : & now
are bringing home again. Snow gone except under
hedges. *Footnote.* Birds do not seem to touch the
berries of the tamus communis [yew] 'tho they look
very red, & inviting : the berries also of the bryonia
alba [white bryony] seem not to be meddled with.
Perhaps they are too acrid. There is a fine crop of
clover of last spring : the frequent showers of last
summer occasioned also a vast growth of grass.

Nov. 30. [Chilgrove] Bright, sunny, soft.

Dec. 1. [Stenning (Steyning, Sussex)] Birds on
the downs are rooks, larks, stone-chats, kites, gulls :
some field-fares, some hawks.

Dec. 2-15. [Ringmer] Not one wheat-ear to be
seen on the downs. The grubs of the scarabaeus
solstitialis abound on the downs : the rooks dig
them out. On what do they feed when they come
forth ? for there are no trees on the South downs.
The county of Sussex abounds in turneps. The
tortoise in M^rs. Snooke's garden went under ground
Novr 21 : came out on the 30th for one day, & retired
to the same hole : lies in a wet border in mud & mire :
with it's back bare. In the late floods the water at
Houghton ran over the clappers :[15] & at Bramber.

Dec. 5. Rooks spend most of their time in mild
weather on their nest-trees : some stares [starlings]
and jack-daws attend them.

Dec. 8. Much wheat still sowing.

Dec. 9. Rooks attend their nest-trees in frost only
morning & evening.

Dec. 11. Flocks of chaffinches : & multitudes of
buntings at the foot of mount Caborn [near Lewes].

Rooks visit their nest-trees every morning just at the dawn of day, being preceeded a few minutes by a flight of daws : & again about sunset. At the close of day they retire into deep woods to roost.

Dec. 15 [Findon] Large gulls on the downs. Some bustards are bred in the parish of Findon, Fieldfares.

Dec. 16. [Chilgrove] They, the shepherds, do not take any wheatears W. of Houghton bridge.

Dec. 17. Chaffinches—many cocks among them. Black rabbits are pretty common on Chilgrove warren. *Footnote.* The parish-well in Findon-village is 200 feet-deep : at Montham on the down the well is full 350 feet. Mr Wood's well at Chilgrove is 156 feet deep ; & yet in some very wet seasons is brimfull : his cellars are some times full.

Dec. 18. [Selborne].

Dec. 25. Yellow water-wagtail frequents the shallow parts of the stream.

Dec. 26. White water wagtail.[16]

Dec. 31. Frost all day.

CHAPTER VII

" Therefore am I still
A lover of the meadows and the woods
And mountains, and of all that we behold
From this green earth."
(WORDSWORTH : Tintern Abbey.)

1774

Jan. 1.　Larks congregate.

Jan. 3.　Thermom^r abroad 22¾. At noon 30. In ye wine vault 43¼.

Jan. 9.　Rain for 24 hours : vast flood. Could not get along down at the pond¹ all day.

Jan. 10.　Wild ducks about.

Jan. 11.　Some snow on the ground. Aurora.

Jan. 14.　Vast rain in the night with thunder. A bittern was shot in shrub-wood. A dog-hunted it on the foot, & sprung it in the covert. On the same day Mr Yalden shot one in a coppice in the parish of Emshot : & about the same time one was killed in the parish of Greatham. These birds are very seldom seen in this district, & are probably driven from their watery haunts by the great floods, & obliged to betake themselves to the uplands. The wings expanded measured just four feet : the tail-feathers shafts & all were just five inches long, & 10 in number. The neck-feathers were very long, & loose like those on the neck of a roost-cock. These birds weighed undrawn, & feathers & all, each 3 p^{ds} & 2 oun : The serrated claw on each middle toe is very curious ! Tho' the colours on the bittern's wings & back are no ways gaudy or radiant, yet are the dark & chestnut streaks so curiously blended & combined, as to give that fowl a surprizing beauty. Both the upper & lower mandible are serrated towards

77

the point, & the upper is emarginated. Two of these birds I dressed, & found the flavour to be like that of the wild duck, or teal, but not so delicate. They were in good case, & their intestines covered with fat. In the crop or gizzard I found nothing that could inform me on what they subsisted : both were quite empty. I found nothing like the flavour of an hare ! The flesh of these birds was very brown. [*Side note.*] It appears since that all these bitterns were killed in Selborne parish, & probably were all of the same family.

Jan. 16.   Jupiter, the moon, & Venus being this evening nearly in a line, & almost at equal distances, make a beautiful appearance. Bat appears : they are out at times every month in the winter : but not in frost.

Jan. 18.   Considerable snow on the ground, frost.

Jan. 20.   Snow remains. Vast halo round the moon.

Jan. 24.   Stone-curlews still appear on Temple farm.

Jan. 28.   Titmouse begins to open [? song]. Red-breast sings.   Song-thrush sings.

Jan. 29.   Snow-drop, wolfs bane,² helleborus foetidus blow.   Gnats appear.   Beetles buz in the evening.

Jan. 31.   The water above the tap in R : Knight's cellar at Faringdon. The land-springs begin to break-out on the downs beyond Andover. A certain token that the rich corn vales must suffer [Cf. D.B., XIX].

Feb. 1.   [Sutton (nr. Winchester)] Considerable snow.

Feb. 2.   The land-springs have not been so high since spring 1764.

Feb. 5.   Frost, sun, yellow evening.

Feb. 6.   Bees play much about their hives.

Feb. 7.   Continual vicissitudes from frost to rain. Land-floods advance.

Feb. 8.   Freezes sharply all day. The cones of laſt year fall very faſt from yᵉ spruce-firs.

Feb. 9.   Jupiter & Venus approximate very faſt. Venus is very bright, & makes ſtrong shadows on the floors, & walls.

Feb. 10.   Weather shifts continually from froſt to rain to the detriment of the wheat & turneps. Wheat looks sadly, & is almoſt heaved out of the ground.

Feb. 13.   Hedge-sparrow sings. Great flock of buntings in the fields towards Faringdon.

Feb. 14.   The ivy, hedera helix, blows in Sept: Oĉtʳ & Novʳ : the berries are now full-grown, & ripen in April : thus fruĉtification goes on in some Inſtances the winter thro'. When the berries are full ripe they are black.

Feb. 16.   Skylarks mount, & essay to sing. House-sparrows get in cluſters, & chirp, & fight. Thrushes whiſtle.

Feb. 17.   Ravens begin to build. Spring-like weather.

Feb. 19.   These great rains retard the preparations for a spring-crop. Grey crows ſtill on the downs.

Feb. 20.   The high wind laſt night blowed down a large old apple-tree in the orchard.

Feb. 21.   Ews³ & lambs look very lank from the drowning rains : the turneps are very washy food.

Feb. 23.   Several muscae appear. Skylarks would sing if the wind would permit.

Feb. 26.   Land-springs rise. The titmouse, which at this time of the year begins to make two quaint, sharp notes, which some people compare to the whetting of a saw, is the marsh-titmouse. It is the great titmouse which sings those three chearful notes which the country people say sounds like " sit ye down " : they call the bird by that name.⁴

Feb. 27.   Ewes die in lambing.

Feb. 28.   Much wheat rotted on the ground in the clays.

March 2.   Venus shadows.

March 4.   Daws resort to churches.

March 5.   Received as a present from M\ Hinton
(to whom it was sent from Exeter, with many more)
one of M\ William Lucombe's new variety of oaks[5] :
it is said to be evergreen, tho' raised at first from an
acorn belonging to a deciduous tree.   They are all
grafted on stocks of common oaks.   My specimen is
a fine young plant, & well-rooted.   The growth of
this sort is said to be wonderful.   Vid : philosoph :
transact : V : 62 : for the year 1772.

March 9-25.   [London] This was the last day of
the wet weather : but the waters were so encreased
by this day's deluge, that most astonishing floods
ensued.   This rain & snow, coming on the back of
such continual deluges, occasioned a flood in the S :
of England beyond anything ever remembred before.
Footnote. In the night between the 8th and 9th a vast
fragment of an hanger in the parish of Hawkley slipped
down ; & at the same time several fields below were
rifted & torn in a wonderful manner : two houses also
& a barn were shattered, a road stopped-up, & some
trees thrown-down.   50 acres of ground were disordered & damaged by this strange accident.   The
turf of some pastures was driven into a sort of waves :
in some places the ground sunk into hollows.[6]   *Footnote*. Thomas kept the journal in my absence.[7]

March 26.   [Selborne] Peaches, nectarines, &
apricots in fine bloom.   No rain since the 9th : stiff
ground still very wet.   Thomas began to mow the
grass plot.   My new-laid turf, where not damaged by
the continual standing of water after the vast rains,
looks well.

March 28.   Hot sun.   Great thunder-shower at
Wintŏn.

March 29.   A fine regular bloom all over the apricot,
peach, & nectarine-trees.   Sheltered the bloom with
some ivy-boughs.

Apr. 1.   No rain since the 9th of March.   *Footnote*.
The saxon word Playstow signifies ludi locus,

gymnasium, palæstra, theatrum. No doubt our little square, called to this day the pleſtor, is of the same import : it is moreover the place to this day where the youths of the village resort for their diversions ; & was moreover no doubt in old days our market-place, when this village was a town.[8]

Apr. 4.   Two swallows appear at Faringdon.

Apr. 5.   The ground harrows, & rakes well.

Apr. 9.   The ring-ouzel appears on it's spring migration. It feeds now on ivy-berries, which juſt begin to ripen. Ivy blossoms in Octob.r In the autumn it feeds on haws, yew berries, &c : also on worms, &c.

Apr. 11.   Shell-snails come out in troops.

Apr. 12.   Nightingale sings. Three swallows appear. Several bank-martins about the verge of the foreſt.

Apr. 13.   Apricots begin to set. Planted seven rows of potatoes. Nightingale in my fields.

Apr. 14.   The blackcap begins to sing in my fields : a moſt punctual bird in it's return. *Footnote.* In the season of nidification the wildeſt birds are comparatively rare. Thus the ringdove breeds in my fields tho' they are continually frequented : & the misselthrush, tho' moſt shy in the autumn & winter, builds in my garden close to a walk where people are passing all day long.

Apr. 17.   The middle willow wren sings with a plaintive, but pleasing note.

Apr. 19.   The titlark begins to sing : A sweet songſter !

Apr. 20.   Turtle cooes.

Apr. 21.   Asparagus begins to sprout.

Apr. 22.   Cuckoo cries.

Apr. 24.   No house-martins appear : they are very backward in coming. One swift seen.

Apr. 25.   Redſtart returns. Wryneck returns, & pipes.

Apr. 26.   No house-martins seen yet, save one by chance.   Apricots, peaches, & nectarines swell : sprinkled the trees with water, & watered the roots.

Apr. 27.   Oaks are felled : the bark runs freely. Many swallows.   Two swifts round the church.

Apr. 28.   Began to cut a little orchard-grass for the horses.   Some few beeches are in leaf.

Apr. 30.   Vine-shoots have been pinched by the frost.   Two house-martins up at Mr Yalden's.   *Footnote.*   If bees, who are much the best setters of cucumbers, do not happen to take kindly to the frames, the best way is to tempt them by a little honey put on the male & female bloom.   When they are once induced to haunt the frames, they set all the fruit, & will hover with impatience round the lights in a morning, 'til the glasses are opened.   *Probatum est.*

May 2.   There is a good bloom on the pear-trees. Great bloom of cherries.

May 3.   The white-throat returns & whistles.

May 4.   Asparagus in plenty.   Orchard-grass cut for the horses.

May 5.   Grass grows, & is forward.   Apple-trees blow.   Plums shew little bloom.

May 6.   The redstart whistles, perching on the tops of tall trees near houses.   *Footnote.*   In some former years, I see, house-martins have not appeared til the beginning of May : the case was the same this Year : & yet they afterwards abounded.   These long delays are more in favour of migration than of a torpid state. [Later note] House-martins afterwards were very plenty.   In Devõn near Exeter Swallows did not arrive 'til April 25 ;   & house-martins not 'til the middle of May.   Swifts were seen in plenty on May 1st.   At Blackburn in Lancashire swifts were seen April 28 :   Swallows April 29.   House-martins May 1st.

May 8.   White-throat warbles softly [this was afterwards crossed out].   Mistake : it was the black-cap : whitethroats are always harsh & unmusical.

May 9.   Chafers have not been plenty since the year 1770.   Eight swifts now appear : they arrive in pairs. Martins encrease.   Hanger almost in full leaf. Chafers in vast numbers.   Regulus non cristatus major, shaking it's wings it makes at intervals a sibilous stammering noise on the tops of the tallest beechen-woods : it abounds in the beechen-woods on the Sussex down where the two other species [willow wren and chiffchaff] are never heard.   It spends it's time on the tops of the tallest trees.   [Cf. T.P.,X.]   The caprimulgus [goatsucker] is the last bird of passage but one : the stoparola [spotted flycatcher] is the last. The house-martin begin to build as early, as when it arrives early.   It came very late.

May 11.   Pulled the first lettuces, brown Dutch, which had stood the winter under the fruit-wall : they begin to loave.⁹

May 12.   Swallows & house-martins begin to collect dirt for building.   The swallow carries straws to mix with it.   Chafers swarm.

May 13.   Crows bring out their young in troops. Horses begin to lie abroad.

May 14.   Swifts are encreased to their *usual* number of about eight or nine pairs.

May 16.   A pair of martins began building their nest against my brew-house.

May 17.   Rooks bring out their young : they & the crows, & daws & ravens, frequent the top of the hanger, & prey on chafers.

May 18.   Thinned the apricots & took off a large basket of fruit.

May 19.   The first leaves of the peaches & nect: sadly blotched, & rivelled¹⁰ [sic].   These leaves seem not to be affected by animals ; but are monstrously distorted.¹¹   Mem: to observe whether the peaches & nect: whose leaves are so blotched, can bear any well-flavoured fruit.   *Later note*.   They bore fine fruit in plenty, considering the wet shady season.

May 20.   Flycatcher appears : the latest summer-bird of passage.   The stoparola is most punctual to the 20th of May ! ! !   This bird, which comes so late, begins building immediately.

May 21.   Apple-trees in fine bloom.   Began to tack some shoots of vines.   Few whitethroats this year.

May 24.   Ophrys nidus avis [bird's nest orchis] Bro: Thomas.   This curious plant was found in bloom in the long Lythe among the dead leaves under the thickest beeches :   & also among some bushes on Dorton.

May 25.   The martins have just finished the shell of a nest left unfinished in some former year under the eaves of my stable.   Apis longicornis bores holes in the grass-walks.

May 26.   Planted one of the ophrys nidus avis with a good root to it in my garden, under a shady hedge. The shell of the martin's nest begun May 16, is about half finished.

May 28.   The crows, rooks, & daws in great numbers continue to devour the chafers on the hanger. Was it not for those birds chafers would destroy everything.   Rooks, now their young are flown, do not roost on their nest-trees, but retire in the evening towards Hartley-woods.   Martins roost in their new nests as soon as ever they are large enough to contain them.

May 30.   The shell of the martin's nest begun May 16 is finished.

May 31.   Pulled off many hundreds of nectarines, which grew in clusters.   The leaves are distempered, & the trees make few shoots.   Vast crop of wall-fruit.

June 1.   Planted vast rows of China-asters in the Garden.

June 2.   Finished tacking the vines for the first time.   Planted-out annuals in the basons down the field.

June 3.   Martins abound : they came late, but appear to be more in number than usual.   Some pairs of martins repair, & inhabit nests of several years standing.

June 4.   The leaves of the mulberry-tree hardly begin to peep.   The vines promise well for bloom. Apis longicornis works at it's nest in the ground only in a morning while the sun shines on the walk.   Earthworms make their casts most in mild weather about March & April : they do not lie torpid in winter, but come forth when there is no frost.

June 5.   The swallows pursue the magpies & buffet them.   Wall-fruit swells.

June 6.   The redstart sits singing on the fane of the may-pole, & on the weather-cock of the tower.

June 7.   Bees swarm & sheep are shorn.   My firs did not blow this year.

June 8.   Bees gather much from the bloom of the buck-thorn, rhamnus catharticus :   & somewhat from the new shoots of the laurel.

June 9.   Chafers are pretty well gone ; they did not deface the hedges this year.   When swifts mute flying, they raise their wings over their backs.

June 10.   Young broods of moor-hens appear. Lettuces that stood the winter are very fine.   Cut my St. foin : it is in full bloom, & a good crop : this is the seventh crop. . . . Swifts have eggs, but do not sit.   Do they lay more than two ?

June 12.   Odd meteorous circle round the sun, which the common people call a mock sun.

June 13.   House-martins gather moss, & grasses for their nests from the Roofs of houses.

June 14.   [Midhurst] Swifts stay-out 'til within 10 minutes of 9.   Ivy-berries are fallen-off.   Young grasshoppers.

June 15.   [Bramshot] There seem to be more hirundines, particularly house-martins, & swifts, about Midhurst than with us.

June 16.   [Selborne] Fern-owl chatters in the hanger.

86 NATURALIST'S JOURNAL 1774

June 18. Variable winds, & clouds flying different ways. Ricked the S^t foin, four jobbs. Rather under made, but not at all damaged by the rain. It was made in swarth [swaths or ridges], & lay 8 days. *Footnote.* Most birds drink sipping a little at a time : but pigeons take a long continued draught like quadrupeds. Some swallows build down the mouths of the chalk draught-holes[12] on Faringdon-common. House-martins retire to rest pretty soon : they roost in their nest as soon as ever it is big enough to contain them. Martins build the shell of a nest frequently, & then forsake it, & build a new one.

June 19. Bees frequent my chimneys : they certainly extract somewhat from the soot, the pitchy part, I suppose.

June 21. [Bp's Sutton (Hants)] Dark, & still, rain.

June 22-*July* 8. [Fyfield] Spiraea filipendula [drop-wort] Valeriana offic : [true valerian]. Quail calls. Young backward rooks just flown. Young nightingales flown. Mayflies abound on the Whorwel streams & are taken by hirundines.

June 23. Nightingales very jealous of y^ir young : & make a jarring harsh noise if you approach them.

June 24. My Bro : [Henry] has brewed a barrel of strong beer with his hordeum nudum.[13] My Brother's hordeum nudum is very large & forward, and has a broad blade like wheat : it is now spindling [growing tall] for ear, & the tops of the beards appear. It will be much forwarder than the common barley. Swifts squeak much. *Footnote.* The swifts that dash round churches, & towers in little parties, squeaking as they go, seem to me to be the cock-birds : they never squeak 'til they come close to the walls or eaves, & possibly then are seranading their females, who are close in their nests attending to the business of incubation. Swifts keep out the latest of any birds, never going to roost in the longest days 'til about a quarter before nine. Just before they retire they squeak & dash & shoot about with wonderful rapidity. They are stirring at least seventeen hours when the days are longest.[14]

June 26. My Brother's vines turn pale on the chalk : the leaves begin to wither.

June 28. Young nestling rooks still. Young partridges, flyers.

June 29. Some swallows this day bring out their broods, which are perchers : they place them on rails that go across a stream, & so take their food up & down the river, feeding their young in exact rotation.

July 1. Swifts, I have just discovered, lay but *two* eggs. They have now naked squab young, & some near half-fledged : so that their broods cannot be out 'til toward the middle or end of July, & therefore they can never breed again before the 20th of August. In laying but two eggs, & breeding but once they differ from all our other hirundines. Scarabaeus solstitialis. The appearance of this insect commences with this month, & ceases about the end of it. These scarabs are the constant food of caprimulgi the month thro'.
Footnote. When Oaks are quite stripped of their leaves by chafers, they are cloathed again soon after mid-summer with a beautiful foliage : but beeches, horse-chest-nuts, & maples, once defaced by those insects, never recover their beauty again for the whole season.

July 4. Fern-owls breed but two young at a time : but breed, I think, twice in a summer.

July 5. Swallows feed their young in the air. Martins, & swallows, that have numerous families, are continually feeding them : while swifts that have but two young to maintain, seem much at their leisure, & do not attend on their nests for hours together, nor appear at all in blowing wet days. Swifts retire to their nests in very heavy showers.

July 1. Farmer Cannings plows with two teams of asses, one in the morning, & one in the afternoon : at night these asses are folded on the fallows ; & in the winter they are kept in a straw-yard where they make dung. [Cf. D.B., XXII.]

July 8. [Sutton] . Showers, sun & clouds, brisk air.

July 9. [Selborne] Young swifts helpless squabs still. Young martins not out. *Footnote.* June 30. I procured a bricklayer to open the tiles in several places of my Bro͏ʳˢ brewhouse in order to examine the state of swift's nests at that season, & the number of their young. This enquiry confirmed my suspicions that they never lay more than *two* eggs at a time : for in several nests which we discovered, there were only *two* squab young apiece. As swifts breed but *once* in a summer, & the other hirundines twice : the latter, who lay from four to six eggs, encrease five times as fast as the former : & therefore it is not to be wondered that swifts are very numerous.

July 12. Martins build nests & forsake them, & now build again. Much hay spoiled : much not cut.

July 13. Martins hover at the mouth of their nests, & feed their young without settling.

July 14. Swifts, at least 30 : at times they seem to come from other villages.

July 15. No young martins out yet. Creeping white mist.

July 16. Swallows strike at owls, & magpies. Cut part of my great mead : grass over-ripe.

July 19. Put part of my meadow-hay in large cock.

July 24. Some young martins come out.

July 22. Hay well made at last. Swifts pursue & drive away an hawk : but do not dart down & strike him with that fury that swallows express on the same occasion. In these attacks they make some noise with their mouths, squeaking a little.

July 24. Young swallows & martins begin to congregate on roofs. These are the first flight.

July 25. Grapes very small & backward for want of sun. qu : if they will ripen. *Later note.* They did in Oĉ͏ᵗʳ.

July 26. Finished my meadow-hay in good order : St foin spoiled.

July 27. Turned out the worst of my S^t foin for thatch for my rick.

July 29. Sowed turnep, & Wrench's radishes. Young swallows continue to come out. Harvest begun on the downs : wheat not ripe. Destroyed a wasps-nest full of young vermiculi [grubs] in their several gradations towards a perfect state. Thermom^r 64 on the stair-case : 59 in the wine-vault. Wrench's radishes come from France, & are said to be a new sort : are called the Nismes radish. Sowed a crop of spinage to stand the winter.

Aug. 1. Wheat cutting near Whorwel : much lodged [laid by storms]. Swifts flie up to the tower, & cling against the walls : qu : are not those young ones that do so.

Aug. 2. Sun, sweet day. A chilly autumnal feel in the mornings & evenings.

Aug. 3. First apricots : first french-beans.

Aug. 4. Swifts began to withdraw [migrate] about this time at Blackburn in Lancashire.

Aug. 6. The trufle-hunter took one pound of trufles at Fyfield. Swifts disappeared at Fyfield on this day. A colony of swifts builds in the tower of London. Swifts in general seemed to withdraw from us on this day. Annuals are stunted, & not likely to blow well.

Aug. 7. Much wheat blighted. Swifts have not appear'd for these two evenings.

Aug. 9. Young martins abound. Wheat-harvest begins with us. The swifts appear again.

Aug. 10. No swifts : are seen no more with us.

Aug. 11. One of my vines looks pale & sickly. Ivy buds for bloom : it blows in Oct^r & Nov^r & the fruit ripens in April.

Aug. 12. Fly-catchers bring out young broods. Mich : daisy blows. Apricots ripen. Some martins, dispossessed of their nests by sparrows, return to them again when their enemies are shot, & breed in them. Several pairs of martins have not yet brought forth their first brood. They meet with interruptions, & leave their nests.

Aug. 15. [Meonstoke] Showers & sun. Meonstoke a sweet district.

Aug. 17. Wheat-harvest general. Large seagulls.

Aug. 18. Two swifts were seen again on this day at Fyfield : none afterwards. Two last swifts seen at Blackburn in Lancashire.

Aug. 20. [Selborne] Vast dew, sweet day. Aster chinensis.

Aug. 21. Sun, sweet day, full moon.

Aug. 23. Missel-thrushes congregate & are very wild. Thistle-down floats. Thompson, who makes this appearance a circumstance attendant on his summer evening,

" Wide o'er the thistly lawn, as swells the breeze,
   A whitening shower of vegetable down
   Amusive floats " . . .

seems to have misapplyed it as to the season : since thistles which do not blow 'til the summer-solstice, cannot shed their down 'til autumn.[15]

Aug. 29. Gathered the first plate of peaches : ripe but not high-flavoured. First bleached endive.

Aug. 30. Pulled the first Wrench's radishes : they are mild & well-flavoured : are long & tap-rooted : bright red above ground, & milk-white under.

Aug. 31. Spitting rain, with wind all day. Wheat begins to grow. Several nectarines rot on the trees. Peaches rot : plums burst & fall off.

Sept. 2. Hop-picking begins at Faringdon. Wasps abound.

Sept. 3. Turned the horses into the great mead : there is good after-grass considering that the field was not mown 'til July 16.

Sept. 4. Wood-owls hoot much.

Sept. 5. Most people at Selborne begin picking their hops. Wheat housed all day.

Sept. 6. Some grapes begin to turn colour. Peaches & Neɗ: much damaged by the wet. Black-caps ſtill.

Sept. 7. Hops brown & small, & not eſteemed very good. Wheat out ſtill.

Sept. 8. Wheat housing. Whitethroats ſtill seen.

Sept. 9. Mushrooms. Hops diſtempered.

Sept. 10. Oats housed all day. Swifts retire usually between the 10th & the 20th of Aug : fly-catchers, ſtoparolae, which are the lateſt summer birds of passage, not appearing 'til the 20 of May, withdraw about the 6th of Septemr.

Sept. 11. Martins do not seem to engage much this year in second broods. Are they discouraged by the cold, wet season?

Sept. 12. Great hail at Wintōn. Wasps abound in woody, wild diſtriɗs far from neighbourhoods : how are they supported there without orchards, or butchers shambles, or grocers shops? *Footnote.* Wasps neſting far from neighbourhoods feed on flowers, & catch flies & caterpillars to carry to their young. Wasps make their neſts with the raspings of sound timber ; hornets with what they gnaw from decayed. These particles of wood are neaded up with a mixture of saliva from their bodies, & moulded into combs.

Sept. 14. Ring-ouzels feed on our haws, & yew-berries in the autumn, & ivy-berries in the spring.

Sept. 15. Ring-ouzels appear on their autumnal migration. Were seen firſt laſt year on the 30th : the year before on the 11th.

Sept. 16. Much barley & oats is housed, but in poor condition. Peaches & neɗ: good, but much eaten by wasps, & honey-bees. Bees are hungry some autumns, & devour the wall-fruit.

Sept. 19. A moor-buzzard with a white head was shot some time ago on Greatham-moor.

Sept. 21. Swallows hawking about very briskly in all the moderate rain. Martins about.

Sept. 22. [Basingstoke; Alton] The oestrus curvicauda is found in Lancashire: probably the kingdom over. It lays it's nits on horses legs, flanks,
&c: each one on a single hair. The maggots when
hatched do not enter the horses skins, but fall to
the ground. On what & how are they supported?
[Cf. 8 Aug. 1770, and *passim*.] *Footnote.* Earth-
worms obtain & encrease in the grass-walks, where
in levelling they were dug down more than 18 inches.
So that they were either left in the soil, deep as it was
removed: or else the eggs or young remained in
the turf. Worms seem to eat earth, or perhaps
rotten vegetables turning to earth; also brick-dust
lying among earth, as appears by their casts. They
delight in slopes, probably to avoid being flooded; &
perhaps supply slopes with mould, as it is washed away
by rains. They draw straws, stalks of vine-leaves, &c.
into their holes, no doubt for the purpose of food.
Without worms perhaps vegetation would go on
but lamely, since they perforate, loosen, & meliorate
the soil, rendering it pervious to rains, the fibres of
plants, &c. Worms come out all the winter in mild
seasons.[16]

Sept. 25. Wood-lark sings.

Sept. 26. Planted numbers of brown Dutch lettuces
under the fruit-wall to stand the winter. *Later note.*
These proved very fine the spring following.

Sept. 27. Mr Yalden mows a field of barley.
Much barley abroad.

Sept. 28. All things in a drowning condition!

Sept. 29. Hops in some places not yet gathered.
Grapes begin to be good: the crop is scanty, & the
branches & berries small.

Sept. 30. Rooks begin to frequent the wall-nut
& carry-off the fruit.

Oct. 3. Gathered in the more choice pears,
Autumn burgamots, chaumontels, &c.; a good crop.

Oct. 5. Mr Yalden houses barley. No hirundines
appear all this day, though the weather is so fine,

& the air full of insects.   [Here follows note on water-
rat : see T.P., XXVI.]   *Footnotes.*   Many ash-trees
bear loads of keys every year ; others never seem to
bear any at all.   The prolific ones are naked of leaves,
and unsightly : those that are steril abound in foliage,
& carry their verdure long while, & are pleasing
objects.   *Later undated note.*   The cones of the
spruce-firs, which were produced Summer 1773 are
all just fallen off : so that they hung two summers
& one winter before they quitted the tree.   My
spruce-firs produced no cones last summer.

Oct. 9.   [Selborne : Lasham] Three swallows at
Faringdon.

Oct. 10.   [Oxford] Dark morning, small showers,
bright afternoon.[17]

Oct. 11.   Began gathering apples, a large crop of
some sorts.   Mr Yalden says he saw a woodcock
this day.

Oct. 12.   Hops sold at Wey-hill fair from £2 : 16 : o
to £4 : 4 : o :   & £5 : o : o to £5 : 10 : o.

Oct. 16.   [Oxford] Great fog, white frost, bright.

Oct. 19.   [Lasham] Dark, & cold, dark, bright.

Oct. 23.   [Selborne] Snipes begin to quit the moors,
& to come up into the wet fallows.

Oct. 25.   Beautiful season for sowing of wheat.
Much wet ground sown.

Oct. 26.   The air swarms with insects, & yet the
hirundines have disappeared for some time : hence
we may infer that want of food alone cannot be the
motive that influences their departure.   *Footnote.*
Many little insects, most of which seem to be tipulae,
[crane fly group] continue still to sport & play about
in the air, not only when the sun shines warm ; but
even in fog & gentle rain, & after sunset.   They
appear at times the winter thro' in mild seasons ;
& even in frost & snow when the sun shines warm.
They retire into trees, especially ever-greens.

Nov. 2.   Rooks gather acorns from the oaks.

Nov. 3.   Great field-fares flock on the down.

Nov. 4.   Grapes now delicate, & in good plenty : they had never ripened, had not Oĉᵗʳ proved a lovely month.   The rallus porzana, [*Rallus aquaticus*] or spotted water-rail, a rare bird, was shot in the sedge of Bean's pond.¹⁸   This was the firſt of the sort that ever I heard-of in these parts.   I sent it to London to be ſtuffed & preserved.   A beautiful bird.

Nov. 11.   Firſt day of winter.   Snow on the ground.

Nov. 12.   Gathered in all the grapes.   Snow on the hills.

Nov. 14.   [Sutton] Froſt, sun.

Nov. 17.   Trimmed, & tacked the vines : pretty good wood towards the S.E. for next year's bearing. The S.W. vines are weak in the wood.

Nov. 21.   [Alton.]

Nov. 22.   [London]¹⁹ *Footnotes*.   When I came to town I found that herrings were out of season : but sprats, which Ray says are undoubtedly young herrings,²⁰ abounded in such quantities, that in these hard times they were a great help to the poor.   Cods & haddocks in plenty : smelts beginning to come in. The public papers here abounded with accounts of moſt severe & early froſts, not only in the more Northern parts of Europe, but on the Rhine, & in Holland.   The news of severe weather usually reaches us some days before the cold arrives ; which moſt times follows soon when we hear of rigorous cold on the Continent.

Dec. 9.   [Selborne] Almoſt continual froſt from Nov. 20 : & some snow frequently falling.   Mergus serratus, the Dun-diver, a very rare bird in these parts, was shot in James Knight's ponds juſt as it was emerging from the waters with a considerable tench in it's Mouth.   It's head, & part of the neck, was of a deep ruſt-colour.   On the back part of the head was a considerable creſt of the same hue.   The sexes in this species, Ray observes, differ so widely, that writers have made two species of them.   It appears from Ray's description that my specimen

with the rust-coloured head was a female, called in
some parts the sparlin-fowl;[21] & is, he supposes,
the female Goosander.

Dec. 13. The frost seems to have done no harm.

Dec. 14. Dark & mild, spitting rain, great rain.
Earth-worms are alert, & throw-up their casts this
mild weather.

Dec. 15. The air abounds with insects dancing
about over the evergreen trees. They seem to be
of the genus of tipula, & empis. Phalaenae come
out in the evening: they seem to be much hardier
than the papiliones, appearing in mild weather all the
winter through. Full moon.

Dec. 17. Mrs Snooke's tortoise, after it had been
buried more than a month, came forth & wandered
round the garden in a disconsolate state, not knowing
where to fix on a spot for it's retreat.

Dec. 18. Rooks resort to their nest-trees.

Dec. 24. Grey & sharp. Vast flight of wild-
fowl haunt Woollmer-pond: the water in some parts
is covered with them. They are probably more
numerous on account of the early severity of the
weather on the continent.

CHAPTER VIII

. . . " *Rapiamus, amici, occasionem de die."* (HORACE :
Epod. xiii., 3, 4.)

1775

Jan. 2. Grey, & white water-wagtails appear
every day : they never leave us in the winter.

Jan. 7. Some ivy-berries half grown.

Jan. 14. The hawk *proinith,* says the new glossary
to Chaucer ; that is picketh, or dresseth her feathers :
from hence the word *preen,* a term in ornithology,
when birds adjust, & oil their feathers.

Jan. 20. Mr Hool's man says that he caught this
day in a lane near Hackwood-park, [near Basingstoke]
many rooks, which attempting to fly fell from the
trees with their wings frozen together by the sleet,
that froze as it fell. There were, he affirms, many
dozens so disabled ! It is certain that Mr H's man
did bring home many rooks & give them to the poor
neighbours.

Jan. 21. Received two bramblings from Mr
Battin of Burkham [Burcombe, near Salisbury].
They are seen but seldom in these parts : are fine
shewy birds.

Jan. 24. Dark & sharp, sun, cutting wind, hard
frost. Icicles. *Footnote.* Chaucer, speaking of Gossamer as a strange phenomenon, says,

" As sore some wonder at the cause of thunder ;
    On ebb, & flode, on *gosomor,* & mist ;
    And on all thing ; 'til that the cause is wist."[1]

Feb. 1. Vast rain, stormy. Much damage was
done by sea & land ; & on the river at London.

Feb. 2. Much damage at Portsmouth by unusual
tides, & at the isle of Wight. *Footnote.* A rook
should be shot weekly the year thro', & it's crop

examined: hence perhaps might be discovered whether in the whole they do more harm or good from the contents at various periods. Tho' this experiment might show that the birds often injure corn, & turneps; yet the continual consumption of grubs, & noxious insects would rather preponderate in their favour.

Feb. 5. Helleborus viridis emerges out of the ground, budding for bloom. Laurustine blooms.

Feb. 8. Many species of muscae come-out. Earthworms lie out at their holes after 'tis dark.

Feb. 9. Many species of Insects are stirring thro' every month in mild winters.

Feb. 10. Mezereon in fine bloom. Peter Wells's well runs over. Spiders, woodlice, lepismae in cupboards, & among sugar, some empedes, gnats, flies of several species, some phalenae in hedges, earth-worms, &c.: are stirring at all times when winters are mild; & are of great service to those soft-billed birds that never leave us.

Feb. 12. Sad accounts from various parts of devastations by storms & inundations. *Footnote.* A spoon-bill, platalea leucorodia Linn: was shot near Yarmouth in Norfolk: it is pretty common in Holland, but very rare indeed in this island.[2] There were several in a flock. They build Willughby says, like Herons in tall trees. Their feet are semipalmated. Those birds in Norfolk must have crossed the German ocean.

Feb. 19. Vast flocks of hen-chaffinches. Honeybees come forth, & gather on the Crocuss.

Feb. 22. Turnep greens run very fast: forward turneps rot.

Feb. 23. Flocks of hen chaffinches, with some bramblings among them. Saw several empty nut-shells with a hole in one side, fix'd in the chinks on the head of a gate-post, as it were in a vice, & pierced, as I suppose, by a nuthatch, sitta europaea.[3] Vid: Willughby's Ornithol:

Feb. 24. Appleshaw river [near Weyhill] runs.

Feb. 25. Honey-bees, & many dipterous insects abound. Frogs croak in ponds. A pair of house-pigeons, which were hatched at Mich: last, now have eggs, & sit. An instance of early fecundity !

Feb. 26. Viola odorata. Ivy-berries begin to turn black.

Feb. 27. Crocuss in great splendour.

Feb. 28. Spiders shoot their webs from clod to clod.

Mar. 3. Rooks begin to build. They began the same day at Fyfield. *Footnote.* Swine & sheep, for such large quadrupeds, become prolific very early ; since a sow at four months old requires the boar : & ram-lambs, which fall in Jan. & Feb. if well kept, will supply the wants of their own dams by the following Octob[r], & beget lambs for the next year. Horses & kine seldom procreate 'til they are two years old.

Mar. 7-18. [Fyfield] Bro[r] Harry's strong beer, which was brewed last Easter monday with the hordeum nudum[4], is now tapped, & incomparably good : it is somewhat deeper-coloured than beer usually is in this country, not from the malt's being higher dryed, but perhaps from the natural colour of the grain. The barrel was by no means new, but old & seasoned. Wheat, it seems, makes also high-coloured beer. Sad season for the sowing of spring-corn. Just such weather this time twelvemonths.

Mar. 10. Rooks are very much engaged in the business of nidification : but they do not roost on their nest-trees 'til some eggs are lain. Rooks are continually fighting & pulling each other's nests to pieces : these proceedings are inconsistent with living in such close community. And yet if a pair offers to build on a single tree, the nest is plundered & demolished at once. Some rooks roost on their nest-trees.

Mar. 11. Vast rain. This rain must occasion great floods. The trufle-hunter came this morning,

& took a few trufles : he complains that those fungi never abound in wet winters, & springs.

Mar. 15. Hard frost, hot sun. Sheltered the fruit-wall bloom with boughs of ivy & yew.

Mar. 16. Ephemerae bisetae come forth.

Mar. 17. Nuthatch brings out & cracks her nuts, & strews the garden-walks with shells. They fix them in a fork of a tree where two boughs meet : on the Orleans plum tree.

Mar. 18. [Selborne] Adoxa moschatellina [fragrant moschatel]. The twigs which the rooks drop in building supply the poor with brush-wood to light their fires.* Some unhappy pairs are not permitted to finish any nest 'til the rest have compleated yir building ; as soon as they get a few sticks together a party comes & demolishes the whole. As soon as rooks have finished their nests, & before they lay, the cocks begin to feed the hens, who receive their bounty with a fondling tremulous voice & fluttering wings, & all the little blandishments that are expressed by the young while in a helpless state. This gallant deportment of the males is continued thro' the whole season of incubation. These birds do not copulate on trees, nor in their nests, but on the ground in open fields.

* Thus did the ravens supply the prophet with necessaries in the wilderness.

Mar. 21. Mrs Snooke's old tortoise came out of the ground, but in a few days buried himself as deep as ever. Earth-worms lie out, & copulate.

Mar. 22. Snake appears : toad comes forth. Frogs spawn. Horse-ants come forth.

Mar. 23. Earthworms travel about in rainy nights, as appears from their sinuous tracks on the soft muddy soil, perhaps in search of food.

Mar. 24. The apricot-bloom, which came out early, seems to be much cut by the late frosts. Peaches & Nect : now in fine bloom.

Mar. 25. Planted a raspberry bed.

Mar. 27. The creeper,[5] a pretty little nimble bird, runs up the bodies & boughs of trees with all the agility of a mouse. It runs also on the lower side of the arms of trees with its back downward. Stays with us all the winter.

Mar. 30. Horse-ants retire under the ground. Wheat-ears appear.

Mar. 31. [Midhurst] Birds eat ivy-berries, which now begin to ripen : they are of great service to the winged race at this season, since most other berries ripen in the autumn. *Footnote.* The shell-less snails, called slugs, are in motion all the winter in mild weather, & commit great depredations on garden-plants, & much injure the green wheat, the loss of which is imputed to earth-worms ; while the shelled snail, the φερεοικος, does not come forth at all 'til about April the tenth ; and not only lays itself up pretty early in the autumn, in places secure from frost ; but also throws-out round the mouth of it's shell a thick operculum formed from it's own saliva ; so that it is perfectly secured, & corked-up as it were, from all inclemencies. Why the naked slug should be so much more able to endure cold than it's housed congener, I cannot pretend to say.

Apr. 1. [Selborne] White frost, sun, dark clouds.

Apr. 2. Bees resort to the hot-beds tempted by some honey spread on the leaves, & blossoms of the cucumbers. When bees do not frequent the frames, the early fruit never sets well : therefore this expedient is very proper for early melons, & cucumbers.

Apr. 7. Prunus spinosa. The black-thorn begins to blow. This tree usually blossoms while cold N.E. winds blow : so that the harsh rugged weather obtaining at this season is called by the country people black-thorn winter.

Apr. 11. Two swallows. Black snail. Some few apricots, which escaped the frost, seem to be set. Some peach & nect : bloom not destroyed. The trees

were struck full of ivy-boughs, which seem to have
been of service against the severe cold.

Apr. 13. The barley-season goes on briskly.
Hops are poling. Curlews clamour. *Footnote.* The
Saxon word *hlithe*, with the h aspirate before it,
signifies *clivus :* hence no doubt two abrupt steep
pasture-fields near this village are called the short,
& long *lithe.* Much such another steep pasture at
about a mile distance is also called the *lithe. Steethe*
in Saxon signifies *ripa*, a perpendicular bank : hence
steethe swalwe, riparia hirundo.

Apr. 17. [Lasham] Mrs Snooke's tortoise came
out of the ground the second time, for the summer.

Apr. 18. [Oxford] Luscinia. Cuculus. Inyx.

Apr. 22. [Lasham] Several beeches in the hanger
begin to leaf. Black snails abound. Worms, when
sick, seem to come out of the Ground to die : under
the same circumstances some amphibiae quit the
water. *Footnote.* Thomas [Hoar] kept a journal of
incidents during my absence.

Apr. 23. [Lasham ; Selborne] Swallows abound ;
but no house-martin, or black-cap. No swift.

Apr. 25. Ivy-berries fall off dead-ripe.

Apr. 27. Early tulips blow. A pair of house-
martins appear, & frequent the nest at the end of the
house ; a single one also wants to go in. These
must be of the family bred there last year. The nest
was built last summer. The martins throw the rubbish out of their nest. Bank-martins abound on
short-heath : they come full as soon as the house-
swallow. Two swans inhabit Oakhanger-ponds :
they came of themselves in the winter with three
more. Pulled down many old House-martins nests ;
they were full of rubbish, & the exuviae of the hippobosca hirundinis in the pupa state. These insects
obtain so much sometimes in y^ir nests, as to render
the place insupportable to the young, & to oblige
them to throw themselves on the ground. The case
is the same sometimes with young swifts.

Apr. 28. Sun, sultry, fierce heat! Midsummer evening. The sun scorched 'til within an hour of setting. Swift appears at Manchester & Fyfield. Apus, one single swift. They usually arrive in pairs. Parhelia, or odd halo round the sun. Described since in Gent: mag.

Apr. 30. Gardens much injured by the heat. White throat appears, & whistles, using odd gesticulations in the air when it mounts above the hedges.

May 1. Sowed two boxes of polyanth-seed from London. Sowed a large bed of carrots, which could not be sowed before on account of the long dry season. Ground still dry & harsh. Some oaks & ashes half in leaf. The beeches on the hanger in full leaf. Trees more than a fortnight forwarder than they have been for some years past.

May 3. One swift at Bramshot; one at Selborne. *Footnote.* At Blackburn in Lancashire swallows first seen April 15: swifts April 28; house-martins May 4th. Cuckow sings April 28: laughing wren [willow wren] sings Apr: 17. Several ponds are dry.

May 5. House-snails abound now: scarce any have appeared before on account of the long drought.

May 9. The long rows of tulips make a gallant shew.

May 12. Fern-owls chatter in the highwood & hanger.

May 13. Papilio Atalanta.[6] This is an autumnal fly, & therefore must appear at this season by accident. Fine rains about the kingdom; but little to the advantage of our district. At Lyndon in Rutland, the first swallow was seen April 14: first swift April 29: first H: martin May 6.

May 14. Two pairs of nightingales in my fields. The country strangely dryed-up. Fine showers about last friday.

May 18. Ponds fail. Watered away hogsheads on the garden, which is burnt to powder.

May 19. No chafers appear as yet : in those seasons that they abound they deface the foliage of the whole country, especially on the downs, where woods & hedges are scarce. Regulus non cristatus stridet voce locustulae [wood wren] : this bird, the latest & largest willow-wren, haunts the tops of the tallest woods, making a stammering noise at intervals, & shivering with it's wings. Bank-martins abound over the ponds in the forest : swifts seldom appear in cold, black days round the church.

May 21. Mr Yalden's tank is dry.

May 23. Dutch-honeysuckles in fine bloom.

May 24. Thrushes now, during this long drought, for want of worms hunt-out shell-snails, & pick them to pieces for their young. My horses begin to lie abroad.

May 26. We are obliged to water the garden continually. Some wells dry.

May 27. No thoro' rain in this district since the 9, 10, & 11 of March. [Here follows a note on upland ponds, referred to in D.B., XXIX.]

May 29. Grass on the common burnt very brown. Tulips decay. No dews for mowing in common.

May 30. House-martins do not build as usual : perhaps are troubled to find wet dirt. Bees swarm. Severe heat in the lanes in the middle of the day.

June 1. Martin begins to build at the end of the brewhouse.

June 3. Hot sun, & brisk gale, sweet even. Dusty beyond comparison. Watered away five hogs-heads of water. Stoparola has five eggs. Rooks live hard : there are no chafers. Barley & oats do not come up; the fields look naked. Some pairs of swifts always build in this village under the low thatched roofs of some of the meanest cottages : & as there fails to be nests in those particular houses, it looks as if some of the same family still returned to the same place.

June 4. Roses begin to blow: pinks bud; fraxinella[7] blows. Garden burnt to powder.

June 5. Diurnal birds suffer for want of worms: thrushes seem to live on snails. *Footnote.* Thrushes do great service in hunting-out the shell-snails in the hedges, & destroying them: the walks are covered with their shells.

June 6. Swifts abound: near 15 pairs: they seem to come from other villages. H: martins now abound, & build briskly.

June 7. Watered the wall-trees well this evening with the engine: the leaves are not blotched & bloated this year, but many shoots are shrivelled, & covered with aphides. Plums & pears abound; moderate crop of apples with me. Vine-shoots very forward.

June 10. Shower in the night. Planted-out vast quantities of annuals both in the borders, & basons; both in the fields, & gardens.

June 11. The autumn-sown brown lettuces, which stood the winter, still continue good. The dry season last friday morning had lasted just 3 months: the 9, 10, & 11 of March were very wet.

June 13. Red kidney beans begin to climb their sticks. Mulberry-tree in full leaf. Snails copulate.

June 14. We had just the skirts of a vast thunder-storm.

June 15. Tremendous thunder, & vast hail yesterday at Bramshot, & Hedley [near Selborne] with prodigious floods. Vast damage done. The hail lay knee-deep. *Footnote.* The shell-snail has hardly appeared at all this season on account of the long dry time. Snails copulate about Midsum[r]; & soon after deposit their eggs in the mould by running their heads & bodies under ground. Hence the way to be rid of them is to kill as many as possible before they begin to breed. In six weeks after wheat is in ear, harvest usually begins; unless delayed by cold, wet, black weather.

June 20.   Meadow-grass very short indeed.

June 21.   Hay makes at a vast rate. Vast crops of plums, currants, & goose-berries. House-martin which laid in an old nest, hatches. House-martins, which breed in an old nest, get the start of those that build in new ones by 10 days, or a fortnight.

June 22.   Pines begin to ripen at Hartley. I have not seen the great species of bat this summer. *Footnote*. Teals breed in Woolmer-forest : jack-snipes breed there also no doubt, since they are to be found there the summer thro'. A person assures me, that Mr Meymot, an old clergyman at Northchappel in Sussex, kept a cuckow in a cage three or four years ; & that he had seen it several times, both winter & summer. It made a little jarring noise, but never cryed *cuckow :* It might perhaps be a hen.[8] He did not remember how it subsisted.

June 25.   Wheat in general out of bloom. After so kind a blowing time we may from the heat of the summer expect an early, & plentiful wheat-harvest.

June 29.   Young minute frogs migrate from the ponds this showery weather, & fill the lanes & paths : they are quite black.

July 1.   On the 28 of June a large quantity of trufles were found near Andover, near two months sooner than the common season. So these roots are in season nine months at least. *Footnote*. House-snails seem to be so checked by the drought, & destroyed by the thrushes, that hardly one annual is eaten or injured. When earth-worms lie-out a nights on the turf, though they extend their bodies a great way, they do not quite leave their holes, but keep the ends of their tails fixed therein ; so that on the least alarm they can retire with precipitation under the earth. Whatever food falls within their reach when thus extended they seem to be content with, such as blades of grass, straws, fallen leaves, the ends of which they often draw into their holes. Even in copulation their hinder parts never quit their holes ; so that no two, except they lie within reach of each others bodies, can have any commerce of that kind ; but as every individual is an hermaphrodite, there is no difficulty in meeting with a mate ; such as would be the case were they of different sexes.

July 4.   Whortle-berries ripe.

July 6.   Wasps begin to come.   Growing weather.

July 10.   Mushrooms begin to appear.

July 11.   Destroyed a wasp's nest which was grown into a considerable bulk, & had many working wasps.

July 12.   Five young kestrils, or windhovers almost fledge are taken in an old magpie's nest.

July 14.   Hay much damaged : many meadows not cut.   This dripping season, which hurts individuals in their hay, does marvelous service to the public, in the spring-corn, after-grass, turneps, fallows, &c. Oats are much recovered, & brought-on.   Wheat begins to change colour ; is not lodged.   *Footnote.* When a person approaches the haunt of fern-owls (caprimulgi) in an evening, they continue flying round the head of the obtruder ; & striking their wings together above their backs, in the manner that the pigeons called smiters are known to do, make a smart snap : perhaps at that time they are jealous for their young ; & their noise & gesture are intended by way of menace.

July 16.   Some of the forwardest birds of some broods of martins are out, the more backward remain in the nest.

July 17.   Some martins are building against Mr Yalden's windows.   Young martins—perchers on the battlements of the tower, where the old ones feed them.   *Footnote.* The young martin becomes a flyer in about sixteen days from the egg : most little birds come to their ἡλικία, or full growth, in about a fortnight : for were they to lie a long time in the nest in a helpless state, few would escape ; some mischief or other would destroy the whole

breed. The more forward pulli are out some days
before the underlings of the same brood.

July 19. Five wasps nests destroyed this evening :
two before.

July 21. Opened the crop of a swift, & found it
filled with the wing-cases & legs, &c. of small
coleoptera.⁹ Hence it is plain the coleoptera soar
high in the air.

July 22. The swifts are so fledge before they
quit the nest that they are not to be distinguished
from their dams on the wing ; yet from the encrease
of their numbers, & from their unusual manner of
clinging to walls & towers one may perceive that
several are now out. And no wonder that they
should begin to bestir themselves, since they will
probably withdraw in a fortnight.

July 23. Birds are much influenced in their choice
of food by colour : for tho' white currans are a much
sweeter fruit than red ; yet they seldom touch the
former 'til they have devoured every bunch of the
latter.¹⁰ The male & female ants of the little yellow
& little black sorts, leaving their nests, fill the air.
The females seem big with eggs. They also run
about on the turf, & seem in great agitation. The
females wander away, & form new colonies when
pregnant.

July 24. Hops throw out good side shoots &
blow. Some few hills have perfect hops. A sea-
lark ¹¹ shot at Newton-pond.

July 28. Ten nests of wasps have been destroy'd
just at hand : they abound & are plowed up every
day.

July 30. By this evenings post I am informed, by
a Gent : who is just come from thence that the hops
all round Canterbury have failed : there are many
hundred acres not worth picking.

July 31. Horses at plow so teized by flies as to be
quite frantic. Horses are never tormented in that
manner 'til after midsumʳ : the people say it is the

nose-fly that diſtraſts them so. I can discover only
such flies as haunt the heads of horses : perhaps at
this season they lay their eggs in the nose & ears.
I never can discern any *oeſtrus* on these days ; only
swarms of small *muscae*.

Aug. 1. [Chilgrove] Small rain, sun & clouds.

Aug. 2. Wheat harveſt is general all about the
downs. When I came juſt beyond Findon I found
wheatear traps which had been open'd about a
week. The shepherds usually begin catching about
the laſt week in July.[12]

Aug. 3-18. [Ringmer] Female viper taken full
of young, 15 in number : gaped & menaced as soon
as they were out of the belly of their dams.[13]

Aug. 4. Little wheat housed. Wheat is very
fine in general. A young cuckow is hatched every
year in some part of Mrs Snooke's out-let, moſt usually
by red-breaſts. No cross-bills this year among
the Scotch-pines. They usually appear about the
beginning of July. No fern-owls.

Aug. 6. Multitudes of swallows of the firſt brood
cluſter on the Scotch-firs. The swifts, or the bulk
of them, departed from Fyfield, about this day.

Aug. 7. Timothy, Mrs Snookes' old tortoise has
been kept full 30 years in her court before the house,
weighs six pounds three quarters, & one ounce.
It was never weighed before, but seems to be much
grown since it came.

Aug. 8. Broods of flycatchers come out.

Aug. 12. Full moon. High tides frequently dis-
compose the weather in places so near the coaſt, even
in the dryeſt, moſt settled seasons, for a day ot two.

Footnote. Cimiccs lincares arc now iħ high copulation on ponds & pools. The females, who vaſtly
exceed the males in bulk, dart & shoot along the
surface of the water with the males on their backs.
When a female chuses to be disengaged, she rears &
jumps & plunges like an unruly colt ; the lover,
thus dismounted, soon finds a new mate.   The females

as fast as their curiosities are satisfied retire to another
part of the lake, perhaps to deposit their foetus in
quiet: hence the sexes are found separate except
where generation is going-on.   From the multitude
of minute young of all gradations of size, these
insects seem without doubt to be viviparous.

Aug. 14.   Two great bats appear.   They feed
high: are very rare in Hants, & Sussex.   Low fog.

Aug. 15.   Dark & still.   Some little farmers have
finished wheat-harvest.

Aug. 16.   Generation seems to be pretty well over
among cimices lineares.   Minute young abroad.

Aug. 17.   [Findon] Rabbits make incomparably
the finest turf, for they not only bite closer than larger
quadrupeds; but they allow no bents to rise: hence
warrens produce much the most delicate turf for
gardens.   Sheep never touch the stalks of grasses.

Aug. 18.   [Chilgrove] Grey.   Sweet afternoon.

Aug. 19.   [Selborne] Wheat-harvest in general
seems to be finished, except where there is turnep
wheat.   Fifteen wasps nests have been destroyed
round the village; yet those plunderers devour the
plum, & eat holes in the peaches & nectarines before
they are ripe; & will soon attack the grapes.   Grapes
begin to turn color: they are forward this year.
Harvest-weather was much finer at Ringmer than
Selborne.   Some wheat a little grown at Newton.

Aug. 20.   Hops are uneven: some grown large,
& some just blown.

Aug. 23.   Sixteen wasps nests destroyed.

Aug. 24.   Wasps abound, & destroy the friut.
Clouds about.   Worms copulate.

Aug. 25.   Sr Simeon Stuart begins to pick his
hops.   Wasps have begun on the Grapes.   Seventeen
wasps nests destroyed.   Peaches are gathered every
day, being injured by the wasps: they are not full
ripe.   *Footnote.   Twaite,* in Saxon is ground cleared
from wood, & plowed; *Woddan* is not a way, but
the verb *to go :* wud is wood in Saxon.

Aug. 27.   8 more wasps nests ; in all 25 have been destroyed round the village.

Aug. 29.   [Meonstoke] showers & sun.

Sept. 1.   [Selborne] Barley begins to be injured. Many fields of barley, green & not mowed.

Sept. 2.   Gathered first grapes : they look well ; & are large ; but not highly flavoured yet. Sad hop-picking : a large crop in this district. Barley in the suds.[14]

Sept. 3.   Great rain. Hops sadly washed. Destroyed the 26th wasps-nest, a vast colony.

Sept. 4.   Linnets congregate. Wasps swarm about the Grapes, tho' so many nests have been destroyed.

Sept. 5.   Grey, spitting, bright & sultry, distant lightening. Wasps swarm.

Sept. 6.   Wasps abound not only in neighbourhoods, but in lone fields, & woods ; how satisfied there ?

Sept. 7.   In the dusk of the evening when beetles begin to buz, partridges begin to call ; these two circumstances are exactly coincident.

Sept. 8.   Wasps abound, & mangle the grapes : we have, I should think, destroyed 50000.

Sept. 9.   Wasps somewhat abated. The day & night insects occupy the annuals alternately : the papilios, muscae, & apes are succeeded at the close of the day by phalenae, earwigs, woodlice, &c. My tallest beech measures in girth at three feet from the ground six feet & four inches. It grows at the S.E. end of Sparrow's hanger, [See map] & appears to be upwards of 70 feet high.

Sept. 11.   Much barley abroad, most of it standing : what is cut lies in a sad way. Hop-picking much interrupted : hops become brown.

Sept. 12.   Put 50 fine bunches of grapes in crape bags to secure them from wasps.

Sept. 13.   Good grapes every day, but not delicate. Bag-ed more grapes.

Sept. 14. Little barley housed towards Winton & Andover. Many crops not ripe.

Sept. 16. Wasps begin to abate. *Footnote.* On friday, Sep[r] 8th, at 10 at night a considerable earthquake was felt at Oxford, Bath, & several other towns.

Sept. 21. [Basingstoke] Showers, rainbow, bright. Barley in a sad condition about Basingstoke. Rams begin to pay court to the ewes.

Sept. 22. [Selborne] Ring-ouzels appear on the common on their autumnal migration. *Footnotes.* The large female-wasps begin to come in a door, & seem as if they were just going to hide, & lay themselves up for the winter. The common wasps are much abated in number. On wednesday the 20 there was a violent storm of thunder & lightening at Fyfield between ten & eleven at night.

Sept. 25. Gathered-in the swan's eggs, & autumn burgamot-pears : a vast crop of the former.

Sept. 26. Gathered in the golden-rennets. Apples are too large from much wet.

Sept. 27. Gathered-in the royal russets, & knobbed russets. Tyed-up endive. Winter garden-crops in vast vigor. *Footnote.* My Arundo donax,[15] which I received from Gibraltar, is grown this year eight or nine feet high : I therefore opened the head of one stalk to see what approaches it had made towards blowing after so hot a summer. When it was cut open we found a long series of leaves enfolded one within the other to a most minute degree, but not the least rudiments of fructification ; so that the plant must have extended itself many feet before it could have attained to it's full stature : & must have required many more weeks of hot weather before it could have brought any seeds to maturity.

Oct. 2. The barometer falls with great precipitation.

Oct. 4. One swallow. What can this bird be doing behind by itself ? Why might not they have all staid, since this individual seems brisk, & vigorous.

Oct. 5. Here & there a straggling swallow. Curlews clamor.

Oct. 6. Just before it was dark a flight of about 12 swallows darted along over my House towards the hill: they seemed as if they settled in the hanger. Now several house-martins appear about the hanger.

Footnote. An oak in Newton-lane near the Cross, by the condensation of the fogs on it's leaves has dripped such quantities for some nights past, that the water stands in puddles, & runs down the ruts. Why this tree should drip so much more than it's neighbours is not easy to say. No doubt this is one of the means by which small upland ponds are still supported with water in the longest droughts; & the reason why they are never dry. What methods of supply upland ponds enjoy, where no trees over-hang, may not be so easy to determine. Perhaps their cool surfaces may attract a fund from the air when it is loaded with fogs & vapors, especially in the night-time. That they have some never-failing stock at hand to counterbalance evaporation & the waste by cattle, is notorious to the meanest observer. For on the chalks no springs are ever seen on the tops or sides of hills, but in the bottoms alone. [Cf. D.B., XXIX.]

Oct. 9. Woodcock returns.

Oct. 10. Woodcock killed this day.

Oct. 11. Lowered the flower-bank in the garden, & laid it on a gentle slope, & filled it with flowers of various sorts.

Oct. 13. Hops were a drug at Weyhill-fair: unusual quantities were exposed for sale.

Oct. 14. Many people sow wheat: the ground works well.

Oct. 15. [Basingstoke] Mr Barker writes word that in Sep. last there fell in the county of Rutland near six inch: & ½ of rain. The beeches on the hanger, & the maples in my fields are now beautifully tinged,

& afford a lovely picturesque scape, very engaging to the imagination.

Oct. 16-20. [Oxford.]

Oct. 17. Turkies get up on the boughs of oaks in pursuit of acorns.

Oct. 19. Vast rain with stormy wind, this storm damaged my trees, & hedges. This storm occasioned much damage at sea, & in the river thames.

Oct. 20. [Basingstoke] One swallow near Wallingford. Strong wind. Acorns abound : the hogs in the lanes & woods seem to be half fat.

Oct. 21. [Selborne] The storm on thursday night tore all the remaining flowers to pieces. *Footnote.* With us the country people call coppice, or brushwood *ris*, or *rice* : now *hris* in Saxon signifies frondes, & is no doubt the word whence our provincial term originates. *Hraed hriz* is frondes celeres : hence probably *Red Rice*, the name of a hunting-seat standing in the midst of a coppice at Andover.[16]

Oct. 22. My autumn crop of spinage this year runs much to seed.

Oct. 25. The arbutus casts it's blossoms & discloses the rudiments of its fruit. In these two instances [ivy and arbutus] fructification [maturation] goes on the winter through. Three martins in the street. Gossamer on every bent. *Footnote.* Bynstede, the name of a parish near us, signifies locus cultus, vel habitatus. This parish abuts on a wild woodland district ,which is a royal forest, & is called the Holt. This parish was probably cultivated when all around were nothing but woodlands, & forests.

Oct. 29. Redwings on the hawthorns. Bat appears.

Oct. 30. Flocks of large fieldfares. Celeri finely blanched.

Oct. 31. Leaves fall very fast. The hangers begin to lose their picturesque beauties.

Nov. 3. Grapes are delicate still; especially those that are not bagged in crape: those that are, are shrivelled, & vapid. *Footnote.* The great month for spring migration is April: tho' the wryneck, one species of willow-wren, & the upland curlew are seen in March: in this month also the winter birds retire. In Sept: most of the short-winged summer birds withdraw; & in Oct: the wood-cock, redwing, & fieldfare return. The hirundines are more irregular in their retreat; for the Swift disappears in the beginning of Aug: the rest of the Genus not 'til Oct: In Nov: the wood-pigeons, & wild fowls return. We have found at times in the parish of Selborne alone about 120 species of birds, which are more than half the number that belong to Great Britain[17] in general; & more than half as many as Linnaeus can produce in the kingdom of Sweden. Mr Pennant enumerates 227 species in Gr: Britain, & Linnaeus about 221 in his native country.

Nov. 13. Charadrius pluvialis. Green whistling plovers [golden plovers] appear: they come in the autumn to us, but do not breed here. They haunt downs.

Nov. 14. Saw yesterday a considerable flock of gulls flying over the hanger to the S.W. Gulls very seldom appear in this district; except sometimes on the forest ponds. *Footnote.* When horses, cows, sheep, deer, &c. feed in wind, & rain, they always keep their heads down the wind, & their tails to the weather; but birds always perch, & chuse to fly, with their heads to the weather to prevent the wings from ruffling their feathers, & the cold & wet from penetrating to their skins.

Nov. 23. The high glass brings no good weather: Barom^rs usually dote,[18] & are mistaken about this time of the year.

Nov. 24. A flight of woodcocks about in the country.

Nov. 25. Many phalaenae appear. Strange that these nocturnal lepidoptera should be so alert, at a season when no day papilios appear, but have long been laid-up for the winter. Trees will not subsist in sharp currents of air : thus after I had opened a vista in the hedge at the E. corner of Baker's hill, no tree that I could plant would grow in that corner : & since I have opened a view from the bottom of the same field into the mead, the ash that grew in the hedge, & now stands naked on the bastion, is dying by inches, & losing all it's boughs. Phalaenae appear about hedges in the night time the winter thro'.

Nov. 26. Very dark season : dark within doors a little after 3 o'clock in the afternoon.

Nov. 27. Arrangement of parts is both *smell* & *color :* thus a sweet, & lovely flower when bruized both stinks & looks ugly. We may add that arrangement of parts is also flavor : since *muddled* liquors & *frozen* meats immediately lose it.

Nov. 29. The grey crow, a bird of winter passage, appears. It is as rare at Selborne, as the carrion crow is in Sweden. This is only the third bird that I have seen in this district. They are common on the downs at Andover, & Wintŏn. The air is unusually damp, with copious condensations on the walls, wainscot, looking-glasses, &c. of houses, in many places running in streams.

Dec. 1. Many species of flies come forth. Bats are out, & preying on phalaenae. The berries of Ivy, which blowed in the end of Sep : now half grown. A noble & providential supply for birds in winter & spring ! for the first severe frost freezes, & spoils all the haws, sometimes by the middle of Novr. Ivy-berries do not seem to freeze. *Footnote.* Large, grey, shell-less cellar-snails lay themselves up about the same time with those that live abroad : hence it is plain that a defect of warmth alone is not the only causes that influences their retreat. The rudiments of the arbŭtus[19] -fruit swell, & grow. Laurustines continue to blow.

Dec. 4. Furze blows. Colds & feverish complaints obtain in this neighbourhood. In London, Portsmouth, & other places colds, & coughs have been general : in Dublin also.

Dec. 8. Grey, gleams of sun, bright. Fogs on the hills : Spring-like : more like Feb : than Dec[r]. *Footnote*. Ravens in their common mode of flying have a peculiarity attending them not unworthy of notice : they turn-over in the air quite on their backs, & that not now & then, but frequently ; often every two or 300 yards. When this odd attitude betides them they fall down several fathoms, uttering a loud croak, & then right themselves again. This strange vacillation seems to be owing to their scratching, when bitten by vermin : the thrusting-out of their leg destroys their aequipoize, & throws their wings out of the true centre of gravity. Ravens spend their leisure-time over some hanging wood in a sort of mock fight, dashing & diving at each other continually ; while their loud croakings make the woody steeps re-echo again [Cf. D.B., XLII, XLIII].

Dec. 10. This epidemic disorder falls heavier on adults than children.

Dec. 13. Ice bears : boys slide.

Dec. 17. People recover from the epidemic disorder.

Dec. 19. 10 or 12 large gulls sailing high in the air over comb-wood pond.

Dec. 21. People fall with colds. Dry weather for near three weeks, 'til the ground was very free from water. *Footnotes*. The country-people, who are abroad in winter-mornings long before sun-rise, talk of much hard frost in some spots, & none in others. The reason of these partial frosts is obvious : for there are, at such times, partial fogs about : where the fog obtains little or no frost appears ; but where the air is clear there it freezes hard. So

the frost takes place either on hill or in dale, where
ever the air happens to be clearest, & freest from vapour.
Hyrn, cornu vel angulus : whence our Faringdon
Hyrn, or hern as we pronounce it, is the corner-
field of the parish. *Heanè*, Humilis : hence perhaps
our honey-lane. Our Gally-hill, is perhaps gallows
hill from *Galga*, crux. Does not domesday book
[Qy. Ancient charters] among other privileges, say
that Priors &c. were allowed Furcas, gallows ? *By*,
habitation : from whence yᵉ adjective *Byn*, as Binsted,
&c. *Deortun*, saltus : hence no doubt our Dorton,
a wild, bushy common just below the village : Deer-
ton, a place where deer are kept. *Eowod*, Ovile :
hence perhaps our field called the Ewel ? *Ymbhanger*
the winding hanger : we have places so named.
Rode, crux : hence our Rode-green near the Priory,
where probably a cross was erected. *Fyrd*, a ford ;
also a *camp :* hence probably our high common-
field to the N.W. is called the fordown. *Ether*,
sepes : the top border that binds down our hedges
& keeps them together is called by our hedgers
ether to this day : the wickering the top along they
call *ethering*. *Gouleins* (Gothic) salutatio : hence
perhaps our word Golly, a sort of jolly kind of oath,
or asseveration much in use among our carters, &
lowest people. *Eorthwicga*, blatta terrena : hence our
absurd word, not peculiar to this district, earwig.

Dec. 27. On every sunny day the winter thro',
clouds of insects usually called gnats (I suppose
tipulae & empedes) appear sporting & dancing over
the tops of the ever-green trees in the shrubbery,
& frisking about as if the business of generation was
still going on. Hence it appears that these diptera
(which by their sizes appear to be of different species)
are not subject to a torpid state in the winter, as
most winged Insects are. At night, & in frosty
weather, & when it rains & blows they seem to retire
into those trees. They often are out in a fog.

CHAPTER IX

"Give me the old road, the same flowers, . . . the old succession of days and garland." (JEFFERIES: *Wild Flowers*.)

1776

Jan. 2. Grey & white wagtails appear every day: they never leave us.

Jan. 11. Some lambs fall.

Jan. 12. A very deep snow. Poor birds begin to be distress'd, & to come in a door: hares do not stir yet. Hares lie-by at first, 'til compelled to beat-out by hunger.

Jan. 13. The snow is drifted-up to the tops of gates, & the lanes are full. Poultry do not stir-out of the hen-houses: they are amazed, & confounded in snow, & would soon perish.

Jan. 14. Rugged, Siberian weather. The narrow lanes are full of snow in some places, which is driven into most romantic, & grotesque shapes. The road-waggons are obliged to stop, & the stage-coaches are much embarrassed. I was obliged to be much abroad on this day, & scarce ever saw its fellow.

Jan. 16. The brambling appears in farm-yards among the chaffinches. It is rare in these parts.

Jan. 17. Rooks come to dunghills close to houses. Sky-larks resort to farm-yards.

Jan. 18. Cats catch all the birds that come in from the cold. Wagtails retire to brooks, & rivu-lets; there they find the aureliae of water-insects. Titmice pull the mosses & lichens off from trees in quest of grubs &c: Nuthatches do the same.

Jan. 19. Snow falling, grey & light, dark & still.

Jan. 20. Fierce frost, sun: none since the snow fell; grey. Clouds fly some from the N.W. some from the E. Hares, compelled by hunger, come into my garden, & eat the pinks. Lambs fall, & are frozen to the ground.

Jan. 21. Flocks of wild-geese pass-over. Snow lies very deep.

Jan. 22. Went to Farnham. Grey & still.

Jan. 23. [London] Therm: in London areas 20. The ground covered with snow & everything frozen up.[1]

Jan. 26. Snow very thick on the roofs & in areas.

Jan. 27. Snow all day, fierce frost at night.

Jan. 28. [S. Lambeth] Fierce frost: ice under people's beds, & cutting winds. Therm^r at Selborne: abroad, 7.

Jan. 29. [Thermom.] 6. As intense frost usually befalls in Jan: our Saxon fore-fathers call'd that month with no small propriety wolf-month; because the severe weather brought down those ravenous beasts out of the woods among the villages.

Jan. 31. Below zero !! 32 deg. below the freezing point. At eleven it rose to 16½. Rime. A most unusual degree of cold for S.E. England.

Feb. 1. Snow now lying on the roofs for 26 days ! Thames frozen above and below bridge: crowds of people running about on the ice. The streets strangely encumbered with snow, which crumbles & treads & looks like bay salt—Carriages run without any noise or clatter. Thaws, I have observed, frequently take place immediately from intense freezing; as men in sickness often begin to mend from a paroxysm.

Feb. 6. Part of the snow gone at Selborne. It lay for many weeks in deep hollow lanes.

Feb. 7. Great rains, & vast floods thro' the month of February.[2]

Mar. 22. [Selborne] Came from London to Selborne. Hot sun: summer-like weather. When I arrived in Hants I found the wheat looking well, & the turneps little injured. My laurels & laurustines somewhat injured; but only those that stood *in hot sunny aspects*. No evergreens quite destroyed; & not half the damage sustained that befell in Jan. 1768. Those laurels that are somewhat scorched on the S. sides, are perfectly untouched on their N. sides. The care I took in ordering the snow to be carefully shaken from the branches wherever it fell, seems greatly to have availed my laurels. Mr Yalden's laurels facing to the N. untouched. Portugal laurels not hurt.

Mar. 26. Flocks of fieldfares remain: no red-wings are seen. No song-thrushes are heard; they seem to be destroyed by the hard weather. Some wagtails survive.

Mar. 28. Hirundo domestica! Hirundo agrestis! Blackbirds are mostly destroyed by shooters. Farmer Tredgold saw five hirundines at Willey-mill next Farnham playing about briskly over the mill-pond: four, he says, were house-swallows, & the fifth an house-martin, with a white rump. These birds are very early! Some few bank-martins haunt round the skirts of London, & frequent the dirty pools in St. George's fields, & near White-chappel: perhaps they build in the scaffold-holes of some deserted house; for steep banks there are none.

Mar. 31. The willows in bloom diversify the coppices, & hedge-rows in a beautiful manner.

Apr. 1. Gossamer floats. Wood-larks hang suspended in the air, & sing all night.

Apr. 2. Peaches & nectarines in bloom make a glorious shew. Thomas began this day mowing the grass-walks.

Apr. 4. No rain since the beginning of March. The ground dry & harsh. *Footnote*. The *Bombylius medius* abounds. It is an hairy insect, like an humble-bee, but with only *two* wings, & a long straight beak, with which it sucks the early flowers, always appearing in March. The female seems to lay it's eggs as it poises on it's wings, by striking it's tail on the ground, & against the grass that stands in it's way in a quick manner for several times together.

Apr. 7.   Linnets, & chaffinches flock still : so are not paired.   Some few sky-larks survive.

Apr. 9.   [Reading] Young geese & ducks.   Four swallows at Alton.

Apr. 10.   [Oxford] One swallow at Wallingford.

Apr. 12.   Oaks are felled : the bark runs.

Apr. 13.   [Selborne] Rain is much wanted.   At Fyfield ; charadrius oedicnemus returns & clamors April 14 ; first swallow, April 16 ; nightingale April 16 : cuckow Ap : 20 ; swifts first seen Apr : 26 : came to their nesting-place May 8.   At Lyndon, in Rutland first swallow Apr : 16 : first swift May 6th

Apr. 18.   Mowed all round the garden.   Cut the first brace of cucumbers : they were well-grown. Nightingale sings.

Apr. 19.   Grass-hopper-lark   whispers.   The bombylius medius is much about in March & the beginning of April, & soon seems to retire.   It is a very early insect.   *Footnote.*   Pulled down the old martins nests against the brew-house & stable : they get foul & full of vermin.   These abounded with fleas, & the cases of Hippoboscae hirundinis. Besides while these birds are *building* they are much more in sight, & are very amusing.

Apr. 22.   Codlings blow.   Hot-beds want rain to make them ferment.

Apr. 24.   Hot-beds never do so well in long dry fits of weather : they do not ferment enough.   The hot dry weather hurries the flowers out of bloom.

Apr. 25.   The wolf-fly[3] appears in windows, & pierces other flies with his rostrum : is of a yellow hue : an asilus of Linn :

Apr. 26.   Pheasants   crow.   Ring-doves   coe. Nest : & peaches swell.   Hops are poling.   *Footnote.* The latest summer birds of passage generally retire the first : this is the case with the hirundo apis, the caprimulgus, & the stoparola.   Birds are never joyous in dry springs : showery seasons are their delight for obvious reasons.

Apr. 30. Birds silent for want of showers. Acer majus in bloom. The sycamore, when in bloom, affords great pabulum for bees, & sends forth an honey-like smell. All the maples have saccharine juices.

May 1. The grass crisp with white frost. Tulips hang their heads in the morning, being pricked with the frost.

May 2. Some missle-thrushes on the down above us : blackbirds & thrushes mostly destroyed.

May 4. Field crickets shrill. Snipes in the forest. The forest quite burnt-up. Small reed-sparrow sings. Young ring-doves fledge. Hay is risen to four pounds per ton : no grass in the fields, & great distress among the cattle.

May 5. Showers all day, with hail, & wind. The ground is pretty well moistened.

May 10. Apis longicornis. This bee appears, but does not bore nests in the ground yet.

May 12. The sycamore or great maple, is in bloom, & at this season makes a beautiful appearance, & affords much pabulum for bees, smelling strongly like honey. The foliage of this tree is very fine, & very ornamental to outlets.

May 14-24. [Fyfield] Spring-corn in a sad state, not half come up.

May 20. Wheat on the downs begins to spindle for ear.

May 21. Medlar blows : this is the most uncouth tree in its growth, the boughs never continuing streight for two feet together.

May 23. Female wasps abound. Young rooks venture-out to the neighbouring trees.

May 24. [Wintŏn] Cold dew, hot sun, soft even.

May 25. [Sutton-Bp's, Selborne] The frost has killed the tops of the wallnut shoots, & ashes; & the annuals where they touched the glass of the

frames ; also many kidney-beans.   The tops of hops,
& potatoes were cut-off by this frost.   Tops of laurels
killed.   The wall-nut trees promised for a vast crop,
'til the shoots were cut off by ye frost.   *Leaf inserted.*
No one that has not attended to such matters, &
taken down remarks, can be aware how much ten
days dripping weather will influence the growth of
grass or corn after a severe dry season.   This present
summer 1776 yields a remarkable instance : for 'til
the 30th of May the fields were burnt-up & naked,
& the barley not half out of the ground ; but now,
June 10th there is an agreeable prospect of plenty.
A very intelligent Clergyman assured me, that hearing
while he was a young student at the University, of
toads being found alive in blocks of stone, & solid
bodies of trees ; he one long vacation took a toad,
& put it in a garden-pot, & laying a tile over the mouth
of the pot, buried it five feet deep in the ground in
his father's garden.   In about 13 months he dug-up
the imprisoned reptile, & found it alive & well,
& considerably grown.   He buried it again as at
first, & on a second visit at about the same period
of time found it circumstanced as before.   He then
deposited the pot as formerly a third time, only
laying the tile so as not quite to cover the whole
of its mouth : but when he came to examine it again
next year, the toad was gone.   He each time trod
the earth down very hard over the pot.

May 26.   Fern-owl first seen : a late summer bird
of passage.

May 29.   Laburnums in beautiful bloom.   Hawthorns blow finely.

May 30.   Strawberries blow well.   The   first
effectual rain after a long dry season.

June 1.   Dames violets, double, blow finely : roses
bud : tulips gone : pinks bud.   Bees begin to swarm.
Tacked the vines the first time.   Began to plant out
annuals in the basons in the field.   Ponds & some
wells begin to be dry.

June 2. Sultry, & heavy clouds. Smell of sulphur in the air. Paid for near 20 wasps : several were breeders ; but some were workers, hatched perhaps this year.

June 3. Soft rain. Grass & corn improved by the rain already. The long-horned bees bore their holes in the walks.

June 5. Boys bring me female-wasps. Apis longicornis bores it's nests & copulates.

June 7. Fly-catcher builds. Farmers cut clover for their Horses.

June 8. Elder begins to blow. Many hundreds of annuals are now planted-out, which have needed no watering. Wheat begins to shoot into ear. Hardly any shell-snails are seen ; they were destroyed, & eaten by the thrushes last summer during the long dry season. This year scarce a thrush, they were killed by the severe winter.

June 9. Forest-fly begins to appear. Grass & corn grow away.

June 12. Drones abound round the mouth of the hive that is expected to swarm. Sheep are shorn.

June 13. Martins begin building at half hour after three in the morning.

June 14. I saw two swifts, entangled with each other, fall out of their nest to the ground, from whence they soon rose & flew away. This accident was probably owing to amorous dalliance. Hence it appears that swifts when down can rise again. Swifts seen only morning & evening : the hens probably are engaged all the day in the business of incubation ; while the cocks are roving after food down to the forest, & lakes. These birds begin to sit about the middle of this month, & have squab young before the month is out.

June 17. Snails begin to engender, & some flew to lay eggs : hence it is matter of consequence to destroy them before midsummer.

June 20. Cut my S^t foin ; a large burden : rather over-blown : the nineth crop. Libellula virgo, sive puella.⁴ Dragon-fly with blue upright wings. *Footnote*. As the way-menders are digging for ſtone in a bank of the ſtreet, they found a large cavern running juſt under the cart-way. This cavity was covered over by a thin ſtratum of rock : so that if the arch had given way under a loaded waggon, considerable damage muſt have ensued.⁵

June 24. Hay makes well. The wind bangs the hedges & flowers about.

June 25-28. [Bramshot] Vine juſt begins to blow : it began laſt year June 7 : in 1774 June 26. Wheat begins to blow. Thomas's bees swarm, & settle on the Balm of Gilead fir.⁶ firſt swarm.

June 26. No young partridges are flyers yet : but by the deportment of the dams it is plain they have chickens hatched ; for they rise & fall before the horses feet, & hobble along as if wounded to draw-off attention from their helpless broods. Sphinx fortè ocellata.⁷ A vaſt insect ; appears after it is dusk, flying with an humming noise, & inserting it's tongue into the bloom of the honey-suckle : it scarcely settles on the plants but feeds on the wing in the manner of humming birds. Omiah, who is gone on board the Resolution, is expected to sail this week for Otaheite with Capt : Cook.⁸

June 28. [Selborne] Flowers in the garden make a gaudy appearance.

June 30. Wheat generally in bloom. The beards of barley begin to peep.

July 1. Full moon. Cherries begin to ripen, but are devoured by sparrows. Began to cut my meadow-hay, a good crop, one 3rd more than laſt year.

July 2. The early brood of swallows are active & adroit, & able to procure their subsiſtence on the wing. Fresh broods come forth daily.

July 3. Black-caps are great thieves among the cherries. The flycatcher is a very harmless & honeſt bird, medling with nothing but inseċts.

July 5. Field-crickets are pretty near silent; they begin their shrilling cry about the middle of May.

July 6. The bees that have not swarmed lie cluſtering round the mouths of the hives. Took-off the frames from the cucum^rs : those under the hand-glasses begin to show fruit. Hay lies in a bad ſtate.

July 8. Second swarm of bees on the same bough of the balm of Gilead fir. Turned the hay-cocks which are in a bad ſtate. Cherries delicate, Mr Grimm,[9] my artiſt, came from London to take some of our fineſt views.

July 9. The bees are very quarrelsome, & ſtung me.

July 10. Some of the little frogs from the ponds ſtroll quite up the hill : they seem to spread in all direċtions.

July 11. Tilia europaea. The lime blows, smells very sweetly, & affords much pabulum for bees. *Footnote.* Bees come & suck the cherries where the birds have broke the skin ; & on some autumns, I remember they attack'd & devoured the peaches & Neċt: where the wasps had once made a beginning.

July 14. Young frogs migrate, & spread around the ponds for more than a furlong : they march about all day long, separating in pursuit of food ; & get to the top of the hill, & into the N. field.

July 16. Bees, when a shower approaches, hurry home. One hive of bees does not swarm ; the bees lie in cluſters at the mouth of the hive.

July 19. Sambucus ebulus. Dwarf elder [Dane's elder] blows. Fungi begin to appear.

July 21. Missle thrushes bring forth their broods, & flock together.

July 22. Bees swarm : the swarm of a swarm, which swarmed itself at the beginning of June.

A neighbour has had nine swarms from four ſtalls
[mother-hives] : two apiece from three of them, &
three from one.

July 23.   Walnuts abound, but are rather small &
spotted.

July 25.   Bees that have not swarmed kill their
drones.

July 26.   Cut the grass in the little meadow.
Hay makes well.   Hops fill their poles, & throw out
lateral shoots.

July 30.   Peacocks begin to moult & caſt their
splendid train.   Total eclipse of the moon.

Aug. 1.   We deſtroyed a ſtrong wasp's neſt,
consiſting of many combs : there were young in all
gradations, from fresh-laid eggs to young wasps
emerging from their aurelia ſtate ; many of which
came forth after we had kept the combs 'til the next
day.   Where a martin's neſt was broken that contained fledge young : the dams immediately repaired
the breach, no doubt with a view to a second brood.

Aug. 5.   Mr Grimm the artiſt left me.   Began to
gather apricots.   Put out two rows of celeri : the
ground dry & harsh.

Aug. 6.   [Meonſtoke] Wheat-harveſt begun at
E. Tiſted & Weſt meon.

Aug. 10.   [Selborne] Hay not housed at Meon-
ſtoke & Warnford.

Aug. 15.   [Chilgrove] Sun, & clouds, sultry,
showers about.

Aug. 16-27.   [Ringmer.]

Aug. 20.   Timothy, the tortoise weighs juſt six
pounds three quarters & two ounces & an half : so
is encreased in weight, since Aug. 1775, juſt one
ounce & an half.

Aug. 26.   While the cows are feeding in moiſt
low paſtures, broods of wagtails, white & grey, run
round them close up to their noses, & under their
very bellies, availing themselves of the flies, & inſects
that settle on their legs, & probably finding worms

& larvae that are roused by the trampling of their feet. Nature is such an oeconomist, that the most incongruous animals can avail themselves of each other! Interest makes strange friendships.

Aug. 27. [Isfield] Grey, sun, sweet day.

Aug. 28. [Ringmer] The tortoise eats voraciously: is particularly fond of kidney-beans. Vast halo round the moon.

Aug. 29. [Findon] Full moon. The rams begin to pay court to the ewes.

Aug. 30. [Chilgrove] Mr Woods of Chilgrove thinks he improves his flock by turning the east-country poll-rams among his horned ewes. The east-country poll sheep have shorter legs, & finer wool; & black faces, & spotted fore legs; & a tuft of wooll in their fore-heads. Much corn of all sorts still abroad. Was wetted thro' on the naked downs near Parham-ash. Some cuckoos remain. N.B. From Lewes to Brighthelmstone, & thence to Beeding-hill, where the wheat-ear traps are frequent no wheat-ears are to be seen: But on the downs west of Beeding, we saw many. A plain proof this, that those traps make a considerable havock among that species of birds.

Aug. 31. [Selborne] Fine harvest day. Some corn housed.

Sept. 1. The barley coming-up unequally is not yet ripe. Hops promise to be very small.

Sept. 2. Vast shower with hail. Turned the horses into the great mead. Much grass. A part of the orchard, where I laid the earth which came out of the garden, was sown in April with rye-grass, & hop-trefoil, & has been mown already three times.

Sept. 3. The season for shooting is come; but scarce any partridges are to be found: the failure of breed is remarkable. The tops of the beeches begin to be tinged with a yellow hue.

Sept. 5. Some wasps on the wall-fruit. Where the wasps gnaw an hole, the honey-bees come &

suck the pulp. All fruits are backward, watry, & bad.

Sept. 6. House-martins do not deal in second broods so much as usual : & yet it should seem that they are not influenced by the cold wet summer, since the swallows seem to be as prolific as ever. Bees injure the wall-fruit in bad autumns ; because they are hindered from gathering honey.

Sept. 8. A sharp, single, crack of thunder at Faringdon : the air was cold, & chilly.

Sept. 12. The wasps, tho' by no means numerous, plunder the hives, & kill the bees, which are weak & feeble, this wet autumn : . . . " asper crabro imparibus se emiscuit armis."[10]

Sept. 13. My muscle-plums are this year are in much more perfection than any other fruit.

Sept. 14. Swallows cluster, & hang about in a particular manner at this season of the year. Honey-bees swarm by thousands, & devour the peaches, & nectarines.

Sept. 15. Swallows catch at walls as they flie about.

Sept. 18. Wagtails join with hirundines, & pursue an hawk high in the air : the former shew great command of wing on the occasion.

Sept. 19. [Basingstoke] Much barley out round Basingstoke ; some standing : & many fields of oats. Green wheat up.

Sept. 20. [Selborne] Peaches, & nect : rot. Wasps are busy still. *Footnote.* Large earth-worms now abound on my grass-plot, where the ground was sunk more than a foot. At first when the earth was removed, none seemed to remain : but whether they were bred from eggs that were concealed in the turf, is hard to say. Worms do not seem to inhabit beneath the vegetable mould.

Sept. 23. Wasps still go into the hives. Gathered-in some of the early pippins : fine baking apples.

Sept. 25. Fine young clover & fine turneps about the country. *Footnote.* The quantities of haws, & sloes this year are prodigious. Those hives of bees that have been taken have proved deficient in wax, & honey. In shady wet summers bees can scarce procure a store sufficient to carry them thro' the winter : if not fed they perish.

Sept. 29. Nothing left abroad but seed-clover, & a few beans.

Oct. 1. Swallows & martins, before they withdraw, not only forsake houses, but do not frequent the villages at all : so that their intercourse with houses is only for the sake of breeding.

Oct. 3. Beautiful wheat season for the wet fallows. The buzzard is a dastardly bird, & beaten not only by the raven, but even by the carrion-crow. Gathered baking-pears.

Oct. 5. Black snails are more sluggish than in summer ; but in sight all day at this season of the year. Saw one hornet.

Oct. 6. Numbers of swallows & martins playing about at Faringdon, & settling on the trees. If hirundines hide in rocks & caverns, how do they, while torpid, avoid being eaten by weasels & other vermin ?

Oct. 7. Gathered some keeping apples. The intercourse between tups [rams] and ewes seems pretty well over. Ewes go, I think, 22 weeks.

Oct. 9. Nuts fall very fast from the hedges.

Oct. 10. Grey, windy, soft & agreeable. Now my grapes are delicate notwithstanding the summer was so wet & shady.

Oct. 11. The red-breast entertains us with his autumnal song.

Oct. 12. The hanging beech-woods begin to be beautifully tinged, & to afford most lovely scapes, very engaging to the eye, & imagination. They afford sweet lights & shades. Maples are also finely tinged. These scenes are worthy the pencil of a Reubens [sic].

Oct. 15. My largest wall-nut tree produced four bushels & a half of nuts : many bunches contained 8, 9, & on to 15 wallnuts each.

Oct. 16. The redbreast's note is very sweet, & pleasing ; did it not carry with it ugly associations of ideas, & put us in mind of the approach of winter.

Oct. 21. A cock pheasant flew over my house, & across the village to the hanger.

Oct. 22. The nuthatches are busy rapping with their bills about the wallnut trees : & as I find wall-nuts fallen down, with holes picked in their shells ; no doubt they are made by those birds.

Oct. 25. One [common, or corn] bunting in the northfield : a rare bird at Selborne. *Footnote.* There is this year a remarkable failure of mushrooms : & the more to be wondered at, since the autumn has been both moist & warm. There is a great failure also of truffes in my Brother's outlet at Fyfield, notwithstanding in simular weather they abounded last year. So that some secret cause influences alike these analogous productions of nature.

Oct. 27. Larks frolick much in the air : when they are in that mood the larkers catch them in nets by means of a twinkling glass : this method they call *daring.*[11]

Oct. 28. [Bp's Sutton] The month of Oct : has been very dry : mill-ponds begin to want water. Sheep frolick.

Oct. 29-*Nov.* 7. [Fyfield] Grey crows return. These are winter birds of passage, & are never seen with us in the summer. The flocks are feeding down the green wheat on the downs, which is very forward, & matted on the ground. They sow wheat on the downs sometimes as soon as the end of July provided the season is showery.

Nov. 1. Four swallows were seen skimming about in a lane below Newton. This circumstance

seems much in favor of hiding, since the hirundines
seemed to be withdrawn for some weeks.   It looks
as if the soft weather had called them out of their
retirement.   My Brother's turkies avail themselves
much of the beech-mast which they find in his grove :
they also delight in acorns, wallnuts, & hasel-nuts :
no wonder therefore that they subsist wild in the
woods of America, where they are supposed to be
indigenous.   They swallow hasel-nuts whole.

Nov. 4.   The trufle-hunter was here this morning :
he did not take more than half a pound, & those
were small.

Nov. 5.   Farmer Cannings has fine weather for his
barley harvest, M*r* Cannings has now 48 acres of
barley abroad either standing or in cock : it was not
sown 'til the rains came in the beginning of June.
Nov*r* 4.   He is now ricking one field ; the other is
standing.   The grain is lank, & the cocks cold, &
damp.

Nov. 6.   Flies abound.   They stay long after
the hirundines are withdrawn.   Tipulae sport in
the air.

Nov. 7.   The great fieldfare [common fieldfare]
returns.   Beetles abound every evening.   Farmer
Cannings's new barley-ricks smoke & ferment like
hotbeds already.

Nov. 8.   [Bp's Sutton] Infinite quantities of haws
& sloes.   Nothing could be more lovely than the
ride from Andover to Alresford over the Hãnts
downs.   The shepherds mow the charlock growing
among the wheat.   I saw no fieldfares[12] all thro'
my Journey.   If they come, as Ray says they do,
" ventis vehementer spirantibus " ;   they can have
had no advantage of that kind ;   for the autumn has
been remarkably still.

Nov. 9.   [Selborne] Grey, clouds & sun, sweet
day.

Nov. 10.   Redwings.   These birds begin to
appear at last.

Nov. 13. Nuthatches rap about on the trees. Crocuss begins to sprout. The leaves of the medlar-tree are now turned of a bright yellow. One of the first trees that becomes naked is the wallnut: the mulberry, & the ash, especially if it bears many keys, and the Horse-chestnut come next. All lopped trees, while their heads are young, carry their leaves a long while. Apple-trees & peaches remain green 'til very late, often 'til the end of Nov^r: young beeches never cast their leaves 'til spring, 'til the new leaves sprout & push them off: in the autumn the beechen-leaves turn of a deep chestnut color. Tall beeches cast their leaves towards the end of Oct^r. *Footnote.* Magpies sometimes, I see, perch on the backs of sheep, & pick the lice & ticks out of their wool; nay, mount on their very heads; while those meek quadrupeds seem pleased, & stand perfectly still, little aware that their eyes are in no small danger; & that their assiduous friends would be glad of an opportunity of picking their bones.

Nov. 19. This afternoon the weather turning suddenly very warm produced an unusual appearance; for the dew on the windows was on the *outside* of the glass, the air being warmer *abroad* than *within*.

Nov. 20. Mrs Snooke's old tortoise at Ringmer went under ground.

Nov. 21. The thatch is torn by the wind.

Nov. 22. The ground was covered with snow at Buxton in Derbyshire.

Nov. 26. A man brought me a common sea-gull alive: three crows had got it down in a field, & were endeavouring to demolish it.

Nov. 28. The nuthatch hunts for nuts in the hedges, & brings them to the forked bough of a certain plum-tree, where it opens them by picking a ragged irregular hole in the small end of the shell. It throws the empty shell on the walk.

Dec. 2. When the therm^r is at 50 flies, & phal-aenae come-out, & bats are often stirring. Beetles flie.

Dec. 3. Worms lie-out very thick on the walks, & grass plot; many in copulation. They are very venereous, & seem to engender all the year.

Dec. 4. Vast condensations on walls & wainscot, which run in streams : these things are colder than the warm wet air.

Dec. 11. Summer-like : the air is full of gossamer, & insects.

Dec. 12. Barom^r 30. Still, dark & spitting, deep fog. When the barom^r gets very high, it is often attended with black spitting weather.

Dec. 12. Missle thrush sings merrily every morning. Song thrush very loud.

Dec. 13. A new magpie's nest, near finished was found in a coppice : this soft season reminds birds of nidification.

Dec. 21. The shortest day : a truly black, & dismal one.

Dec. 23. Strong N. auroras tho' the moon was very bright.

Dec. 31. A grosbeak[13] was shot near the village. They sometimes come to us in the winter.

CHAPTER X

" While, with poring eye
I gazed, myself creating what I saw."
(COWPER : *The Winter Evening.*)

1777

Jan. 1. [Lasham] Steady frost, snow on the ground.
Jan. 4. [Selborne] Dark & thawing, frost, snow
on the ground. Larks congregate : roads hard, &
beaten.
Jan. 8. Bottles of water frozen in chambers.
Haws frozen on the hedges & spoiled so as to be
no longer of any service to the birds.
Jan. 10. Thaw. Tipulae are playing about as
if there had been no frost at all.
Jan. 16. Grey sun, sweet day. Bees, & flies
moving : air full of insects : spiders shoot their
webs : butter-fly out.
Feb. 3. The Planet Mercury is now to be seen
every evening : it is nearer to the horizon than Venus,
& more to the right hand, setting somewhat S. of
the W. about six in the evening. Will be visible
about six days longer.
Feb. 9. Very harsh day. The fieldfares now feed
on sloes, which abound on the hedges. 'Til now
I never observed that any birds touched the sloes.
Leaf inserted. About the beginning of July, a
species of Fly (Musca) obtains, which proves very
tormenting to horses, trying still to enter their
nostrils, and ears, & actually laying their eggs in the
latter & perhaps in both of those organs. When
these abound, horses in wood-land districts become
very impatient at their work, continually tossing their
heads, & rubbing their noses on each other, regardless
of the driver : so that accidents often ensue. In

the heat of the day, men are often obliged to desist
from plowing : saddle-horses are also very troublesome at such seasons. Country-people call this
insect the *nose fly*.[1] In the decline of the year, when
the mornings & evenings become chilly, many species
of flies (muscae) retire into houses, & swarm in the
windows. At first they are very brisk & alert :
but as they grow more torpid, one cannot help
observing that they move with difficulty, & are
scarce able to lift their legs, which seem as if glued to
the glass : and by degrees many do actually stick
on till they die in the place.[2] Now as flies have flat
skinny palms, or soles to their feet, which enable
them to walk on glass & other smooth bodies by
means of the pressure of the atmosphere ; may not
this pressure be the means of their embarrassment
as they grow more feeble ; 'til at last their powers
become quite inadequate to the weight of the in-
cumbent air bearing hard upon their more languid
feet ; & so at last they stick to the walls & windows,
where they remain, & are found dead.[3]

Feb. 19. [London.][4]

Mar. 26 & 27. Two sultry days ;[5] Mrs Snooke's
tortoise came forth out of the ground ; but retired
again to it's hybernaculum in a day or two, & did
not appear any more for near a fortnight. Swallows
appeared also on the same days, & withdrew again :
a strong proof this of their hiding.

Mar. 27. A swarm of bees came forth at Kingsley
[Hants], and were hived. From that day to April
the 10th harsh, severe weather obtained with frequent
frosts & ice, & cutting winds. How are these bees
to subsist so early in an empty hive ? On March 26
& 27, two, sunny, sultry days, swallows were seen
at Cobham, in Surry. Therm^{rs} were at that time
in London up at 66 in the shade.

Apr. 11. Returned from London to Selborne.

Apr. 18. The golden-crested wren frequents the
fir-trees, & probably builds in them. *Footnote.*

Tho' the spring has been remarkably harsh & drying, yet the ground crumbles, & dresses very well for the spring-crops. The reason is, the driness of the winter : since the ground bakes hardest after it has been most drenched with water.

Apr. 20. The house-snail, φεριοικος, begins to appear : the naked black-snail comes forth much sooner. Slugs, which are covered with slime, as whales are with blubber, are moving all the winter in mild weather.

Apr. 24. The cock green-finch begins to toy, & hang about on the wing in a very peculiar manner. These gestures proceed from amorous propensities.

Apr. 25. The titlark rises, & sings sweetly in its descent. The Ring-dove hangs on its wings, & toys in the air.

Apr. 27. Notwithstanding the dry winter & spring, the pond on the common [Wood pond] is brim full.

Apr. 29. The bark of oak now runs ; & I am felling some trees. When trees are sawn-off, & thrown, a rushing sound is heard from the but, often attended with a little frothing & bubbling-out of the sap. This rushing or hissing is occasioned by the motion of the air escaping thro' the vessels of the wood.

May 10. The scenes round the village are beautifully diversifyed by the bloom of the pear-trees, plums, & cherries. A great flood on the Thames in consequence of the rain on friday night [May 9].

May 17. Sun, fine day, showers. Most vivid rainbow.

May 19. Swallows begin to collect dirt from the road, & to carry it into chimneys for the business of nidification.

May 26. The grasshopper-lark whispers at night.

May 27. Field crickets begin their shrilling summer sound. My horses began to be turned-out a nights.

May 28. Clouds flie different ways. Distant thunder.

June 2. Began some alterations previous to the building my parlor.[6]

June 3. The foliage of the nectarines is much blotched & shrivelled, so that the trees look poorly.

June 5. The caterpillars of some phalaenae abound on the foliage of the apricots, which they tye together with their webs, & gnaw & deface in a bad manner. We wash the trees with the garden-engine.

June 6. Began to build the walls of my parlor, which is 23 feet & half by 18 feet; & 12 feet high & 3 inch:

June 7. The bees gather earnestly from the flowers of the buck-thorn. Tho' we are exempt from chafers this season round this district; yet between Winchester & Southampton they swarm so as to devour everything; the country stinks of them.

June 10. The ground chops [cracks, cleaves] & bakes very hard.

June 11. From the egg-shells flung-out it appears that young martins are hatched in a nest built last year. The circumstance of the ready-built nest makes the brood so much the forwarder.

June 17. My building is interrupted by the rain.

June 20. Tremella nostoc[7] abounds in the field-walks; a sign that the earth is drenched with water.

June 21. Wheat begins to come into ear. A pair of martins began a nest this day over the garden-door. The brick-burner has received great damage among his ware that was drying by the continual rains.

June 22. Swallows are hawking after food for their young 'til near nine o' the clock. They take true pains to support their family.

June 24. Kidney-beans look miserably. A poor cold solstice for tender plants. Wheat looks yellow. My bees when swarming settle every year on the boughs of the Balm of Gilead Fir. Yesterday they settled at first in two swarms, which soon coalesced into one. To a thinking mind few phenomena are more striking than the clustering of bees on some bough where they remain in order, as it were, to be ready for hiving :

> . . . " arbore summâ
> Confluere, & lentis uvam demittere ramis."[8]

June 26. Began to cut my S^t foin ; large & much lodged, & full of wild grasses. The tenth crop.

June 27. Boys bring me female wasps, & hornets. Ophrys nidus avis [bird's nest orchis].

June 30. The pair of martins that began their nest near the stair-case window on June the 21 : finished the shell this day.

July 1. Some labourers digging for stone found in an hole in the rock a red-breast's nest containing one young cuckow half-fledged. The wonder was how the old cuckow could discover a nest in so secret, & sequestered a place.

July 4. New moon. The vines begin to blow. They blowed in 1774 June 26 : & in 1775 June 7 : & in 1776 June 25.

July 6. My S^t foin lies in a rotting state. Birds are very voracious in their squab state, as appears from the consequences of eating which they eject from their nests in marvelous quantities : as they arrive so rapidly at their full ἡλικια, much nutrition must necessarily be wanted.

July 7. Winter-like : we are obliged to keep fires.

July 8. Rain, rain, rain. Bees cluster round the mouth of one hive ; but cannot swarm. Bees must be starved soon, having no weather fit for gathering honey : no sun, nor dry days. A swarm of bees, which had waited many days for an opportunity, came-out in a short gleam of sunshine just before an heavy shower, between 3 & 4 in the afternoon, & settled on the balm of Gilead-fir. When an hive was fixed over them they went into it of themselves.

The young swallows that come-out are shivering, & ready to starve.

July 10.   A swarm of bees has hung-out in a torpid state for many days.

July 11.   Bees swarm by heaps.   31 swifts appear: so that if near half of them are not strangers the young broods are out.

July 12.   Ricked the St foin : it lay 12 days washed with continual showers, & yet is not quite spoiled.

July 13.   The backward wheat is in beautiful bloom : the fields look quite white with blossoms. The forward wheat is out of bloom, & therefore from the late weather not likely to be so good.

July 15.   Rye, which blows early, in a bad state ; no promise of a crop.

July 18.   Swifts dash & frolick about, & seem to be teaching their young the use of their wings. Thatched my rick of meadow-hay with the damaged St foin instead of straw.   Bees begin gathering at three o'clock in the morning : Swallows are stirring at half hour after two.

July 21.   My building is interrupted by the rain.

July 28.   Lime trees in full bloom : on these the bees gather much honey.

July 29.   This morning more than 50 swifts sailed slowly over the village towards the S : there were almost double the number that belong to this place ; & were probably actuated by some tendencies towards their retreat, which is now near at hand.

July 30.   Pond-heads are blown-up : & roads torn by the torrents.   Great flood at Gracious street [Selborne].   Several   mills   are   damaged.   Hay drowned.   Finished the walls of my new parlor.

Aug. 1.   Reared the roof of my new building. *Inserted leaf.*   On July 29 such vast rains fell about Iping [near Midhurst], Bramshot, Haselmere, &c. that they tore vast holes in the turnpike-roads, covered several meadows with sand, & silt, blowed-up the heads of several ponds, carryed away part of the country-bridge at Iping, & the garden-walls of the paper mill, & endangered the mill & house. A paper-mill near Haselmere was ruined, & many 100 ae [librae = £] damage sustained. Much hay was swept away down the rivers, & some lives were lost. A post-boy was drowned near Haselmere, & an other as he was passing from Farnham to Alton : the Gent : in the chaise saved himself by swimming. These torrents were local ; for at Lewes, which lies about the middle of the county of Sussex, they had a very wet time, but experienced none of these devastations. After ewes & lambs are shorn there is great confusion & bleating, neither the dams nor the young being able to distinguish one another as before. This embarrassment seems not so much to arise from the loss of the fleece, which may occasion an alteration in their appearance, as from the defect of that *notus odor*, discriminating each individual personally : which also is confounded by the strong scent of the pitch & tar wherewith they are newly marked ; for the brute creation recognize each other more from the *smell* than the *sight* ; & in matters of *Identity* & *Diversity* appeal much more to their *noses* than to their *eyes*.

Aug. 7. Finished the chimney of my parlor : it measures 30 feet from the hearth to the top.

Aug. 8. Flocks of lap-wings migrate to the downs & uplands.

Aug. 15. Male & female ants come forth & migrate in vast troops : every ant-hill is in strange commotion, & hurry. The pair of martins which began to build on June 21 brought-out their brood this day in part : [*added note*] the rest remain in the nest, Aug. 17.

Aug. 17. White butter-flies settle on wet mud in crowds. *Footnote.* No swift was seen after Aug. 14 : so punctual are they in their migrations, or retreat ! The latest swift I ever saw was only once on Aug. 21, but they often withdraw by the 10.

Aug. 26. A spotted water-hen [moorhen] shot in the forest.

Aug. 27. The large winged female ants, after they have wandered from their nests lose their wings & settle new colonies : are in their flying state food for birds, particularly hirundines. No wasps : & if there were, there is no fruit for them.

Aug. 30. Finished tiling the new parlor in good dry condition just before the rain came. The walls & timbers will be in much better order for this circumstance. *Footnotes.* The pair of martins brought-out all their young Aug. 26 : they still roost in the nest. The nest was begun June 21. Woolmer-forest produces young teals, & young large snipes ; but never, that we can find, any young jack-snipes.[9]

Aug. 31. 'Til now the whole month of Aug : has been dry & pleasant. The evenings begin to feel chilly.

Sept. 1. Cold, white dew, sun, brisk air, clouds about, sun breaks out. Destroyed a small wasp's-nest : the combs were few, but full of young.

Sept. 3. The working-wasps are very small, perhaps half starved in their larva-state for want of pears, plums, &c.

Sept. 5. [Shopwick, near Chichester] Sultry & gloomy. Wasps abound.

Sept. 6-17. [Ringmer.]

Sept. 6. Wheatears (birds) continue to be taken : are esteemed an elegant dish. Horse-ants travel home to their nests laden with flies, which they have caught ; & the aureliae of smaller ants, which they seize by violence.

Sept. 7. Swallows & house-martins dip much in ponds. Vast Northern Aurora.

Sept. 9. Fern-owls haunt Mrs Snooke's orchard in autumn.

Sept. 11. Mrs Snooke's tortoise devours kidney-beans & cucumbers in a most voracious manner : swallows it's food almost whole. *Footnote.* Timothy

the tortoise weighed six pounds 3 quarters, 2 oun:
& an half : so is not at all encreased in weight since
this time last year.   The scales were not very exact.

Sept. 14.   Black cluster-grapes begin to turn color.
A tremendous & awful earthquake at Manchester,
& the district round.   The earthquake happened a
little before eleven o' the clock in the forenoon,
when many of the inhabitants were assembled at
their respective places of worship.

Sept. 17.   The sky this evening, being what they
call a mackerel sky, was most beautiful, & much
admired in many parts of the country.   *Footnote.*
As the beautiful mackerel sky was remarked &
admired at Ringmer, near Lewes, London, & Selborne
at the same time ; it is a plain proof that those fleecy
clouds were very high in ye atmosphere.   These
places lie in a triangle whose shortest base is more
than 50 miles.   Italian skies !   Full moon.   The
creeping fogs in the pastures are very picturesque
& amusing [interesting] & represent arms of the sea,
rivers, & lakes.

Sept. 18.   [Findon] Deep, wet fog.   Sweet day.

Sept. 19.   [Chilgrove] Ring-ousels on the downs
on their autumnal visit.   Lapwings about on the
downs attended by starlings : few stone-curlews.
Sweet Italian skies.   The foliage of the beeches
remarkably decayed & rusty.

Sept. 20.   Some corn abroad : a vast burden of
straw, & many ricks.

Sept. 24.   The walks begin to be strewed with
leaves.   Vivid Northern Aurora.

Sept. 27.   Distant lightening.   We had but little
rain, only the skirts of the storm.   The dry weather,
which was of infinite service to the country after so
wet a summer, might fairly be said to last eight
weeks ; three of which had no rain at all, & much
sun-shine.

Oct. 1.   Bright stars.   This day, Mr Richardson
of Bramshot shot a wood-cock : it was large & plump

& a female : it lay in a moorish [boggy] piece of ground. This bird was sent to London, where as the porter carryed it along the streets he was offered a guinea for it.

Oct. 3. What becomes of those massy clouds that often incumber the atmosphere in the day, & yet disappear in the evening. Do they melt down into dew ? *Footnote.* Some of the store wethers on this down now prove fat, & weigh 15 pounds a quarter. This incident never befals but in long dry seasons ; & then the mutton has a delicate flavour.

Oct. 8. [Bramshot place] Fine autumnal weather. Mr Richardson's nectarines & peaches still in perfection.

Oct. 10. [Selborne] Vast fog, sweet day. Gossamer abounds.

Oct. 11. Found the Sphinx atropos, or death's head-moth, a noble insect, of vast size : it lays it's eggs on the Jasmine. When handled, it makes a little, stridulous noise. A squirrel in my hedges. Insects retreat into the roof of my new building.

Oct. 13. [Worting] Red-wings appear.

Oct. 14. [Whitchurch] Vast shower.

Oct. 15-22. [Oxford.][10]

Oct. 23. [Reading.]

Oct. 24. [Lassam.]

Oct. 25. [Selborne] Hogs are put-up in their fatting pens. The hanging woods are beautifully tinged.

Oct. 30. Gluts of rain, much thunder. The trees & hedges are much broken, & the thatch is torn. Much damage done to the shipping : chimneys, & some houses blown down in London.

Nov. 2. Ring-ouzels still on the downs near Alresford. They have left these parts some time.

Nov. 3. Sea-gulls, winter-mews, haunt the fallows. Beetles flie.

Nov. 4. 21 house-martins appeared playing about under the hanger. The air was full of insects. Others

that saw the martins in an other part of the hanger
say there were more than 150! This was a mistake.
Added note. No martins have been observed since
Oct. 7th 'til this day, when more than 20 were playing
about & catching their food over my fields, & along
the side of the hanger. It is remarkable that tho'
this species of Hirundines usually withdraws pretty
early in Oct. yet a flight has for many years been
seen again *for one day* on or about the 4th of Novr.
Farther it is worthy of notice, that when the Thermr
rises above 50, the bat awakens, & comes forth to
feed of an evening in every winter-month. These
circumstances favour the notion of a torpid state
in birds; & are against the migration of swallows
in this kingdom.

Nov. 7. Put the sashes into my new room.

Nov. 8. Put the first coat of plaster on the battin
work & ceiling of my new room.

Nov. 14. Thatched roofs smoke in the sun:
when this appearance happens rain seldom ensues
that day. This morning they send-up vast volumes
of reek.

Nov. 17. Large field-fares abound: vast clouds
of them on the common.

Nov. 20. Bees come-out much from their hives,
& are very alert.

Nov. 21. Planted a number of small beeches"
in the tall hedges.

Nov. 22. Beeches love to grow in crouded situations, & will insinuate themselves thro' the thickest
covert, so as to surmount it all. Are therefore
proper to mend thin places in tall hedges. Strong
N. aurora, extending to the W. and S.W.: some
streaks of fiery red.

Nov. 24. Gathered in all the grapes for fear of
the frost. We have now enjoyed a dry good season,
with no more rain than has been useful, ever since
the first week in August.

Nov. 25. Men stack their turneps, a new fashion that prevails all at once; & sow the ground with wheat. They dung the fields in summer as for wheat.[12]

Nov. 27. Began planing the floor-battins for my new parlor: they are very fine, & without knots; 500 feet.

Dec. 1. The brick-layers began to lay on the second coat of plaster in my new parlor.

Dec. 2. There is now in this district a considerable flight of woodcocks.

Dec. 3. [Newton] Vast N. Auroras.

Dec. 4. [Selborne] Cold wind, rain with snow. N. Auroras.

Dec. 9. [Bp's Sutton] Grey, sunny, & soft.

Dec. 11. [Selborne] The plasterer began the cornice of my new parlor.

Dec. 16. One black rat was killed at Shalden [near Alton] some months ago, & esteemed a great curiosity. The Norway rats destroy all the indigenous ones.[13]

Dec. 20. Finished plowing-up the Ewel-close, a wheat-stubble, to prepare it for barley, & grass-seeds: it must be plowed thrice. The ground is pretty dry, but tough & heavy, requiring naturally much meliorating. This week Wolmer-pond was fished; & out of it was taken, an eye-witness tells me, a *pike* that weighed 30 pounds.

Dec. 22. For want of rain the millers are much in want of water. Carried out many loads of dung from the cucumber beds on the great meadow. Finished the cornice of the great parlor.

Dec. 24. This day the plasterers put a finishing hand to the ceiling, cornice, and side-plaster-work of my great parlor. The latter is done on battin-work standing-out 3 inches from the walls.

Dec. 26. A fox ran up the street at noon-day.

Footnotes. No birds love to fly down the wind, which protrudes them too fast & hurries them out of their poise : besides it blows-up their feathers, & exposes them to the cold.[14]   All birds love to perch as well as to fly with their heads to the windward.   The christenings at Faringdon near Alton, Hants from the year 1760 to 1777 inclusive were 152 : the burials at the same place in the same period were 124.   So that the births exceed the deaths by 28.   I have buried many very old people there : yet of late several young folks have dyed of a decline.

CHAPTER XI

" Four seasons fill the measure of the year;
There are four seasons in the mind of man."
KEATS : *The Human Seasons.*

1778

Jan. 1. Fires are made every day in my new parlour : the walls sweat much.

Jan. 2. There is reason to fear that the plasterer has done a mischief to the last coat of my battin-plaster that should carry the paper of my room by improvidently mixing *wood ashes* with the morter ; because the *alcaline salts* of the wood will be very long before they will be dry at all, & will be apt to relax & turn moist again when foggy damp weather returns. If any ashes at all he should have used *sea-coal*, & not *vegetable ashes* ; but a mixture of loam & horses dung would have been best.

Jan. 14. The wind very still, for so low a barometer. *Footnote.* Foxes abound in the neighbourhood, & are very mischievous among the farm-yards, & hen-roosts. The fox-hounds have lately harrassed Harteley-woods, & have driven them out of those strong coverts, & thickets.

Jan. 26. Snow on the ground, which is icy, & slippery.

Jan. 28. Frost comes in a doors. Little shining particles of ice, appear on the ceiling, cornice, & walls of my great parlor : the vapor condensed on the plaster is frozen in spite of frequent fires in the chimney. I now set a chafing dish of clear-burnt charcoal in the room on the floor.

Feb. 6.[1] Ravens carry over materials & seem to be building. *Footnote.* Foxes begin now to be

very rank, & to smell so high, that as one rides along
of a morning it is easy to distinguish where they have
been the night before.   At this season the intercourse between the sexes commences ; & the females
intimate their wants to the males by three or four
little sharp yelpings or barkings frequently repeated.
This anecdote[2] I learned by living formerly by an
house opposite to a neighbour that kept a tame
bitch fox, which every spring about candlemas began
her amorous serenade as soon as it grew dark, &
continued it nightly thro' ye months of Feb. & March.
The wheat this year looks very weak & poor ; last
winter it was proud & gay ; & yet after a cold wet
summer the crop was very indifferent.   Farmer
Lassam feeds his early lambs, & their ewes with oats
& bran : the lambs are large & fat.

Feb. 15.   The sun at setting shines into the E.
corner of my great parlor.

Feb. 19.   The dry air crisps my plaster in the new
parlor.

Mar. 2.   [Farnham] Snow in the night, sun, & mild.

Mar. 3-12.   [South Lambeth] Turkey-cock struts
& gobbles.

Mar. 7.   Rain, harsh & dark, much London
smoke.

Mar. 10.   Titlarks in cages essay to sing.   For
want of sun hot-beds languish.   Every matter in
field, & garden is very backward.

Mar. 12.   [London] Dark & windy.   Bright.

Mar. 13-25.   [S. Lambeth] Ice, bright sun.   Full
moon.

Mar. 14.   The green wood-pecker laughs in the
fields of Vauxhall.   Owl hoots at Vauxhall.

Mar. 22.   Frogs spawn in ditches.

Mar. 26.   [London.]

Mar. 27-*Apr.* 1.   [London.]

Mar. 31.   Three or four bank-martins were seen
over Oakhanger-pond.   Flies abound in the pastry-
cooks shops.

Apr. 2-4. [London.]

Apr. 4. [London, Selborne] A swallow was seen this morning near Ripley.   Young geese.

Apr. 10. Three bernacle-geese³ on a pond at Bramshot : one was shot & sent to me.

Apr. 11. The plaster of my great parlor now dries very fast.

Apr. 12. Like Midsummer !

Apr. 13. One beech is in full leaf in the Lythe.

Apr. 16. Planted three beds of asparagus. Planted potatoes.   No swallow.

Apr. 19. The little laughing yellow wren whistles.

Apr. 20. [Lasham] Sun, showers of hail & sleet.

Apr. 21. [Caversham (Oxon)] Frost, snow-storm.

Apr. 22-24. [Oxford.]

Apr. 25. [Alton] The Lathraea squammaria,⁴ a rare plant, is just discovered in bloom in the Litton-coppice at Selborne, just below the church, near the foot-bridge.

Apr. 26. [Selborne] *Inserted leaf.* A day or two before any house-martins had been observed, Thomas Hoar distinctly heard pretty late one evening the twittering notes of those birds from under the eaves of my brewhouse, between the ceiling & the thatch.   Now the quere is, whether those birds had harboured there the winter thro', and were just awakening from their slumbers, or whether they had only just taken possession of that place unnoticed, & were lately arrived from some distant district. If the former was the case, they went not far to seek for an Hybernaculum, since they nestle every year along the eaves of that building.   Mr Derham wrote word to the R. Society " that some time before any Swifts had been seen, (I think before the month of March was out), he heard them squeaking behind the weather-tiles on the front of his parsonage-house."   It is pity that so curious a Naturalist did not proceed to the taking-down some of the tiles, that he might have satisfyed his eyes as well as his hearing. As a notion had prevailed that Hirundines at first coming were lean & emaciated, I procured an H. martin to be shot as soon as it appeared : but the bird, when it come to be opened, was fat & fleshy. It's stomach was full of the legs & wings of small coleoptera. There can be no doubt that the *Horn* shown to me at Mr Lever's museum,[5] vast as it was, belonged to the *Genus* of *Bos :* for it was *concavum, antrorsùm versum lunatum laeve :* whereas had it related to the *Genus* of *Capra,* it would have been *concavum, sursùm versum, erectum, scabrum.* Neither can it by any means belong to the *genus* of *Cervus,* for then it would have been *solidum, corio hirto tectum, apice crescens :* nor for as strong reasons to the *Genus* of *Cvis ;* for then it would have been *concavum, retrorsum versum, intortum, rugosum.* It must therefore of course have belonged to the *Genus* of *Bos.*[6]

May 4. The king & queen are this day at Portsmouth to see the fleet at Spithead. There were five general running firings, which shook my house & made the windows jar. The firings were at ten, twelve, one, four.

May 8. Sowed the Ewel-close, now barley, with 12 pounds of white clover, two bushels of Rye-grass, & a quarter of meadow-grass seeds from a farmer's hay-loft. The ground is too wet, & will not harrow well. Strong wheat-land.

May 16. Nightingales visit my fields & sing awhile : but withdraw, & travel on : some years they breed with me.

May 18. The wind damages the flowers, & beats-off the blossoms from the apple & pear-trees.

May 19. Blowing & cold. In such weather as this the swifts seldom appear. Bees suffer, & get weak.

May 21. The rooks bring their young out, after the chafers.

May 27. The missel-thrush sings much : his song is loud, & clear, but without any variety ; consisting of only two or three wild notes.

May 30. Barn-owls are out in the day, taking their prey in the sunshine about noon.

June 6. Snake gorges a toad much larger than itself. When full it is very sluggish, & helpless, & easily taken.

June 7. The cucumbers abate in their bearing ; & always do at this time of the year.

June 8. Bees go into their hives covered all over with a yellow farina, so that they look like wasps.

June 10. Full moon. Sweet summer's day. The laburnums are in bloom, & high beauty. Wheat begins to push a few ears.

June 11. Cut my S^t foin, the 11th crop. Weeds obtain much, & the crop grows thinner every year.

June 13. Finished laying the floor of my great parlor.

June 14. White butter-flies unnumerable : woe to the cabbages !

June 19. My garden is much bound up, & chopped. Annuals languish for want of moisture.

June 20. The elders, water-elders [guelder roses], fox-gloves, & other solstitial plants begin to be in bloom. Blue dragon-flies appear. Cucumbers, which had stopped for a time bear again. *Footnote*. My favourite old Galloway, who is touched in his wind, was allowed to taste no water for 21 days ; by which means his infirmity grew much less troublesome. He was turned to grass every night, and became fat & hearty, and moved with ease. During this abstinence he staled less than usual, & his dung was harder & dryer than what usually fall from grass-horses. After refraining a while he shewed little propensity for drink. A good lesson this to people, who by perpetual guzzling create a perpetual thirst. When permitted to drink he shewed no eagerness for water.

June 23. Began to cut my meadow. A good crop, especially where the ground was dunged.

June 24. Strawberries ripen. Notwithstanding the vast bloom there are no plums nor many pears; a moderate share of apples : few currans, & gooseberries. Few cherries. Great crop of medlars. Tempest at Farnham.

June 26. Ricked the meadow-hay, six jobbs, & an half in most delicate order. The hay this year fine, & free from weeds. Did not mow the little mead.

June 30. Finished-off my great parlor, & hung the door. The ceiling, & sides are perfectly dry.

July 1. The meadow-rick sinks fast.

July 3. Thatched the hay-ricks : delicate hay.

July 5. We have had no thunder-shower all this summer, tho' many have fallen in sight all around us. Much mischief by this thunder in distant parts.

July 6. The thunder-clouds sunk all away in the night; & we have had no rain. My well sinks very fast. Watered the garden, which is much scorched.

July 11. Finished cutting the hedges. Watered the garden. Many ponds are dry. Much hay ricked.

Footnotes. The young martins that were hatched June 11th began to come-out of their nest July 7th, so that they arrive at their ἡλικία in somewhat less than a month. A colony of black ants comes forth every midsummer from under my stair-case, which stands in the middle of the house; & as soon as the males & females (which fill all the windows & rooms) are flown away, the workers retire under the stairs & are seen no more. It does not appear how this nest can have any communication with the garden or yard; & if not, how can these ants subsist in perpetual darkness & confinement !

July 13. Bestowed great waterings in the garden.

July 14. The little pond on our common has still plenty of water ! ponds in bottoms are dry.

July 18.   We have never had rain enough to lay the dust since saturday June 13 : now five weeks. By watering the fruit-trees we have procured much young wood.   The thermometer belonging to my brother Thomas White of South Lambeth was in the most shady part of his garden on July 5th & July 14 : *up at 88*, a degree of heat not very common even at Gibraltar ! ! July 5 : Therm^r at Lyndon in Rutland 85.

July 20.   Much thunder.   Some people in the village were struck down by the storm, but not hurt. The stroke seemed to them like a violent push or shove.   The ground is well-soaked.   Wheat much lodged.   Frogs migrate from ponds.

July 22.   Sowed first endive.   Planted-out Savoys, choux de Milan, cabbages, &c.   The ground works well, & falls very fine.   Sowed parsley, which has failed before.   Planted out more annuals.

July 25.   The water shines in the fallows.   Much damage done about London by lightening on July 20.

July 27.   Few turnips are yet sown : they were prevented first by the dry weather, & then by the rain.

July 28.   Wallnuts & hazel-nuts abound.   One bank-martin at Combwood-pond : the only one I ever saw so far from the forest.

July 29.   The fruit of the wild merry-trees⁷ being now ripe, diverts the thrushes &c : from eating the currans, goose-berries, &c : therefore useful in outlets.

Aug. 8.   Full moon.   The pair of martins which build by the stair-case window, where their first brood came-out on July 7 : are now hatching a second brood, as appears by some egg-shells thrown-out.

Aug. 12.   My well sinks very much.

Aug. 13.   There is this year the greatest crop of wheat in the North-field that ever was remembered.

Aug. 23.   Flies torment the horses in a most unusual manner.

Aug. 26.   The failure of turnips this year is very great.

Aug. 27.   Selborne people begin hop-picking. The tops of beeches begin to turn yellow.

Sept. 4.   Ladies-traces[8] blow, & abound in the long Lithe. A rare plant. *Footnote.* The young house-martins of the first flight are often very troublesome by attempting to get into the nest among the second callow broods ; while their dams are as earnest to keep them out, & drive them away.

Sept. 11.   Martins congregate in vast flocks, & frequent trees, & seem to roost in them. The second brood of Martins near the stair-case window, which were hatched Aug. 8 : came-out Septemʳ 5ᵗʰ. So that the building a nest, & rearing two broods take up much about four months, May, June, July, & August; during September they congregate, & retire in October.

Sept. 15.   Just at the close of day several teams of ducks fly over the common from the forest : they go probably to the streams about Alresford.

Sept. 16.   [Meonstoke] Many ponds are dry a second time.

Sept. 19.   [Selborne] A lime-avenue in Rotherfield-park has shed all it's leaves. Many ponds are dry a second time.

Sept. 21.   Gathered-in the large white pippins. There are now some wasps.

Sept. 22.   Bee-stalls are very heavy this year : this hot dry summer has proved advantageous to bees. Vast N. Aurora, very red, & coping over in the zenith.

Sept. 23.   [Chilgrove] Ring-ouzels appear on their autumnal visit.

Sept. 24.   [Shopwick (near Chichester)] No stonecurlews congregate this autumn at Chilgrove.

Sept. 26-*Oct.* 7.   [Ringmer] Mʳˢ Snooke has gathered-in all her apples, & pears : her fruit is finely flavoured in such hot years. Mʳˢ Snooke's black grapes begin to ripen. No wasps here. The distress in this place for want of water is very great :

they have few wells in this deep loam; & the little pits & ponds are all dry; so that the neighbours all come for water to Mrs Snooke's ponds.

Sept. 29. Herrings come into season. The after-grass in this grazing-country is very short, & scanty.

Oct. 2. Timothy, the old tortoise, weighed six pounds, & eleven ounces averdupoise.

Oct. 3. White low fogs over the brooks.

Oct. 5. Whitings in season still. Many martins, & some swallows hover about the cliffs [chalk quarries] near Lewes.

Oct. 8. [Findon] Not one wheatear to be seen on all the downs. Swallows abound between Bright-helmstone & Beeding. Not one ring-ouzel to be seen on the downs either coming or going.

Oct. 9. [Chilgrove] Many martins near Houghton-bridge. Some swallows all the way.

Oct. 10. [Selborne] My crop of apples is large; pears are but few; medlars in abundance; wallnuts many, but not very good. One apple-tree produced ten bushels.

Oct. 11. Redwings begin to appear on their winter visit. Some ring-ouzels still about. When redwings come, woodcocks are near at hand.

Oct. 13. Near 40 ravens have been playing about over the hanger all day.

Oct. 14. The hanger & my hedges are faintly tinged with a variety of shades & colours. Ravens play over the hanger.

Oct. 16. The rooks carry-off the wallnuts, & acorns from the trees. One house-martin appears: by it's air & manner it seemed to be a young one: it scouted along as if pinched with the cold.

Oct. 17. Gathered-in the berberries,[9] a great crop.

Oct. 19. The vines are naked, & the grapes exposed to the frost. The crop is very large. The farmers complain that the ground is too dry for sowing.

Oct. 20. Planted long rows of tulips in the garden, & field. Linnets flock.

Oct. My well is so low that if much water is wanted it soon becomes foul & turbid. This inconvenience has never happened before since my Father sunk the well about 40 years ago. But then it must be remembered, that we have had no rains to influence the springs since July twelvemonth. The single oak in the meadow has born this year about 6 & ½ bushels of acorns.

Oct. 24. Farmers put-up their fatting hogs.

Oct. 25. Truly winter-weather. Red-wings abound.

Oct. 29. The bat is out. Beetles hum.

Oct. 31. Great field-fares abound. My well rises.

Nov. 4. Full moon. Tit-mice creep into the martins nests, & probably eat the pupae of the hippoboscae hirundinis.

Nov. 6. Planted six proliferous fiery lily-bulbs[10] from Hambledon in the flower-borders.

Nov. 7. My Chaumontelle-pears now come into eating, & are very delicate.

Nov. 11. Planted in the borders some ferrugineous foxgloves.

Nov. 12. The vast yew-tree at Prior's-dean [near Selborne] is a female : males in general grow to the largest bulk. The yew-tree of East-Tisted is a female. The great yew-tree at Selborne, & two very large ones at Faringdon are all males.[11]

Nov. 17. Phalaenae flie in abundance about my hedges : those & some others, such as spiders, wood-lice, slippery jacks,[12] & some gnats, & tipulae come forth all the winter in mild weather.

Nov. 27. Finished trimming & tacking my vines : the wood is pretty well ripened for next year. Notwithstanding the vehemence of last summer, & the lasting heat, yet my grapes were not so early nor so well ripened as in some moderate years. In particular in 1775 my crops began to be gathered the first

week in Septem^r : & were in high perfection all the
autumn : whereas this year we could not gather at
all 'til Octob^r : & then the flavour was not delicate :
& many clusters never ripened at all. A proof this
that somewhat more is requisite in the production of
fine fruits than mere heat. My peaches & nectarines
also this summer were not in such perfection as in
some former seasons.

Dec. 1. Planted an old Newington-peach, & a
Roman nectarine.

Dec. 3. My well is risen very much.

Dec. 9. Warm fog, small rain. Vast condensations : the trees on the down, & hanger run in
streams down their bodies. Walls sweat. The dew
this morning was on the *outsides* of the windows :
a token that the air was colder within than without.

Dec. 13. Peter Wells's well at Gracious street
begins to run over. The lavants rise at Faringdon.[13]

Dec. 21. Vast flocks of fieldfares. Are these
prognostic of hard weather?

Dec. 23. Wheat grows much. Grass grows.

Dec. 24. Several little black ants appear about
the kitchen-hearth. These must be the same that are
seen annually in hot weather on the stairs, with which
some how they have a communication thro' a thick
wall, or under the pavement, into the middle of the
house.

CHAPTER XII

" Happy the man, whose wish and care
A few paternal acres bound,
Content to breathe his native air
In his own ground."

(POPE : *Ode on Solitude.*)

1779

Jan. 1.   Storm all night.   The may-pole is blown-down.   Thatch & tiles damaged.   Great damage is done both by sea & land.

Jan. 4.   Water froze in my chamber-window.

Jan. 10.   My therm^r is broken.[1]

Jan. 16.   Sowed the great mead in part, & all Berriman's field (laid down last year with grass-seeds) with good peat-ashes.

Jan. 17.   Ice on ponds is very thick.

Jan. 22.   Bees come-out, & gather on the snow-drops.   Many gnats in the air.

Jan. 29.   Out of the wind there is frost ; but none where the S. wind blows.

Jan. 30.   Tulips begin to peep.

Feb. 7.   Lambs come very fast. Bats appear. Field-pease are sowing.

Feb. 9.   The garden works well : sowed pease, & planted beans.   Crocus's blow.

Feb. 12.   The dry season lasted from Dec. 14 to Feb : 12.   This whole time was very still ; & even the frosty part of it very moderate : the barom^r was up at 30 great part of the time.   We have experienced no such winter since the year 1750.

Feb. 14.   The hazels are finely illuminated with their male-bloom.

Feb. 15.   A vivid Aurora : a red belt from East to West.

Feb. 16.   Crocus's blow-out. When the vernal crocus blows, the autumnal crocus peeps out of the ground. Bees gather on the crocuss.

Feb. 17.   Bees rob each other, & fight.

Feb. 20.   Field-crickets have opened their holes, & ſtand in the mouths of them basking in the sun ! They do not usually appear 'til March.

Feb. 23.   [Selborne ; South Lambeth] Drivers use the summer track.² Roads duſty.

Feb. 23-*Mar.* 19.   [South Lambeth.]

Feb. 26.   Pilewort [lesser celandine] Summer-like.

Feb. 27.   The gardener begins to mow my Brother's grass-walks.

Feb. 28.   Gossamer abound. Frogs swarm in the ditches. Spawn.

Mar. 6.   Radishes pulled in the cold ground.

Mar. 13.   The roads in a moſt duſty, smothering condition.

Mar. 14.   Small rain. Quick-set hedges begin to leaf. Duſt is laid.

Mar. 17.   Tussilago farfara [coltsfoot]. Stellaria holoſtia [greater ſtitchwort].

Mar. 20.   [Selborne.]

Mar. 25.   Piƈturesque, partial fogs, looking like seas, islands, rivers, harbours, &c. ! ! Vivid Auroras.

Mar. 26.   Made an asparagus bed : that which was made laſt spring was spoiled for want of rain. Planted potatoes ; sowed carrots.

Mar. 30.   *Bombylius medius :* many appear down the long Lithe. *Field-crickets* bask at the mouths of their holes : they seem to be yet in their pupa-ſtate ; as yet they show no wings.

Apr. 2.   Efts appear.

Apr. 5.   [Worting.]

Apr. 6.   [Caversham & Reading.]

Apr. 7.   [Oxford.]

Apr. 10.   [Selborne] The beeches on the hanger begin to show leaves.

Apr. 11. Ivy-berries are ripe : the birds eat them, & stain the walks with their dung.

Apr. 14. Two cuckows appear in my outlet. The mole-cricket jars. The wry-neck appears, & pipes.

Apr. 15. Thunder-like clouds in the W. at break of day. Nightingale sings in my outlet. Black-cap sings. Dark clouds to the W. & N.W. Lightening.

Apr. 17. Rain greatly wanted. No spring corn comes up. The dry weather has now lasted four months ; from the 15th of Decemr 1778. Apple-trees blow this year a full month sooner than last year. The hanger is pretty well in full leaf : last year not 'til May 15. Musca meridiana.

Apr. 18. Some young grass-hoppers appear : they are very minute.[3]

Apr. 21. Lathraea squammaria, in the Church-litten coppice near the bridge among the hasel-stems, is out of bloom.

Apr. 23. The caterpillars of some phalaenae attack the foliage of the apricots again.

Apr. 24. Hail, stormy, strong wind. The wind broke-off the great elm in the churchyard short in two : the head of which injured the yew-tree. The garden is much damaged by the wind. Many tulips & other flowers are injured by the hail. *Footnotes*. The lightening on friday morning shivered the masts of the Terrible man of war in Portsmouth harbour. The field-crickets in the short Lithe have cast their skins, are much encreased in bulk, show their wings, being now arrived at their ἡλικια. 'Til this altera-tion they are in their pupa-state, but are alert, & eat ; yet cannot chirp, nor propagate their kind.

Apr. 26. Opened the leaves of the Apricot-trees, & killed many hundreds of caterpillars which infest their foliage. These insects would lay the tree bare. They roll the leaves up in a kind of web.[4] N.B. By care & attention the leaves were saved this year.

Apr. 28. Five *long-legged plovers, charadrius himantopus*, were shot at Frinsham-pond. There were three brace in all. These are the most rare of all British birds. Their legs are marvellously long for the bulk of their bodies. To be in proportion of weight for inches the legs of the *Flamingo* should be more than 10 feet in length.[5]

Apr. 30. Two swifts seen at Puttenham in Surrey. Bank-martins on the heaths all the way to London.

May 1. A pair of Creepers (Certhia) build at one end of the parsonage-house at Greatham, [near Selborne] behind some loose plaster. It is very amusing to see them run creeping up the walls with the agility of a mouse. They take great delight in climbing up steep surfaces, & support themselves in their progress with their tails, which are long, & stiff & inclined downwards.

May 3. Shower of snow. The snow lay but a small time. Began to turn my horses into my field lain down last year with rye-grass & dutch-clover. Wheat looks wretchedly.

May 5. The swifts which dashed-by on saturday last have not appeared since; & were therefore probably on their passage.[6]

May 8. A good crop of rye-grass in the field sown last year; but the white clover takes only in patches. Sowed 4 pounds more of white clover, & a willow basket of hay-seeds. [*Later.*] The white clover since is spread all over the field.

May 22. Nightingales have eggs. They build a very inartificial nest with dead leaves, & dry stalks. Their eggs are of a dull olive colour. A boy took my nest with five eggs: but the cock continues to sing: so probably they will build again.

May 24. Fiery lily blows: orange lily blows.[7]

May 26. The nightingale continues to sing; & therefore is probably building again.

May 28. Young pheasants !

May 31. Cut my *Saint foin*, the 12th crop. The smoke lies low over the fields. Glow-worms begin to appear.

June 1. [Bramshot-place] In Mr Richardson's garden ripe scarlet strawberries every day; large

SAINFOIN.
White's fodder crop.

artichokes, pease, radishes, beans just at hand. Bramshot soil is a warm, sandy loam.[8] Small cauliflowers Wheat shoots into ear. Barley & pease are good on the sands. The sands by liming, & turniping produce as good corn as the clays. *Footnote.* Many

large edible chestnut-trees which grew on the turn-
pike road near Bramshot-place were cut this spring
for repairs : but they are miserably shaky, & make
wretched timber. They are not only shaky, but
what the workmen call *cup-shaky*, coming apart in
great plugs, & round pieces as big as a man's leg.
The timber is grained like oak, but much softer.

June 6. [Selborne] Sparrows take possession of
the martins nests. When we shot the cock, the hen
soon found another male ; & when we killed the
hen, the cock soon procured another mate ; & so on
for three or four times.

June 19. Farmer Turner cut my great meadow.
He bought the crop. Wood-strawberries begin to
ripen.

June 22. Farmer Turner housed his hay : it
should, I think, have lain a day longer.

June 23. Golden-crowned wrens, & creepers
bring-out their broods.

June 24. Things in the garden do not grow. Clap
of thunder. Vine-bloom smells fragrantly.

June 26. Cold black solstice.

July 3. Hops are remarkably bad, covered with
aphides, & honey-dews.

July 7. Vipers are big with young.

July 9. A surprizing humming of bees all over the
common, tho' none can be seen ! This is frequently
the case in hot weather.[9]

July 11. By the number of swifts round the church
which seem to be encreased to more than 30, their
young ones must be come out.

July 12. Apricots, the young tree, ripen. Mossed
[mulched] the hills of the white cucumbers to keep them
moist.

July 13. Therm^r 79 ! The grass-mowers complain of the heat.

July 14. Dwarf elder blows. Red martagons
begin to blow. Large kidney beans bud for bloom.
Grapes swell.

July 25. Puff-balls come up in my grass-plot, &
walks : they came from the common in the turf.
There are many fairy-rings in my walks, in these the
puff-balls thrive best.   The fairy-rings alter & vary
in their shape.

July 27.   Planted out in trenches four rows of
celeri.

Aug. 7.   Rain, rain, rain.   Wheat under the
hedges begins to grow.

Aug. 10.   Wheat lies in a bad way.   Peaches, &
plums rot.   Wheat grows.   Flying ants swarm in
millions.   Inches of rain 3 [hundredths].[10]

Aug. 17.   Much wheat housed.   Drank tea at the
hermitage.[11]

Aug. 21.   Sun, brisk air, sweet even.   Many
people have finished wheat-harvest.

Aug. 23.   Sun, clouds, thunder shower, red even :
Great blackness.

Aug. 27.   Full moon.   My well is shallow & the
water foul.

Aug. 29.   House-crickets are heard in all the
gardens, & court-yards.   One came to my kitchen-
hearth.

Aug. 30-*Sept.* 8.   [Fyfield] Harvest is nearly at
an end on the downs.

Aug. 31.   The grass burns.

Sept. 2.   Partridges innumerable.   Barley-harvest
finished here.

Sept. 6.   The trufle-hunter trie's my brother's
groves ; but finds few trufles, & those very small.
They want moisture.

Sept. 7.   No mushrooms for want of more moisture.

Sept. 9.   [Worting] The greens of the turneps in
light shallow land are quite withered away for want
of moisture.

Sept. 10.   [Ufton in Berks ; Worting] Gloomy,
still, sultry, soft showers.

Sept. 11. [Selborne] The rain that fell in my absence was 73 [hundredths].

Sept. 13. Gathered-in the filberts, a large crop.

Sept. 25. Full moon. No mushrooms have appeared all this month. I find that the best crop is usually in Aug : & if they are not taken then, the season for catchup [ketchup] is lost. Many other fungi.

Sept. 27. Gathered-in the pears. The Cardillac-tree bore five bushels. Apples are few ; & the crop of grapes small.

Sept. 28. Grapes are rich, & sweet.

Oct. 3. Began lighting fires in the parlor.

Oct. 4. Mushrooms abound. Made catchup.

Oct. 12. [Funtingdon (Funtington, near Chichester)] Bad fevers near Chichester.

Oct. 13. [Shopwick] Small showers, sun.

Oct. 15-27. [Ringmer] Heavy clouds, rain. One martin at Lewes.

Oct. 16. Vast rain with strong wind for near 24 hours. *Footnote*. Thomas Hoar kept the measure of the rain [at Selborne] during my absence.[12]

Oct. 18. Mrs Snooke's black grapes are very fine. Ponds are very low.

Oct. 20. Many libellulae in copulation. Rooks frequent their nest-trees very much.

Oct. 23. Whitings, & herrings. Timothy, the old tortoise at this house, weighs 6 pounds 9 ounces & an half averdupoise. It weighed last year, Oct. 2, an ounce & an half more. But perhaps the abstemious life that it lives at this season may have reduced it's bulk : for tortoises seem to eat nothing for some weeks before they lay-up. However this enquiry shews, that these reptiles do not, as some have imagined, continue the grow as long as they live. This poor being has been very torpid for some time ; but it does not usually retire under ground 'til the middle of next month.

Oct. 24. Many hornets about the vines.

Oct. 28. [Arundel] Sheep-ponds on the downs are all filled by the rains. A great beast-market this day at Arundel.

Oct. 29. [Selborne.]

Nov. 2. My well is risen very fast.

Nov. 5. On this day many house-martins were seen playing about the side of the hanger. See Novemʳ 4 : 1777.

Nov. 18. [Upton in Berks] Frost, ice, snow, grey, bright.

Nov. 20. [Selborne.]

Nov. 23. An eclipse of the moon, total but not central.

Nov. 25. Mʳˢ Snooke's old tortoise retired under the ground.

Nov. 28. The ground is glutted with water.

Nov. 29. Snow was halfshoe-deep on the hill. Distant lightening.

Dec. 2. Vast condensations in the great parlor : the grate, the marble-jams, the tables, the chairs, the walls are covered with dew. [*Later.*] This inconvenience may be prevented by keeping the window-shutters, & the door close shut in such moist seasons.

Dec. 13. Much thunder. Great hail at Faringdon.

Dec. 22. Ground covered with snow.

Dec. 25. Vast rime, strong frost, bright, & still, fog. The hanging woods when covered with a copious rime appear most beautiful & grotesque.

Dec. 26. Most beautiful rimes.

CHAPTER XIII

" Here at my feet what wonders pass !
What endless, active life is here ! "
(M. ARNOLD : *Lines Written in Kensington Gardens.*)

1780

Jan. 1. Ice is very thick.

Jan. 2. Vast condensations, & drippings from the trees.

Jan. 8. Hard frost. On this day Sr George Rodney took a large Spanish convoy off cape Finister.

Jan. 12. For want of snow to cover, the garden-things suffer ; cabbages, spinage, celeri, &c.

Jan. 16. On this day Sr George Rodney defeated the Spanish fleet off Cadiz. A precipitate thaw. The rain last night tore-up the N : field lane very badly.

Jan. 18. The ground very dirty, & hollow from the frost.

Jan. 23. Vast rime on the trees all day. The naked turnips suffer, especially where the fields incline to the sun : they are frozen, & then thawed, & so rot. Those farmers are best off this winter, who pulled their turnips, & stacked them up in buildings, & under hedges. Lambs fall very fast.

Jan. 27. Hares, driven by the severity of the weather, crop the pinks in the garden.

Jan. 31. The cabbage-plants that were to have stood the winter seem to be killed : lettuces under the fruit-wall are damaged. [*Later*] but they all recovered. Ground hard & dusty.

Feb. 4. Wag-tails appear about the streams thro' all the severe weather.

Feb. 11. Eruptive fevers, & sore throats lurk about the parish : some die of this disorder.

Feb. 12. The farmers begin to plow after the frost.

Feb. 13. Turnips are all rotten.

Feb. 16. Sowed a crop of spinage, the autumn-sown being much diminished by the frost. Ground works well. Planted broad beans.

Feb. 23. Ivy-berries, now near ripe, are coddled [softened and partially fermented] with the frost. Heath fires.

Feb. 26. The ground is covered with snow. People that were abroad early say the cold was very intense. The ground as hard as iron.

Feb. 29-*Mar.* 11. [S. Lambeth]

Feb. 29. [S. Lambeth] Remarkable vivid Aurora borealis.

Mar. 6. [London] Sky-larks mount & sing.

Mar. 8. Mrs Snooke dyed, aged 86.

Mar. 12. [Ringmer][1] No turnips to be seen on the road.

Mar. 14. Chaffinches sing but in a shorter way than in Hants.[2]

Mar. 15. Mrs Snooke was buried.

Mar. 17. [Dorking] Brought away Mrs Snooke's old tortoise, Timothy, which she valued much, & had treated kindly for near 40 years. When dug out of it's hybernaculum, it resented the Insult by hissing.

Mar. 18. [Selborne] No turnips to be seen on the road. Green plovers [lapwings] on the common. The uncrested wren, the smallest species, called in this place the *Chif-chaf*, is very loud in the Lythe. This is the earliest summer bird of passage, & the harbinger of spring. It has only two piercing notes. *Footnote.* Thomas kept the quantity of rain in my absence. [Only two entries made.]

Mar. 20. We took the tortoise out of it's box, & buried it in the garden: but the weather being warm it heaved up the mould, & walked twice down to the bottom of the long walk to survey the premises.

Mar. 21. The tortoise is quite awake, & came-out all day long : towards the evening it buried itself in part.

Mar. 25. Sowed carrots, parsneps, planted potatoes. Ground works well. Tortoise sleeps.

Mar. 28. The tortoise put-out his head in the morning.

Mar. 30. The tortoise keeps close.

. *Apr.* 5. The frost injured the bloom of the walltrees : covered the bloom with boughs of ivy.

Apr. 7. Tortoise keeps still in it's hole.

Apr. 10. Planted two more beds of asparagus.

Apr. 15. Cucumbers swell. Tortoise sleeps on. Radishes are drawn.

Apr. 17. On this day Sir G : B : Rodney defeated the French fleet off Martinique.

Apr. 21. The tortoise heaves up the earth, & puts out it's head.

Apr. 22. Tortoise comes-forth & walks round his coop : will not eat lettuce yet : goes to sleep at four o'clock p : m : In the hot weather last summer a flight of house-crickets were dispersed about the village : one got from the garden into my kitchen-chimney, & continued there all the winter. There is now a considerable encrease & many young appear in the evening running about, & hunting for crumbs. From this circumstance it should seem that the impregnated females migrate. This is the case with ants.

Apr. 30. A sprig of *Antirrhinum cymbalaria, the ivy-leaved Toadsflax*, which was planted last year on a shady water-table [ledge or projection at top of plinth] of the wall of my house, grew at a vast rate, & extended itself full *nine* feet ; & was in perpetual bloom 'til the hard frost came. In the severity of the winter it seemed to die : but it now revives again with vigor, & shows the rudiments of flowers. When in perfection it is a lovely plant. *Lathraea squammaria* blows in the coppice below the church-litten near the foot-bridge over the stream.

May 2. Tortoise marches about :. eat part of a piece of cucumber-paring.

May 6. Made an hot-bed for the hand glasses. I opened a hen swift, which a cat had caught, & found she was in high condition, very plump & fat : in her body were the rudiments of several eggs, two of which were larger than the rest, & would probably have been produced this season. Cats often catch swifts as they stoop to go up under the eaves of low houses. The cock red-breast is a gallant bird, & feeds his hen as they hop about on the walks, who receives his bounty with great pleasure, shivering with her wings, & expressing much complacency. [*Inserted leaf.*] The quantity of rain that fell at Selborne³ between May 1st, 1779, & May 1st, 1780.

		inch : hund :
In May 1779		2 : 71
„	June	2 : 0
„	July	5 : 35
„	August	2 : 12
„	September	3 : 22
„	October	4 : 03
„	November	2 : 66
„	December	6 : 28
„	January, 1780	1 : 80
„	February	1 : 03
„	March	1 : 92
„	April	3 : 57
		36 : 69

May 7. Wild cherries in bloom make a fine show in my hedges.

May 8. The *Lathraea squammaria* grows also on the bank of Trimming's orchard, just above the dry wall, opposite Grange-yard.

May 10. Stormy all night. Tortoise scarce moves during this wet time. *Tremella nostoc* abounds on the grass walks.

May 11.   Tortoise moves about, but does not feed yet.

May 12.   The missel-thrush drives the mag-pies, & Jays from the garden.   Lettuces that stood the winter come into use.   Hops are poled, but make weak shoots.

May 13.   Vines are backward in their shoots, but show rudiments of fruit.   The cores of the spruce-firs, produced last year, now fall.   After a fast of 7, or 8 months, the tortoise which in Oct. 1779 weighed six pounds 9 oun : & ½ averdupoise, weighs now only 6 pounds 4 ounces.   Timothy began to break his fast May 17 on the globe-thistle, & American willow-herb ; his favourite food is lettuce, & dande-lion, cucumber, & kidney-beans.

May 16.   Wheat looks somewhat yellow.   Men sow barley : but the ground is cold, & cloddy.

May 18.   Field-crickets in their pupa-state lie-out before their holes.   Magpies tear the missel-thrushes nest to pieces, & swallow the eggs.

May 19.   *Helleborus viridis* sheds it's seeds in my garden, & produces many young plants.

May 27.   Large blue flag iris blows.   Flesh-flies abound.   Timothy the tortoise possesses a much greater share of discernment than I was aware of : & . . . " Is much too wise to go into a well ; " for when he arrives at the haha, he distinguishes the fall of the ground, & retires with caution, or marches carefully along the edge : he delights in crawling up the flower-bank, & walking along it's verge.

May 29.   The tortoise shunned the heat, it was so intense.

May 30.   Columbines, a fine variegated sort, blow.

May 31.   Master Etty[4] went on board the Van-sittart India-man at Spithead.   Thunderstorm in the night with a fine shower.

June 1.   Distant clouds, sultry, thunder-clouds. Sulphurous smell in the air.   Sweet even, small shower.   Strawberries blow well.   Medlar shows

much bloom.   Honey-suckles blow.   Fern-owl
chatter : chur-worm [mole cricket] jars.   The tortoise shuns the intense heat by covering itself with
dead grass ; & does not eat 'til the afternoon.   Terrible storms in Oxfordshire, & Wilts.

June 2.   Finished papering my great parlor.

June 3.   The phalaena called the swift night-
hawk [humming-bird hawk-moth] appears.

June 5.   Tortoise does not move.   Tulips fade.
Cinnamon-roses blow.

June 6.   Red valerian blows.   [*Later note.*] Terrible riots in London : & unpresidented burnings,
& devastations by the mob.⁵

June 8.   The limes show their bracteal leaves, &
rudiments of blossoms.   Sweet Williams begin to
blow.

June 9.   Hoed the quick-sets at the bottom of
the hanger.

June 10.   Fraxinells begin to blow.   Wheat-ears
begin to shoot-out.

June 11.   Field-pease look well.   All the young
rooks have not left their nest-trees.   Glow-worms
appear.

June 12.   Dragon-flies.   Bees swarm.   Sheep are
shorn.

June 15.   Vivid Aurora to the W.

June 19.   [S. Lambeth] Dust well-laid on the road.
Barley in ear on the sands.   Much upland-hay
mowed near London.

June 20.   Early pease abound.   Strawberries, &
cherries ill-ripened, & very small.   Much wall-
fruit.   Roses blow.

June 22.   Gloomy & moist, rain.   Sold my Sᵗ
foin the 13ᵗʰ crop.⁶   Lighted a fire in the dining
room.   Rain at S. Lambeth 32 [hundredths of an inch].

June 23.   Swifts stay-out 'til a quarter before nine
o' the clock.

June 24.   Cistus ledon [gum cistus] blows.   *Footnote.*   Thomas kept the rain account at Selborne.

June 25-29. [S. Lambeth.]

June 27. Swallows feed their young on the ground in Mr Curtis's botanic garden in George's fields.[7]

June 29. Jane[8] White was married to Mr Clement.

June 30. [Selborne] The portugal-laurel blows in a beautiful manner.

July 1. The red valerians, roses, iris's, corn-flags, honey-suckles, lilies, &c., make a gallant shew. Most of the pinks were destroyed in the winter by the hares. We put Timothy into a tub of water, & found that he sunk gradually, & walked on the bottom of the tub : he seemed quite out of his element, & was much dismayed. This species seems not at all amphibious. Timothy seems to be the *Testudo Graeca* of Linnaeus.[9] Dr Chandler[10] who saw the operation, says there is a species of tortoise in the Levant that at times frequents ponds & lakes : and my Bro : John White, affirms the same of a sort in Andalusia.

July 3. The tortoise weighs six pounds & three quarters averdupoise ; six pds. 12 oun :

July 4. Female ants, big with egg, come-out from under the stairs.

July 5. Began to cut the tall hedges. Put a bed of moss round the white cucumbers. Young part-ridges run.

July 8. The excrement of the tortoise is hard & solid : but when that creature urines, as it often does plentifully, it voids after the water a soft white matter, much like the dung of birds of prey, which dries away into a sort of chalk-like substance.

July 10. Timothy eats voraciously ; but picks out the hearts & stems of Coss-lettuce, holding the outer leaves back with his feet.

July 11. Finished my great parlor, by hanging curtains, & fixing the looking-glass.

July 14. Seeds of lathraea squammaria ripen.

July 17. White Jasmine begins to blow. The solstitial chafer now flies : this insect is the food of fern-owls thro' this month.

July 19.   Puff-balls appear in my grass-plot.

July 21.   The late orange, & the white lilies blowing together, make a fine show.

July 24.   Tortoise eats endive & poppies.

July 26.   Vast fog at sea, over the Sussex-downs.

July 27.   Tortoise eats goose-berries.

July 28.   Vast crops of cow-grass.   Much hay made.   Vast lights in the air from all quarters. Crickets swarm in my kitchen-chimney, & have bored thro' into the common parlor.   *Footnote.* The flies, called by our people *Nose flies*, torment the horses at plow.   They lay their eggs in the ears as well as the noses of cattle.   Some of our farmer's work their teams with little baskets tyed-on over the horses noses.   These flies seem to prevail only in Italy.[11]   Round the eaves of the Priory farm-house are 40 martins-nests, which have sent forth their first brood in swarms,   At 4 young to a nest only, the first brood will produce 160 ; & the second the same, which together make 320 : add to these the 40 pairs of old ones, which make in all 400 ; a vast flight for one house !!   The first, when congregating on the tiles, covers one side of the roof !

July 30.   Young snipes were seen at the Bishop of Winchester's table at Farnham-castle on this day : they are bred on all the moory-heaths of this neighbourhood.

July 31.   Dined at Bramshot.   Turnips flourish on the sands.   Mr Richardson's garden at Bramshot-place abounds with fruit.

Aug. 1.   Much latter-grass in delicate order. Wheat turns very fast.   Old wheat rises in price.

Aug. 2.   *Papilio Machaon alis caudatis*, concoloribus, flavis, limbo fusco, lunulis flavis, angulo ani fulvo,[12] appears in my garden, being the first specimen of that species that ever I saw in this district.   In Essex & Sussex they are more common.   A person brought me a young snipe from the forest.

Aug. 4.   Several broods of blackbirds & thrushes devour the currans, &c: 'til the wild cherries are eaten they do not annoy the garden.

Aug. 5.   My pendent pantry, made of deal & fine fly-wire, & suspended in the great wallnut tree, proves an incomparable preservative for meat against flesh-flies. The flesh by hanging in a brisk current of air becomes dry on the surface, & keeps 'til it is tender without tainting.

Aug. 10.   Sowed a crop of spinage for the winter, and spring, & trod the seed well in.

Aug. 11.   The *Papilio Machaon* never appeared but once in my garden.

Aug. 12.   Dust flies. Gardens suffer from want of rain. Much wheat bound. Timothy, in the beginning of May, after fasting all the winter, weighed only *six pounds* & *four ounces* averdupoise; is now encreased to *six pounds* & 15 *ounces*, averdupoise.

Aug. 13.   Ponds are very low.

Aug. 14.   Sope-wort [*Saponaria officinalis*] blows. Dwarf elder continues in bloom.

Aug. 16.   Lord Cornwallis gained a signal victory over General Gates in South Carolina, near Camden.

Aug. 17.   Fell-wort [*Gentiana Amarella*] blows on the hanger.

Aug. 22.   Timothy is sluggish, & scarce moves.

Aug. 25.   The thermom^r which on the stair-case stood at 67, in the wine-vault became $60\frac{3}{4}$; & again when it was 66 on the stair-case, by being plunged into a bucket of water, fresh-drawn, it fell to 51.

Aug. 28.   [Winton: Fyfield] There were at the King's house at Wintŏn 1600 Spanish prisoners; rather small men, & some very swarthy: here & there a fairish lad.

Aug. 29.   On this day the people at Selborne were to begin picking of hops: the crop of hops is prodigious.

Aug. 31.   The season is so dry that no trufle-hunter has yet tryed my brother's grove [at Fyfield].

Sept. 3-7. [Fyfield.]

Sept. 4. The trufle-man came & hunted my brother's grove for the first time : but found only half a pound of trufles, & those shrivelled & decayed, for want of moisture.

Sept. 8. [Winton : Selborne] Barley-harvest on the downs.

Sept. 9. My kidney-beans are much withered for want of rain : cucumbers bear : peaches begin to come : endives large, & tyed-up. Gathered-in the Burgamot-pears ; they easily part from their stems. Hop-picking partly ended. Myriads of flying ants, of the small pale, yellow sort, fly from their nests & fill the air.

Sept. 10. The motions of Timothy the tortoise are much circumscribed : he has taken to the border under the fruit-wall, & makes very short excursions : he sleeps under a Marvel of Peru. Lapwings frequent the upland fallows.

Sept. 12. Timothy still feeds a little. *Ophrys spiralis*, *Ladies-traces*, blows plentifully in the long lithe, & on the common near the beechen-grove.

Sept. 16. The *Antirrhinum cymbalaria* is grown to an enormous size, extending itself side-ways 15 or 16 feet, & 7 or 8 in height ! ! it grows on the water-table of a N.W. wall of my house, & runs up among the shoots of a Jasmine.

Sept. 17. When we call loudly thro' the speaking-trumpet to Timothy, he does not seem to regard the noise.

Sept. 18. Timothy eats heartily.

Sept. 19. Hornets settle on mellow fruit among the honey-bees & carry them off [Quotation of 12 Sept. 1776 repeated].

Sept. 25. [Seleburne][13] When people walk in a deep white fog by night with a lanthorn, If they will turn their backs to the light they will see their shades impressed on the fog in rude, gigantic proportions. This phenomenon seems not to have been attended to; but implies the great density of the meteor at that juncture.

Sept. 26.   Moles live in the middle of the hanger.

Sept. 27.   Finished a Bostal, or sloping path up the hanger from the foot of the zigzag to the corner of the Wadden, in length 414 yards.[14]   A fine romantic walk, shady & beautiful.   In digging along the hanger the labourers found many pyrites[15] perfectly round, lying in the clay; & in the chalk below several large cornua Ammonis [ammonites].

Sept. 28.   The China hollyhocks in my strong soil grow too tall, & are just beginning to blow.   Began to light fires in the parlor.

Oct. 2.   Cleaned-out the zigzag.   The spinage sown in Aug: now in perfection.

Oct. 3.   No ring-ouzels seen this autumn yet. Timothy very dull.

Oct. 11.   A tremendous storm in the Leeward islands, which occasioned vast damage among the shipping, &c.   Vast halo round the moon.

Oct. 12.   Spinage grown very large: a vast crop.

Oct. 13.   The tortoise scarcely moves.

Oct. 14.   Dug up the carrots, & potatoes.

Oct. 13, 14.   On these two days many house-martins were feeding & flying along the hanger as usual, 'til a quarter past five in the afternoon, when they all scudded away in great haste to the S.E. and darted down among the low beechen oaken shrubs above the cottages at the end of the hill.   After making this observation I waited 'til it was quite dusk, but saw them no more; & returned home well pleased with the incident, hoping that at this late season it might lead to some useful discovery, & point out their winter retreat.   Since that, I have only seen two on Oct.r 22 in the morning.   These circumstances put together make it look very suspicious that this late flock at least will not withdraw into warmer climes, but that they will lie dormant within 300 yards of the village.

THE PLESTOR
(see p. 80)

THE ZIGZAG, SELBORNE HANGER

[face p. 178

Oct. 15.   The cause, occasion, call it what you will, of *fairy-rings*, subsists in the turf, & is conveyable with it : for the turf of my garden-walks, brought from the down above, abounds with those appearances, which vary their shape, & shift situation continually, discovering themselves now in circles, now in segments, & sometimes in irregular patches, & spots.   Wherever they obtain, puff-balls abound ; the seeds of which were doubtless also brought in the turf.[16]   Hunter's moon.   Much lightening, & distant thunder.   [*Later.*]   This storm did much damage at Hammersmith, Putney, Wandsworth, &c. in the Isle of wight, & at Plymouth, &c. &c.   [*Another note.*]   This storm did great damage in the isle of Wight, Lancashire, & at Torbay where our fleet of observation lay, & over on the coast of France near Brest :   & was part of that hurricane which occasioned such horrible devastations in the W. Indies.

Oct. 16.   Grapes improve in the morning.

Oct. 18.   Jet-ants still in motion.

Oct. 25.   No beech-mast : many acorns.   Many wall-nuts.   My great tree produced, when measured in their husks, eight bushels.

Oct. 26.   Planted *two rows* of small *lettuces* under the fruit-wall to stand the winter : the ground works very fine.   The rows reach the whole length of the wall.

Oct. 28.   Bees begin gathering honey on the bloom of the crocus's, & finish with the bloom of Ivy.

Oct. 29.   Men put their hogs up a fatting. Timothy the tortoise, who in May last, after fasting all the winter, weighed only 6 pds. & four ounces : & in Aug. when full feed weighed 6 pds. & 15 ounces : weighs now 6 pds. 9 oun. & $\frac{1}{2}$: & so he did last Octr at Ringmer.   Thus his weight fluctuates, according as he fasts or abstains.

Nov. 2.   Leaves fall very fast.   My hedges shew beautiful lights, & shades : the yellow of the tall maples makes a fine contrast against the green hazels.

Nov. 3. Timothy, who is placed under a hen-coop near the fruit-wall, scarce moves at all.

Nov. 4. Planted out some slips of red pinks : & set some rows of Tulips.

Nov. 6. The tortoise begins to dig mould for his winter-retreat : he has much moss in his coop, under which he conceals himself.

Nov. 7. Some snow on the ground. Many trees were stripped last night : vine-leaves begin to fall. Winter-weather. Gathered the barberries, a vast crop.

Nov. 8. Gathered-in a great many grapes, because the vines cast their leaves. The crop of grapes is prodigious : perhaps the greatest I ever had.

Nov. 9. Timothy does not stir.

Nov. 11. Several wood-cocks seen this day : stone-curlews are not yet gone.

Nov. 12. Wells, streams, & ponds are very low.

Nov. 13. Wheat-stubbles plow-up in fine order ; green wheat comes up well. *Tortoise* goes under ground : over him is thrown a coat of moss. The border being very light & mellow, the tortoise has thrown the mould entirely over his shell, leaving only a small breatheing hole near his head. Timothy lies in the border under the fruit-wall, in an aspect where he will enjoy the warmth of the sun, & where no wet can annoy him : a hen-coop over his back protects him from dogs, &c.

Nov. 17. Somewhat eats the pinks in the garden.

Nov. 18. Crickets in the chimney cry but faintly & do not appear much. The cats kill all they can see.

Footnote. The severity of the weather quickened Timothy's retreat : he used to stay above ground 'til about the 20th. At Ringmer he used to lay himself up in a wet swampy border : indeed he had no choice.

Nov. 20. The ponds fill very fast.

Nov. 22. Some animal eats off the pinks.

Nov. 23. Multitudes of Starlings appear at New-ton, & run feeding about in the grass-fields. No number is known to breed in these parts. This is therefore an emigration from some other district.

Nov. 28. Timothy lies very snug but does not get any deeper.

Nov. 29. Rear Adm. Sr Samuel Hood sailed with 8 ships of the line.

Nov. 30. Hares eat all the pinks.

Dec. 2. The well is risen two rounds [turns of the axle of the windlass]. Planted about a doz. of the roots of the *Spiraea filipendula* [dropwort] sent me from Lyndon. Sr George Baker[17] directed Mrs Barker to take these roots powdered for the gravel. This plant does not grow with us, but is common at Salisbury plain, & the downs about Winton, & Andover, appearing among the bushes, & flowering about midsummer.

Dec. 6. Planted out Sweet-Williams, vine, & goose-berry cuttings, honey-suckle cuttings; & several crab stocks grafted from a curious & valuable green apple growing at South Lambeth in Surrey.

Dec. 11. The air is full of gnats, & tipulae. Gossamer about. The paths are dry like summer.

Dec. 12. Bees play out from their hives. Spring-like. The Barometer at S. Lambeth was this day at 30-6-10: a sure token that S: Lambeth is much lower than Selborne [Selborne 30-1-10].

Dec. 16. A plant of missel-toe grows on a bough of the medlar: it abounds in my hedges on the maple. The air is full of Insects. Turkies strut & gobble. Many young lambs at the Priory [Farm].

Dec. 17. My well has risen lately, notwithstanding the long dry season.

Dec. 19. At two o'clock in the shortest days the shades of my kitchen & hall-chimnies fall just on the middle of the window of J. Carpenter's workshop; & the shade of the chimney of my great parlor just in the midst of John Hall's side-front.

Hard frost all day. Covered the artichokes, & young asparagus with litter.

Dec. 21. Level snow: no drifting. Cutting wind.

Dec. 22. Antir. Cymbalaria still in bloom. Two hares in the garden last night. Snow wastes: eaves drip. Cocks crow.

Dec. 23. No rain has fallen since the 24th of Novr. The millers complain for want of water.

Dec. 29. Snow totally gone.

Dec. 30. No shower since the 24th of Novemr.

CHAPTER XIV

" To look Nature in the face." (Motto of the late Prof.
H. G. SEELEY.)

1781

Jan. 3. Some snow on the ground. Vast halo
round the moon.

Jan. 6. In the church-yard at Faringdon are two
male-yew-trees, the largest of which measures 30 feet
in girth.

Jan. 14. The ground is as hard as a rock. The
roads & fallows are dusty.

Jan. 15. Millers complain for want of water.
Footnotes. At the end of the new parlor a box-tree
is nearly killed by the current of air ; while a laurel
in the same circumstances seems not to be affected
at all. From Novr 25 to Jan. 18 there fell only 69
of rain, & snow ! less than ¾ of an inch.

Jan. 21. At the corner of my great parlor there
is such a current of air that it has half killed a box-
tree planted for a screen ; while a laurel planted in
the same draught remains unhurt. [*Later.*] This
laurel continues to flourish ; Octr 1782.

Jan. 24. Flood at Gracious street [Selborne].

Jan. 26. My *Heliotrope*,[1] which is J. Carpenter's
workshop, shows plainly that the days are lengthened
considerably : for on the shortest day the shades of
my two old chimneys fall exactly in the middle of
the great window of that edifice at half an hour after
two P.M., but now they are shifted into the quick-
set hedge, many yards to the S.E.

Feb. 10. The nuthatch brings his nuts almost
every day to the alcove, & fixing them in one corner
of the pediment drills holes in their sides, & after he has
picked out the kernels, throws the shells to the ground.

Feb. 12. Sea-gulls appear : in stormy weather they leave the sea.

Feb. 13. Stormy all night. Much thatch blown-off, & some trees thrown down.

Feb. 14. A pair of ravens build in the hanger.

Feb. 15. Strong N. aurora : very red in the N.E.

Feb. 16. A row of crocus's in the field in bloom.
Footnote. The storms in the beginning of the week did great damage by sea, & by land.

Feb. 21. The snow melts as it falls, except on the hills.

Feb. 27. [Alton] Vast storm. Had the duration of this storm been equal to it's strength, nothing could have withstood it's fury. As it was, it did prodigious damage. The tiles were blown from the roof of Newton church with such violence, that shivers from them broke the windows of the great farm-house at near 30 yards distance. This storm blew the alcove back into the hedge, & threw down the stone dial-post.

Feb. 28-*Mar.* 18. [S. Lambeth.]

Feb. 5. Wheat looks miserably in Battersea-field.[2]

Feb. 11. The tortoise came-forth & continued to be alert 'til the 25th, & then eat some lettuce ; when the weather turning very harsh he retired under the straw in his coop. About this time the shell-snails at this place began to move.

Feb. 15. On this day Lord Cornwallis gained a considerable victory 'over Gen. Greene at Guildford in N. Carolina.

Feb. 19, 20. [London.]

Feb. 21-28. [S. Lambeth.]

Feb. 24. The dust on the roads is become a terrible nuisance.

Feb. 29. [Alton.]

Feb. 30. [Selborne.]

Feb. 31. Cucumber-plants show fruit ; but are rather drawn, from the long dry season, which never is favourable to hot-beds.

Apr. 1. The tortoise came-out for two hours.

Apr. 2. Tortoise out. Timothy weighs 6 lbs. 8¾ oz. The beginning of last May he weighed only 6 lbs. 4 oz.

Apr. 3. Timothy eats heartily. The wry-neck appears & pipes. *Bombylius medius* still: bobs his tail in flight against the grass, as if in the act of laying eggs.

Apr. 5. Searched the S.E. end of the hanger for house-martins, but without any success, tho' many young men assisted. They examined the beechen-shrubs & holes in the steep hanger.

Apr. 11. While two labourers were examining the shrubs & cavities at the S.E. end of the hanger, a house-martin came down the street & flew into a nest under Benham's eaves. This appearance is rather early for that bird. Quae: whether it was disturbed by the two men on the hill?

Apr. 16. [Worting.]

Apr. 17. [Caversham: Reading.]

Apr. 18. [Oxford] Some bank-martins at Walling-ford-bridge.

Apr. 21. [Alton] White-throat on the road.

Apr. 22. [Selborne.]

Apr. 26. A pair of Nightingales haunt my fields: the cock sings nightly in the Portugal-laurel, & balm of Gilead fir.

Apr. 30. Men pole their hops. Dragon-fly & musca meridiana. Ponds begin to be dry.

May 2. Field-crickets *crink*: this note is very summer-like, & chearful.

May 5. No rain enough to measure.

May 7. Vast bloom among the apples: the crop but small. No rain to measure since April 12.

May 8. Timothy lies close this cold weather.

May 10. A small sort of caterpillar annoys the goose-berry trees.

May 11. *Fern-owl* chatters. When this bird is heard, summer is usually established.

May 12. My well sinks very fast : indeed much water has been drawn lately for watering. Mr Yalden's tank is almost dry.

May 13. The rain last night broke the stems of several of the tulips, which are in full bloom. The rain from the S.

May 14. The hops wanted rain, & began to be annoyed by the aphides. The ground finely refreshed. Vast rocklike, distant clouds.

May 15. Killed some hundreds of shell-less snails about the garden. The boys every day kill some large wasps, that feed on the sycamore-bloom, on the Plestor. Several small wasps appear as well as large breeders [foundress wasps].

May 20. Wheat very gross in some fields.

June 1. Grass-walks burn very much. Ground chops. Roses begin to blow. Wheat spindles for ear.

June 2. Tulips are gone. The heat injures the flowers in bloom. S^t foin in full bloom. Fly-catcher has five eggs.

June 3. Wheat-ears begin to burst-out. Boys bring hornets. The planet Venus is just become an evening star : but being now in the descending signs ; that is, the end of Virgo, where it now is, being a lower part of the Zodiac than the end of Leo, where the sun is ; Venus does not continue up an hour after the sun, & therefore must be always in a strong twilight. It sets at present N. of the west ; but will be in the S.W. but not set an hour after the sun 'til Octr from which time it will make a good figure 'til March in the S.W., W., & a little to the N. of the W.

June 9. A pair of swallows hawk for flies 'til within a quarter of nine o'clock ; they probably have young hatched.

June 13. The house-martins which build in old nests begin to hatch, as may be seen by their throwing-out the egg-shells.

June 14. We have planted-out a vast show of annuals, which will want no watering.

June 16. My garden in nice order, & full of flowers in bloom. Lilies, roses, fraxinellas, red valerians, Iris's, &c. &c. now make a gaudy show.

June 18. The S^t foin is in a bad way about the neighbourhood.

June 19. A strange swarm of bees came & settled on my balm of Gilead fir.

June 20. Much thunder, & vast showers to the westward. Vast storm & rain at Wintŏn. These storms were very terrible at Sarūm, & in the vale of white-horse³ ; &c.

June 21. Finished cutting the St foin, which has stood full long. The 14th crop. Sold it to John Hale. In some parts a good burden.

June 24. [Here follows an account of the honey-buzzards reproduced in T.P., XLIII.] *Additional note.* The male honey-buzzard still haunts about the hanger, & on sunny mornings soars above the hill to inhale the coolness of the upper air. There are this summer about this church 11, or 12 pairs of swifts.

June 25. Our fields of pease are in a sad lousy state.

June 26. Young redstarts come abroad.

June 27. The honey-buzzard sits hard.

June 30. About nine in the evening a large shining meteor appeared falling from the S. towards the E. in a inclination of about 45 degrees, & parting in two before I lost sight of it. I was in Baker's Hill in the shrubbery, having a very bad horizon ; & therefore could not see how & where it fell.

July 1. Wheel round the sun.

July 2. Made my rick of meadow-hay, which contains six jobbs, without one drop of rain. Some part of it would have been better, I think, had there been some sun on the day of making.

July 4. The bloom of the lime hangs in beautiful golden tassels.

July 6. Brisk gale. The wheat, in large fields, undulates before the gale in a most amusing [interesting] manner.

July 7. Timothy the tortoise, who weighed April 2 : after fasting all the winter only *six pounds* 8 oun. & ¾ : weighs now *seven pounds*, & *one ounce* : weighed last august *six pounds*, & 15 ounces. From the encreased number of the *Swifts*, it seems as if they had brought out many of their young. About eight in the evening, Swifts get together in a large party, & course round the environs of the church, as if teaching their broods the art of flying. As yet they do not retire 'til three quarters after 8 o' the clock ; & before they withdraw, the *bats* come forth : so that day & night animals take each others places in a curious succession! All the swifts that play around the church do not seem to roost under it's eaves. Some pairs, I know, reside under some of the cottage roofs. Three or four pairs of lapwings hatched their broods this summer on the common : the young, which run long before they can flie, sculk among the fern. They usually affect low, moist situations.

July 11. Trenched-out celeriac, & some of the new-advertized large celeri. Planted out some endive. A pair of house-martins, that built under the eaves of my stable, lost their nest in part by a drip, just as most of the young were flown. They are now repairing their habitation in order to rear a second brood. [Here follows the account of the sparrow-hawks, T.P., XLIII ; with this added note.] They [the sparrow hawks] carried-off also from a farm-yard a young duck larger than themselves.

July 14. The hay that is down is now entirely spoiled. These soft rains sop & drench every thing. A young man brought me a live specimen of a *Papilio Machaon*, taken below Temple. The first specimen that ever I saw of that species in these parts was in my own garden in last Augt 2nd.

July 15.   The farmers complain of smut in their wheat.

July 16.   Wheat-harvest begins at Headley.

July 17.   The sparrow-hawks continue their depredations.

July 18.   [Bramshot-place] Lapwings haunt the uplands still.   Farmers complain that their wheat is blited.   At Bramshot-place, the house of Mr Richardson, in the wilderness near the stream, grows wild, & in plenty, *Sorbus aucuparia* [= *Pyrus aucuparia*], the *quicken-tree*, or *mountain-ash*, *Rhamnus frangula, berry-bearing alder*; & *Teucrium scorodonia, wood-sage*, & whortle-berries.   The soil is sandy.   In the garden at Dowland's, [Bramshott parish] the seat, lately, of Mr Kent, stands a large *Liriodendrum tulipifera*, or *tulip-tree*, which was in flower.   The soil is poor sand; but produces beautiful pendulous *Larches*. M^r R's garden, tho' a sand, abounds in fruit, & in all manner of good & forward kitchen-crops.   Many China-asters this spring seeded themselves there, & were forward; some cucumber-plants also grew-up of themselves from the seeds of a rejected cucumber thrown aside last autumn.   The well at Dowland's is 130 feet deep; at Bramshot-place . . . [gap in original].   M^r R's garden is at an average a fortnight before mine.

July 19.   House-martins abound at Lipock [Liphook.]

July 21.   [Selborne] The planet Mars figures every evening & makes a golden & splendid shew.   This planet being in opposition to the sun, is now near us, & consequently bright.

July 22.   All the first meadow-hay about us was spoiled : all the latter was ricked in delicate order. Late in the evening the swifts course round with their young high in the air.   They are some times so numerous that one might suspect they are joined by parties from other villages.   The fly-catchers have quite forsaken my house & garden : they never breed twice.

July 23. Of those China hollyhocks that stood the winter the tall ones are plain & single : the stunted ones are double & variegated.

July 25. The crop on my largest Apricot-tree is still prodigious, tho' in May I pulled off 30, or 40 dozen.

July 26. The blackbirds & thrushes, that have devoured all the wild cherries in the meadow, now begin to plunder the garden.

July 27. Cran-berries ripen.

July 28. Gleaners bring home bundles of corn. The black-birds, & thrushes come from the woods in troops to plunder my garden. We shot 30 black-birds, & thrushes. The white-throats are bold thieves ; nor are the red-breasts at all honest with respect to currans. Birds are guided by colour, & do not touch any white fruits 'til they have cleared all the red ; they eat the red grapes, rasps, currans, & goose berries first. [Cf. 1770, Note 10.]

July 29. Timothy comes-out but little, while the weather is so hot : he skulks among the carrots, & cabbages. Red-breasts eat the berries of the honey-suckle.

July 30. The ants, male, female, & workers, come forth from under my stairs by thousands.

Aug. 1. The honey-bees suck the goose-berries, where the birds have broke the skin.

Aug. 3. Now the ants under the stairs have sent-out their males, & females, they no longer appear.

Aug. 5. Small scuds of rain. No rain to measure since July 14. On this day a bloody & obstinate engagement happened between Admiral Hyde Parker, & a Dutch fleet off the Dogger-bank.

Aug. 6. Every ant-hill is in a strange hurry & confusion ; & all the winged ants, agitated by some violent impulse, are leaving their homes ; &, bent on emigration, swarm by myriads in the air, to the great emolument of the hirundines, which fare luxuriously. Those that escape the swallows return no more to their nests, but looking out for new retreats, lay a foundation for future colonies. All the females at these times are pregnant.

Aug. 8. We have shot 31 black-birds, & saved our gooseberries.

Aug. 9. One swift, perhaps a pair, going in & out of the eaves of the church. Why do these linger behind the rest, which have withdrawn some days ? have they a backward brood delayed by some accident ?

Aug. 11. Ponds & streams fail. People in many parts in great want of water. The reapers were never interrupted by rain one hour the harvest thro'.

Aug. 13. The pond on Selborne down has still some good water in it ; Newton pond is all mud. Many annuals are shrivelled-up for want of moisture. The drought is very great. Hops are injured for want of rain.

Aug. 14. The bank-martins at the sand-pit on Short-heath are now busy about their second brood, & have thrown out their egg-shells from their holes. The dams & first broods make a large flight. When we approached their caverns, they seemed anxious, & uttered a little wailing note. My well is low in water ; but a constant spring bubbles up from the bottom. Some neighbouring wells are dry. My well is 63 feet deep.

Aug. 16. Sowed a crop of winter-spinach, & pressed the ground close with the garden-roller. The ground turned-up very dry, & harsh.

Aug. 17. The small pond in Newton great farm field, near the verge of the common,⁴ is full nearly of good clear water ! while ponds in vales are empty. One swift ! The crevice thro' which the swift goes up under the eaves of the church is so narrow as not to admit a person's hand.

Aug. 18. Some wasps at the butcher's shop.

Aug. 19. Mʳ Pink's turnips are infested with black caterpillars ;⁵ he turned 80 ducks into the field, hoping they would have destroyed them; but they did not seem much to relish the sort of food. I have known whole broods of ducks destroyed by their eating too freely of hairy caterpillars.

Aug. 21.   No wasps; but several hornets, which devour the nectarines. The wasps are probably kept down by the numbers of breeders that the boys destroyed for me in the spring.

Aug. 23.   Caught 8 hornets with a twig tipped with bird-lime. No wasps in my garden, nor at the grocer's, or butcher's shop. Five or six hornets will carry off a whole nectarine in the space of a day.

Aug. 24.   Tho' white butterflies abound, & lay many eggs on the cabbages; yet thro' over-heat, & want of moisture, they do not hatch and turn to palmers [caterpillars]; but dry & shrivel to nothing. One swift still frequents the eaves of the church; & moreover has, I discover, two *young* nearly fledged, which show their white chins at the mouth of the crevice. This incident of so late a brood of swifts is an exception to the whole of my observations ever since I have bestowed any attention on that species of hirundines!

Aug. 30.   Between nine & ten at night a thunder-storm with much vivid lightening began to grow up from the N.W. & W: but it took a circuit round to the S. & E. & so missed us. We had only the skirts of the tempest, & a little heavy rain for a short time. Ten miles off to the southward there were vast rains.

Aug. 31.   Began to use endive, which is large & well-blanched. *No swifts.* We searched the eaves to no purpose. In searching the eaves for the young swifts, we found in a nest two callow dead swifts, on which had been formed a second nest. These nests were full of the black shining cases of the hippoboscae hirundinis.

Sept. 1.   We have caught about 20 hornets with a twig tipped with bird-lime.

Sept. 4. Gathered one bunch of black *grapes*, which was ripe & well-flavoured. It grew close to the wall, pressed down by a bough.

Sept. 7. Dined at Bramshot-place.

Sept. 9. Red-breasts whistle agreeably on the tops of hop-poles, &c. but are prognostic of autumn. Young fern-owl.

Sept. 10. Red-breasts feed on elder-berries, enter rooms, & spoil the furniture. Bro^r T. and M.[6] come to Selborne. Timothy, whose appetite is now on the decline, weighs only 7 pounds & ¾ of an ounce : at Midsumm^r he weighed 7 ae 1 oun. [ae used for lbs.]

Sept. 11. Bean-harvest & vetch-harvest.

Sept. 13. Beans heavy.

Sept. 14. Timothy the tortoise dull & torpid.

Sept. 15. Thunder & lightening in all quarters round. The spring called *Well-head*[7] sends forth now, after a severe hot dry summer, & dry spring & winter preceding, *nine* gallons of water in a minute ; which is 540 in an hour ; & 12960, or 216 hogsh : in 24 hours, or one natural day. At this time the wells are very low, & all the ponds in the vales dry.

Sept. 16. The boys destroyed a hornets nest : it was but small. *Ophris spiralis, ladies traces*, seed.

Sept. 18. Fly-catchers seem to be gone : they breed but once.

Sept. 20. [Meonstoke] The Well is now so low, that Thomas [Hoar] found some difficulty in getting water sufficient to Brew with.

Sept. 21. Hooker's-hill mended by Tom Prior: the ditch below which was made about fifty years ago, is now open'd and cleaned.[8]

Sept. 22. [Selborne] The well at Filmer-hill[9] is 60 yards deep : at Privet, on the top of the hill, they have no wells, & have been greatly distressed for water the summer thro'. The Warnford, & Meon-stoke stream as full, & bright, as if there had been no drought.

Sept. 23. Began to light fires in the parlor. Aurora.

Sept. 24. The wind blows down apples & pears. Vivid aurora.

Sept. 25. Gathered swan's egg pears, a large crop. Surprising Auroras, very red in the W!!! The young swarms of bees this summer are light; the old stocks are heavy.

Sept. 26. Dug up potatoes : earthed up celeri. Gathered knobbed russetings, a large crop. Our building-sand[10] from Wolmer-forest seems pure from dirt ; but examined thro' a microscope proves not to be sharp, & angular, but smooth as from collision. It is of a yellow colour. " The amazing number of swallows* that at this time are flying in London, is a very uncommon appearance. They seem greatly affected by the severe cold weather we have experienced for some days past, since the wind has been northerly ; they fly in at windows, & are so tamed or numbed, that boys beat them down, as they fly in the streets." The Gazetteer. [*White's note.* * Probably house-martins, & swallows.]

Sept. 27. Gathered Cadilliac-pears, dearlings,[11] & royal russets.

Sept. 28. Dug up potatoes, & carrots.

Sept. 29. My well has now only three feet in water : it has never been so low, since my father sunk it, more than forty years ago.

Sept. 30. Men put-up their hogs to fat. House-flies muscae domesticae, now croud about the fire-place, run on hearths, & sport in the chimney-corner.

Oct. 1. [*Note, apparently by Mary White*] Good riding round Woolmere Pond, within it's banks, on the sand. Cleaned my well by drawing out about 100 buckets of muddy water : there was little rubbish at bottom. There were two good springs, one at the bottom, & one about three feet above. Nothing had been done to this well for about 40 years.

[Cf. 22 Oct. 1778.] The man at bottom in the cleaning brought up several marbles & taws that we had thrown down when children.

Oct. 3.   Bought a bay-Welch Galloway mare. Out of the horses that were offered to me to try, there were ten mares to one gelding.

Oct. 4.   No h: martins, nor swallows in the village, nor sand-martins about the forest. Ld Stawel[12] was fishing Wolmer-pond with a long net drawn by ten men.

Oct. 5.   No h: martins, nor swallows about the villages, nor sand-martins at the pit on Short-heath.[13] The white-sand in the pit above, observed thro' a microscope, appears more sharp, & angular than the yellow sand of the forest. Gathered in the nonparels, & royal russets.   Much gossamer flying.

Oct. 6.   Several herons at Wolmer-pond, & a *tringa octrophus*, or white-rumped sand-piper, [green sandpiper] Cranmer-pond in Wolmer-forest is quite dry.

Oct. 8.   Several women & children have eruptions on their hands, &c., is this owing to the lowness of the water in the wells, &c? It seems this often befalls after they have been employed in hop-picking.

Oct. 9.   The grass was covered with cob-webs, which being loaded with dew, looked like white frost. A *grey hen*[14] was lately killed on that part of Hind-head which is called the Devil's punch-bowl. This solitary bird has haunted those parts for some time.

Oct. 10.   My well rises. My hedges are beautifully tinged. Wood-larks sing sweetly thro' this soft weather.

Oct. 11.   A brood of swallows over Oakhanger-pond!

Oct. 12.   Farmer Parsons fetches a waggon-load of water from Dorton[15] for brewing! Wells fail.

Oct. 13.   On frequented roads the dust is very troublesome.

Oɛt. 14. The greens of turnips wither, & look rusty. The distress in these parts for want of water is very uncommon. The well at the Grange-farm is dry ; & so are many in the villages round : & even the well at *Old-place* in the parish of E. Tisted, tho' 270 feet, or 45 fathoms deep, will not afford water for a brewing. All the while the little pond on Selborne down[16] has still some water, tho' it is very low : & the little pond just over the hedge in Newton great farm abates but little. The ponds in the vales are now dry a third time, Most of the wells in Selborne-street are empty ; & mine has only three feet of water. The people at Medsted, Bentworth,[17] & those upland parts are in great want. *Well-head* sends-out a considerable stream still, not apparently abated since we measured it last. The last wet month was Decem[r] 1779, during which fell 6 inch : 28 hund : of rain : since which the quantity of water has been very little.

Oɛt. 16. The mill at Hawkley cannot work one-tenth of the time for want of water.

Oɛt. 17. Greatham-mill can work but 3 hours in the day.

Oɛt. 19. On this ill-fated day Lord Cornwallis, & all his army surrendered themselves prisoners of war to the united forces of France & America at York-town in Virginia.

Oɛt. 21. The distress for water in many places is great. A notion has always obtained, that in England hot summers are productive of fine crops of wheat : yet in the years 1780, and 1781, tho' the heat was intense, the wheat was much mildewed, & the crop light. Quaere, Does not severe heat, while the straw is milky, occasion it's juices to exsude, which being extravasated, occasion spots, discolour the stems & blades, & injure the health of the plants ? The heat of the two last summers has scalded & scorched the stems of the wall-fruit trees, & has fetched-off the bark.

Oct. 22. Men continue to fetch peat from the forest.

Oct. 23. Farmer Eaves of *Old place* [E. Tisted, 3 miles] fetches his water from *Well-head*. Jupiter is now very low in the S.W. at sun-set.

Oct. 24. The tortoise is very torpid, but does not bury itself.

Oct. 25. Acorns abound, & help poor men's hogs. " There has lately been felt in diverse parts of Hungary so extraordinary a heat, that the husband-men could only work in the night. All the snow that has covered the Carpathian mountains for more than a century is entirely melted." St James's Chronicle.

Oct. 26. Men sow their wheat in absolute dust. Bro : T. and M. went away.

Oct. 27. My well sinks & is very low. The tortoise begins to dig into the ground. Mr Yalden[18] fetches water from Well-head. The bat is out this warm evening.

Oct. 28. The planet Venus, which became an evening star in June, but was not visible 'til lately, now makes a resplendent appearance. On Sel-borne down are many oblong tumuli, some what resembling graves but larger, supposed by the country people to be the earth of saw-pits. But as they mostly lie one way from S.E. to N.W. & are many of them very near to each other, it is most probable that they were occasioned by some purpose of a different kind. My bro : Tho : ordered two to be dug across ; one of which produced nothing extra-ordinary ; while in the other was found a blackish substance : but how, & in what quantity it lay, & whether it consisted of ashes & cinders, or of *humus animalis*, we had no opportunity to examine from the precipitancy of the labourer, who filled up the trench he had opened without giving proper notice of the occurrence.[19]

Oct. 29. From the scantiness of the grass I have given for sometime 9ᵈ pr. pd. for butter; a price here not known before.

Oct. 30. The *tortoise* retires under ground, within his coop.

Oct. 31. The water is so scanty in the streams, that the millers cannot grind barley sufficient for mens hogs. Dairy-farms cannot fill the butter-pots of their customers.

Nov. 1. Much wheat-land not sown yet; because men are afraid to sow their corn in the dust. Some water still in the pond on Selborne down; & the pond on Newton-farm, over the hedge, is half full. No drought, equal to the present, has been known since autumn 1740, which being preceeded by a dry summer & spring, & the terrible long frost of winter 1739, exhausted most of the wells & ponds, & distressed the country greatly.

Nov. 3. Hogs, in eating acorns, chew them very small, & reject all the husks. The plenty of acorns this year avails the hogs of poor men & brings them forward without corn.

Nov. 4. The wintry & huge constellation, Orion, begins now to make his appearance in the evening, exhibiting his enormous figure in the E. Tho' my grapes ripen in the most disadvantageous years: yet from the concurring circumstances of a hot summer, & a failure of wasps, I think my crop was never so delicate before, nor ever supplyed my table so long.

Nov. 6. The walls of the hall & entry are all in a float with condensation.

Nov. 7. John Hale has cleansed the pond on the down, & carried out a large quantity of mud. Husbandry seems to be much improved at Selborne within these 20 years, & their crops of wheat are generally better: not that they plough oftener, or perhaps manure more than they did formerly; but from the more frequent harrowings & draggings

now in use, which pulverize our strong soil, & render
it more fertile than any other expedient yet in practice.

Nov. 8. The *tortoise* came out of his coop, & has
buried himself in the laurel-hedge. When my great
parlor is kept close shut-up, it is not at all affected
by condensations on the wainscot or paper, tho' the
hall & entry are all in a float.

Nov. 10. Hares eat down the pinks, & cloves in
the garden: & yet sportsmen complain that the
breed this year is very small; alledging that dry
summers, tho' kindly for partridges, are detrimental
to hares.

Nov. 11. The house-martins have disappointed
us again, as they did last year, with respect to their
Novemʳ visit for one day. On Nov. 5ᵗʰ 1779, &
Nov. 4ᵗʰ 1777, they showed themselves all day along
the hanger in considerable numbers, after they had
withdrawn for some weeks : when, had they been
properly watched, their place of retreat in the evening,
I make no doubt, might have been easily discovered.
Once in a few years they make us a visit of this sort,
some time in the first week in November.

Nov. 15. Vast rain in the night, with strong
wind. My well is risen near six feet. Thomas
begins to dress the vines. The crop of grapes is
just over, having lasted in perfection more than ten
weeks.

Nov. 19. Planted two Cypresses in the garden.
They came from S. Lambeth.

Nov. 21. Finished dressing the vines. The new
wood was small, & not highly ripened; so it was
laid-in the shorter. It covers the walls regularly.

Nov. 24. Cascades fall from the fields into the
hollow lanes.

Nov. 25. Fog, wh. frost. As the fog cleared
away, the warm sun occasioned a prodigious reek,
or steam to arise from the thatched roofs. In the
evening picturesque partial fogs come rolling-in
up the Lithe from the forest.

Nov. 26. The planet Venus now is visible at Selborne over the hanger. Planted against the fruit-wall, three (*sic*) well-trained trees that are to begin to bear fruit next year : viz : 1 Peterboro' nectarine : N.E. end. 1 Montaban peach. 1 Red Magdalen peach, against the scullery. 1 Elrouge Nectarine. These trees came from Mr Shiell's nursery at Lambeth & cost 7s. 6d. apiece on the spot. They have healthy wood, & well-trained heads. Planted a Virginian creeper against the wall of my house next the garden.

Nov. 27. Began to use some of the advertised Celeri, which, I think, is crisper & finer flavoured than any sort that I have met with.

Nov. 28. This month proved a very wet one, more so than any month since Decem^r 1779 : in which fell 6 in. 28, in this 6. 18.

Nov. 29-*Dec.* 6. [Oxford] Dark & hazy, deep fog. [A break in the diary until Dec. 7. "Thomas measured the rain."]

Dec. 7. [Alton] The weather was dark, still, & foggy all the time that I was absent, & the wind mostly N.E. & E.

Dec. 8. [Selborne] Rain.

Dec. 9. George Tanner's bullfinch, a cock bird of this year, began from it's first moulting to look dingy ; & is now quite black on the back, rump & all ; & very dusky on the breast. This bird has lived chiefly on hemp-seed. But T. Dewey's, & —— Horley's two bull-finches, both of the same age with the former, & also of the same sex, retain their natural colours, which are glossy & vivid, tho' they both have been supported by hemp-seed. Hence the notion that hemp seed blackens bull-finches does not hold good in all instances ; or at least not in the first year.[20]

Dec. 11. Trenched some ground in the meadow for carrots.

Dec. 12. Larby now digs & double trenches a weedy spot in the great meadow. The ground is black, & mellow, & fit for carrots, for which it is intended.

Dec. 14.   Some of my friends have sported lately in the forest : they beat the moors & morasses, & found some jack & whole snipes [common snipes], most of which they killed, with a teal, a pheasant, & some partridges.   The flight of snipes is but small yet.   There is now, I hear, a flight of woodcocks in the upland coverts.

Dec. 17.   Heard's well is now dry : it is of a vast depth.

Dec. 21.   Furze is in bloom.   Several young lambs at the Priory.   Shortest day.

Dec. 25.   Sun, bright & pleasant.   A gardener in this village has lately cut several large cauliflowers, growing without any glasses.   The boys are playing in their shirts.   On this day Admiral Kempenfelt fell in with a large convoy from Brest, & took a number of French transports.

CHAPTER XV

"All trivial fond records." (*Hamlet*, Act 1, Sc. 5.)

1782

Jan. 1. Winter aconites blow.

Jan. 3. Mezereon blows. Viola tricolor, red lamium, & grounsel blow. Hazel catkins open.

Jan. 6. Wells now rise considerably. Bees come out of their hives. Gnats play about.

Jan. 9. The wind blowed-down the rain-measurer. Wells rise very fast, & are now up to their usual pitch.

Jan. 10. The earth is well-drenched; streams run; & torrents fall from the fields into the hollow lanes.[1]

Jan. 14. On this day Sᵣ G. Rodney, after having been wind-bound for several weeks, sailed to the W. Indies from Torbay.

Jan. 28. Water-cresses come in. The χείμαρρος, or winter-spring, or lavant, runs from the hanger into gracious street.

Feb. 1. The wheatfields so hard that they carry a waggon & horses.

Feb. 5. Knee-holly, or butcher's broom, blows.

Feb. 8. Venus *shadows* very stongly, showing the bars of the windows on the floors & walls.

Feb. 9. Venus sheds again her silvery light on the walls of my chamber, &c. & *shadows* very strongly.

Feb. 13. Things froze in the pantry. Shallow snow : ground very hard.

Feb. 17. Last night the moon was midway between Mars & Venus : she is this evening just above, close above the former, at the seeming distance of two feet.

Feb. 18. [Alton.]

Feb. 19-*Mar.* 15. [S. Lambeth] Thomas kept an account of the rain in my absence.

Mar. 3. [London] Daffodil opens. Misselthrush sings.

Mar. 6. Almond tree in bloom.

Mar. 16. [Selborne] Frost, ice, small flights of snow. Peaches, & Nect. forwarder in bloom than apricots.

Mar. 19. Cleaned-up the alleys, & borders of the k : garden.

Mar. 20. The wheat-ear is seen on our down.

Mar. 21. Vast flocks of large Fieldfares appear : they are probably intent on the business of migration.

Mar. 23. A farmer tells me he foresaw this extraordinary weather by the prognostic deportment of his flock; which, when turned-out on a down two or three mornings ago, gamboled & frolicked about like so many lambs.

Mar. 28. Poor Timothy was flooded in his hybernaculum amidst the laurel-hedge; & might have been drowned, had not his friend Thomas come to his assistance & taken him away.

Mar. 30. Apricots shew hardly any bloom : they exhausted themselves with bearing last year. Peaches & nect: abound with blossoms just opening; as do the new, trained trees planted last Nov^r.

Mar. 31. The earth is glutted with water.

Apr. 1. Vast rains ! 91 [hundredths].

Apr. 3. The prospect at Newton was most lovely ; as usually is the case after much rain, &c.

Apr. 6. Many hail-storms about. Sunny evening, pleasant. The wind veered about to every black storm. On this day the *lavants* began to break at Chawton, two fields above the church.

Apr. 7. Peter Wells's well runs over. Strong *lavant* thro' Cobb's court-yard.

Apr. 11. Forked the asparagus-beds, & planted some firs in the outlet [prospect in grounds].

Apr. 12. A person thought he heard a black-cap. On this day Sʳ George Rodney in the W. Indies obtained a great victory over Admiral de Grasse, whom he took in the Ville de Paris. He also captured four more French ships of the line, & sunk one. On the 19ᵗʰ Admiral Hood, who was detatched with his division in pursuit of the flying enemy, took two more line of battle ships, & two frigates in the passage of Mono, between Porto Rico, & Domingo.

Apr. 17. Several *Black-caps* are heard to sing.

Apr. 20. On this day Admiral Barrington discovered a convoy newly from Brest. His fleet took 12 or thirteen transports; a French man of war of 74 guns, called the Pagasus; & a 64 gun ship, named the Actionnaire, armée en flute [flûte.]

Apr. 27. Hyacinths in full bloom. My sort is very fine.

May 1. One beech in the Lith, which is always forward, in full leaf. The beeches in the hanger are naked, but the buds are opening. Nightingale sings in my fields; but no cuckow is heard. Bees keep within their hives. No gnats appear. Redstart is heard. Sowed the dug plot in the great mead with carrots.

May 2. Two swifts at Nore hill passed by me at a steady rate towards this village as if they were just arrived.²

May 4. Vegetation is at a stand, & Timothy the tortoise fast asleep. The trees are still naked.

May 10. The tortoise weighs 6ae [*librae*] 11oun. 2dr. He weighed Spring 1781, 6 : 8 : 4 & May 1780, 6 : 4 : 0. [Here follow notes on *pettichaps* and on swallows perching near water (D.B., LVII).]

May 12. A pair of white owls breed under Mʳ Yalden's roof of his house : they get in thro' the leaden-gutter that conveys the water from the roof to the tank. They have eggs. About this time many swallows were found dead at & about Fyfield :

some were fallen down the chimnies, & some were lying on the banks of brooks, & streams.

May 14. Tortoise eats the leaves of poppies.

May 22. Men pole their hops, which are backward, but strong. Some hail.

May 27. Men have not been able to sow all their barley.

May 31. From Jan. 1, 1782 to May 31 D° inclusive, the quantity of rain at this place is 24 inch. 7 hund. This is after the rate of about 58 inch. for the whole year. This evening *Chafers* begin to fly in great abundance. They suit their appearance to the coming-out of the young foliage, which in kindly seasons would have been much earlier.

June 2. Mr Pink is obliged to leave 26 acres of barley-ground unsown. Feverish colds begin to be very frequent in this neighbourhood, & indeed the country over. Within the bills of mortality this disorder is quite epidemic, so that hardly an individual escapes. This complaint seems to have originated in Russia, & to have extended all over Europe.[3] The great inclemency of the spring may best account for this universal malady.

June 4. Kidney-beans in a poor way : they have all been in danger of rotting.

June 5. My Bro^r Thomas White nailed-up several large 'scallop shells under the eaves of his house at South Lambeth, to see if the house-martins would build in them. These conveniences had not been fixed half an hour, before several pairs settled upon them ; &, expressing great complacency, began to build immediately The shells were nailed on horizontally with the hollow side upward ; & should, I think, have a hole drilled in their bottoms to let-off moisture from driving rains.

June 9. When the servants have been gone to bed some time, & the kitchen left dark, the hearth swarms with young crickets about the size of ants : there is an other set among them of larger growth :

so that it appears that two broods have hatched this spring.

June 11. Standard honey-suckles, having lost their first shoots by the frosts, will produce little bloom this summer.

June 13. A house-martin drowned in the water-tub : this accident seems to have been owing to fighting.

June 14. Ephemerae, may-flies, appear, playing over the streams : their motions are very peculiar, up & down for many yards, almost in a perpendicular line.

June 15. Hung-out my pendent meat-safe. The martins over the garden-door have thrown-out two eggs : they had not been sat on. A pair of part-ridges haunt Baker's hill, & dust themselves along the verge of the brick-walk. Many people droop with this feverish cold : not only women & children, but robust labourers. In general the disorder does not last long, neither does it prove at all mortal in these parts.

June 16. This hot weather makes the tortoise so alert that he traverses all the garden by six o'clock in the morning. When the sun grows very powerful he retires under a garden-mat, or the shelter of some cabbage ; not loving to be about in vehement heat. In such weather he eats greedily.

June 18. *Apis longicornis* swarms in my walk down Baker's hill, & bores the ground full of holes, both in the grass, & brick-walk. Peat begins to be brought in. On this day there was a great thunder-storm in London. Probably much rain fell this day at some distance to the S.W. While the thunder was about, the stone pavement in some parts of the entry & kitchen sweated & stood in drops of water. The farmers say, that the chafers, wch abound in some parts, fall off the hedges, & trees on the sheeps backs, where being entangled in the wooll they die, & being blown by flies, fill the sheep with maggots.

The epidemic disorder rages in an alarming manner in our fleet. Sr John L. Ross has left the N. sea, & is returned to the downs, not being able to continue his cruize on account of the general sickness of his crews.

June 20. The smoke from lime-kilns hangs along the forest in level tracts for miles.

June 23. Jupiter makes, & has made for some weeks past a beautiful & resplendent appearance every evening to the S.E. Saturn, who is very near, is much obscured by the brilliancy of the former. The sun at setting shines along the hanger in these long days, & tinges the stems of the tall beeches of a golden colour in a most picturesque, & amusing manner !! Just at the summer solstice the sun at setting shines directly up my broad walk against the urn, & tall fir. Fox-gloves, thistles, butterfly-orchis, blow in the high wood. In the garden roses, corn-flags, Iris's, red valerian, lychnis's, &c. blow.

June 24. The disorder seems rather to abate in these parts. Some few sufferers have had relapses.

June 26. *Serapias latifolia* [helleborine] blossoms in the Hanger, & high wood.

June 29. Louring, with cool gale, sun, mist on the hills, golden stripe in the W. Double catch-flies [*Lychnis*] make a lovely show.

June 30. Neither veal nor lamb is so fat this summer as usual : the reason is, because the cows, & ewes were much reduced by the coldness & wetness of the last very ungenial spring : We have had no rain since June 13. The ground is bound as hard as iron, & chopped & cracked in a strange manner. Gardens languish for want of moisture, & the spring-corn looks sadly. The ears of wheat in general are very small. The wetter the spring is, the more our ground binds in summer.

July 1. The crop of carrots in the great meadow will be good.

July 2.   Cut my S^t foin, & sold it to John Carpenter. This is the 15th crop. It continues as good as it has been for some years.

July 4.   My flower-bank now in high beauty.

July 5.   [Worldham (near Selborne).]

July 6.   [Selborne] Several titlarks nests were mowed-out in the S^t foin.

July 8.   [Bramshot Place] Rode to Fir-grove in the parish of Bramshot, & saw the house & garden. The south wall of the kitchen-garden is covered with a range of vines of the sort called the *millers-grape*. Each vine was trained within a very narrow space, & their boughs upright: yet they had fine wood, & promised for much fruit, & were almost in full bloom. Mr Richardson's vines, my sort, did not blow then: but Fir-grove is much more sheltered than Bramshot-place. The soils are the same, a warm sandy loam [Hythe Beds]. When we came to Evely-corner [in Wolmer Forest] a hen-partridge came out of a ditch, & ran along shivering with her wings, & crying out as if wounded, & unable to get from us. While the dam acted this distress, the boy who attended me, saw her brood, that was small & unable to fly, run for shelter into an old fox-earth under the bank. So wonderful a power is instinct.

July 12.   [Selborne] A great flock of lap-wings passes over us from the uplands to the forest. They frequent the uplands at this season.

July 14.   Rain. This weather will occasion much after-grass. Field-pease, & spring corn thrive.

July 16.   A covey of young partridges frequents my out-let. Hops do not cover their poles, nor throw-out any side-shoots.

July 17.   The great Portugal-laurel in most beautiful bloom. Tremella nostoc abounds.

July 21.   Artichokes   come   in.   Wood-straw-berries over.

July 23.   Will: Tanner shot a sparrow-hawk, which had infested the village for some time. It

had lately made havock among the young swallows,
& h. martins, which are slow & inactive : the dams
insult all hawks with impunity.   [Cf. D.B., XVIII.]

July 24.   Whortle-berries ripen.   Bought an aged
brown Galloway of M͏ͬ Bradley of Alton.

July 27.   Vast rain.   Swallows-nests with their
young washed down the chimney.

July 29.   Fine   wood-straw-berries   again.   A
strong stream of water runs in Norton mead among
the hay-cocks.

Aug. 1.   Timothy the tortoise weighed seven
pounds & three ounces.

Aug. 7.   Vast rains in the night.   The quantity
of rain since Jan. 1ˢᵗ 1782, is 36 in. 1 h. ! ! !

Aug. 9.   Hops are injured by the late winds.
Swifts about Reading.

Aug. 11.   Five swifts at E. Tisted.

Aug. 12.   Swifts about Windsor.

Aug. 13.   Bro : Tho : White⁴ & daughter came.

Aug. 14.   The lavant runs by the side of Cobb's
court-yard.   Swifts about⁵ High-Wycombe.

Aug. 15.   Potatoes for the first time.   Thirteen
swifts over the Lythe : seven at Harteley.   Do they
not at this season move from village to village ?
The hay is all badly spoiled ; & men begin to fear that
the wheat will grow as it stands.   That which is
lodged is in much danger.   A fledged young swift
was found alive on the ground in the church-yard :
it was full of hippoboscae.   We gave it two or three
flies, & tossed it up on the church.   Gathered one
handful of kidney beans.   The ground is quite
glutted with rain.

Aug. 17.   Cranberries, but not ripe.

Aug. 18.   Linnets congregate & therefore have
probably done breeding.   Saw a *Papilio Machaon* in
my garden : this is only the third of this species
that ever I have seen in this district.   It was alert,
& wild.   It is the only swallow-tailed fly in this
island.

Aug. 21.   Hay, of cow-grass, is housing.   Wheat-sheaves are bound in single bands.[6]

Aug. 22.   Goody Hammond[7] goes off from the garden to glean wheat.   The quantity of rain from Jan. 1st 1782 to Aug. 23rd is 40 inc. 52 h.

Aug. 23.   The pond on Selborne-down is brim-full, & has run over.

Aug. 24.   Newton great pond runs over.

Aug. 25.   Clays pond runs over.   Not one wasp or hornet to be seen: nor if there were, is there any fruit to support them.   On such a summer, it seems quite a wonder that the whole race is not extinct.   In great plum-years, that fruit seems to be their principal food.

Aug. 28.   Wheat grows as it lies : & the lodged wheat uncut is in a bad state.

Aug. 29.   The store-sheep on the down are in good case.   Some men began to house wheat before the shower.   Were forced to light a fire in the parlor. On this day the Royal George, a 100 gun ship, was unfortunately over-set at Spithead, as she was heaving down.   Admiral Kempenfelt, & about 900 people, men, women, & children, were lost.   The lower port-holes being open the ship filled, & sunk in about 8 minutes.

Aug. 30.   The air is chilly, & has an October-feel.

Aug. 31.   Began to turn the horses into the great meadow : there is a fine head of grass.   In the month of Augst there fell 8 inch, 28 h. of rain !

Sept. 2.   Contrary to all rule the wheat this year is heavy, & the straw short : last year, tho' so much heat prevailed, the wheat was light & the straw was long.

Sept. 3.   Neps Thomas Holt White & Henry Holt White came from Fyfield.

Sept. 4.   Began to cut the first endive : finely blanched.   Curlews clamor.

Sept. 5.   The air is full of flying ants, & the hirundines live luxuriously.

Sept. 6.   Sister Barker & Nieces Mary & Eliz: Barker came from Lyndon in Rutland.   Planted

several Ladies-traces & Serapias's from the long
Lithe, in the bank near the alcove.

Sept. 7.   Many Selborne farmers finished wheat-
harvest.   The latter housings are in delicate order :
the early housed will be cold, & damp.   The swifts
left Lyndon in the county of Rutland, for the most
part, about Aug^st 23.   Some continued 'till Aug^st
29 : & one 'till Septem^r 3rd ! ! In all our observation
Mr Barker & I never saw or heard of a swift in
Septem^r, tho' we have remarked them for more than
40 years.   All nature this summer seems to keep
pace with the backwardness of the season.

Sept. 8.   People   complain   of   harvest-bugs.
Therm^r in the sun 110.   On this day Mrs Brown, of
Uppingham in the County of Rutland, eldest daughter
of my Sister Barker, was brought to bed of a daughter,
her third child.   My nephews & nieces living are
now 17 nephews :   15 nieces :   2 grand nephews :
2 grand nieces :   2 nephews by marriage :   total 38.
[*Later.*] One Niece since, 39.   8 nephews & nieces
dead.

Sept. 11.   Goody Hammond returned to weed in
the garden.   Got-in two loads of wheat at last in
good order.   The perfoliated yellow centaury in
seed on the bank above Tull's cottage.   Chlora
perfoliata.   On this day Lord Howe sailed from
Spithead with 34 ships of the line, as is supposed for
the relief of Gibraltar.

Sept. 13.   Some few orleans-plums.   Ravens about
the hill.   All the Selborne wheat in, except some
turnip-wheat at the priory [farm].

Sept. 15.   [Long passage quoted from Dr Hux-
ham ; see D.B., LX.]   My brother Thomas White
opened two of the most promising tumuli on the
down above my house.   In the first, at about three
feet under ground, was found a blackish matter that
had some appearance of an *humus animalis* ; in the
second, at the same depth, a greyish matter that
looked like the first rudiments of a bed of *pyrites*.[8]

Sept. 16.   Hops are very small; but the haulm is clean & free from insects.

Sept. 18.   [Newton] The woods & hangers still look very green : the tops of the beeches are scarcely tinged.

Sept. 19.   [Basingstoke ; Selborne] Barley mowing about the country.

Sept. 20.   One little starveling wasp.

Sept. 23.   Many swarms of bees have dyed this summer : the badness of the weather has prevented their thriving.

Sept. 25.   Sad hop, & harvest weather.

Sept. 26.   Barley grows in swarth [ridge of mown corn].   [Long quotation from a letter on rainfall, received from Thomas Barker : See T.P., V, note.]

Sept. 27.   Bro : Thomas White, his daughter, & two sons left Selborne.

Sept. 29.   It is remarkable that this wet cold weather produces no good mushrooms.   A great plenty of the pale, coarse sort [horse mushrooms] appeared early in the autumn, but I have seen none with the salmon-coloured laminae, which are the only edible sort.

Sept. 30.   Many wasps at Lyndon in Rutland, tho' none in the great heats of autumn 1781.   So there is some mystery in their breeding that we do not understand.   *Note.*   At the autumnal aequinox,[9] the evenings are remarkably dark, because the sun at that time sets more in a right angle to the horizon, than at any other season.   But of late these uncomfortable glooms have been much softened by frequent N. Auroras.   This circumstance of autumnal darkness did not escape *the poet of nature :* who says,

" Now black, & deep the night begins to fall,
    A shade immense.   Sunk in the quenching gloom
    Magnificent & vast are heaven & earth ;
    Order confounded lies ;   all beauty void ;
    Distinction lost ;   & gay variety
    One universal blot :   such the fair power
    Of light, to kindle, & create the whole."
                                        Thomson's Autumn.

Oct. 3.   Rain 78 in my garden, on the tower 59.

Oct. 4.   Numbers of pheasants at Inne down-coppice.

Oct. 6.   Wood-cock returns, & is seen in the Hanger.   Young martins in the nest at little World-ham ; probably Ward-le-ham. [Cf. T.P., I.]

Oct. 8.   Sad weather for the barley.   Barley housed at Bramshot; & other places, being green & damp, has heated violently, & endangered the firing of barns.[10]   All the hops of this parish this year are carried to Wey-hill [near Andover] in two waggons : good crops require four or five.   Gathered two or three bunches of grapes : they have some colour, but are crude & sour.—By the evening being so light, there must be great N. Auroras.   Mr Yalden finished mowing his barley.

Oct. 10.   We make tarts, & puddings with the crude unripened grapes.   Gathered-in the Virgol-euse, & Chaumontelle pears, a good crop : somewhat has gnawn many of the former like wasps or hornets.

Oct. 11.   Lord Howe arrived in the straits of Gibraltar.

Oct. 12.   The paths are dry & crisp.   Men housed barley 'till between ten & eleven at night.   Pleasant starlight.

Oct. 13.   The great farmer at Newton has 105 acres of barley abroad.   Mr Pink still has 40 acres of barley abroad.

Oct. 14.   Sister Barker & her two daughters left Selborne.

Oct. 16.   Gathered-in my apples.   Knobbed russetings, & nonpariels, a few.   Near four bushels of dearlings on the meadow-tree : fruit small.

Oct. 17.   No baking pears.   Gathered-in medlars. Dug up carrots, a good crop, but small in size.   The tortoise not only gets into the sun under the fruit-wall ; but he tilts one edge of his shell against the wall, so as to incline his back to it's rays : by which contrivance he obtains more heat than if he lay in his natural position. And yet this poor reptile has never read, that planes inclining to the horizon receive more heat from the sun than any other elevation! At four P.M. he retires to bed under the broad foliage of a holyhock. He has ceased to eat for some time.

Oct. 19. Lord Howe compleated the relief of Gibraltar.

Oct. 20. No corn abroad but a few vetches. Lord Howe had a skirmish with the combined fleet, in which he had 68 killed, & 208 wounded.

Oct. 23-31. [Fyfield] My brother's children & plantations strangely grown in two years. On the downs the green wheat looked very chearful & pleasant. Some wheat & turnips covered with charlock in full bloom. This proves, it seems, but a poor trufle-year at Fyfield; so they fail some times in very wet years, as well as in very dry ones.

Oct. 24. Grapes at this place eatable: the sort, black cluster, came from Selborne.

Oct. 25. The gold & silver-fish lie sleeping all day in their silver-bowl towards the surface of the water: people that have attended to them suppose this circumstance prognostic of rain. Jupiter & Saturn approach to each other very fast.

Oct. 27. [Long entry on gold-fish reproduced in D.B., LIV.]

Nov. 1. [Selborne] Some flocks of starlings on the wide downs between Andover, & Winton. Several martins were playing about over the chalk-bank at the E. end of Whorwel village. Can any one suppose but that they came out of that bank that morning to enjoy the warm sunshine, & would retire into it again before night?

Nov. 4. I watched the S.E. end of the hanger, hoping to have seen some house-martins, as they sometimes appear about this day but was disappointed.

Nov. 8. Men are interrupted in their wheat-sowing in the mornings by hard frost.

Nov. 11. Planted 50 tulips, which I bought of Dan : Wheeler, in the border opposite to the great parlor-windows. They are, I think, good flowers.

Nov. 14. Lord Howe arrived at Portsmouth with 16 men of war. He was absent just nine weeks. If a frost happens, even when the ground is considerably dry, as soon as a thaw takes place, the paths & fields are all in a batter. Country people say that the frost *draws* moisture. But the true philosophy is, that the steam & vapours continually ascending from the earth, are bound-in by the frost, & not suffered to escape, 'till released by the thaw. No wonder then that the surface is all in a float ; since the quantity of moisture by evaporation that arises daily from every acre of ground is astonishing. Dr Watson, by experiment, found it to be from 1600 to 1900 gallons in 12 hours, according to the degree of heat in the earth, & the quantity of rain newly fallen.— See Watson's Chem : essays : Vol : 3 : p : 55 : 56 .

Nov. 15. The torrent down the stoney-lane as you go towards Rood has run all the spring, summer, & autumn, joining Well-head stream at the bridge.

Nov. 18. No hogs have annoyed us this year in my outlet. They usually force-in after the acorns, nuts, beech & maple mast ; & occasion much trouble.

Nov. 19. One way or other we have used most of the grapes.

Nov. 21. The conjunction of Jupiter & Saturn is over ; & the former, which lately was just below the latter, is now to the E. of him, & in a line parallel with the horizon. These planets are so near the sun at setting as to be visible but a small time : & are so low as not to be seen at all at Selborne, because of the hill.

Nov. 24. [Long passage quoted from Gassendus, with White's remarks thereon; see D.B., LVI, where the following note is in part, omitted.] [Pleasure of music.] In particular this autumn my Nieces M. & E. Barker played many elegant lessons in a very masterly manner from Niccolai & others, which still teize my imagination, & recurr irresistably to my memory at seasons ; & even when I am desirous of thinking of other matters.

Nov. 26. The woods, & hedges are beautifully fringed with snow. Ordered Thomas carefully to beat-off the snow that lodges on the South side of the laurels & laurustines.

Nov. 27. Fierce frost. Rime hangs all day on the hanger. The hares, press'd by hunger, haunt the gardens & devour the pinks, cabbages, parsley, &c. Cats catch the red-breasts. Timothy the tortoise sleeps in the fruit-border under the wall, covered with a hen-coop, in which is a good armfull of straw. Here he will lie warm, secure, & dry. His back is partly covered with mould.

Dec. 1. [Long note on the peregrine falcon, which forms half of letter, D.B., LVII.]

Dec. 4. Farmer Lassam's lambs begins to fall.

Dec. 9. Rime on the hill.

Dec. 16. A hare frequents the garden, & eats the celeri-tops, the spinage, young cabbages, pinks, scabious's, &c.

Dec. 25. The boys at Faringdon play in the churchyard in their shirts. They did so this day twelvemonth.

Dec. 26. Crocus's shoot. Feb-like weather. The Plestor, & street dry & clean.

Dec. 27. Large flock of wood-pigeons in the N. field.

Dec. 28. Boys play at Marbles on the Plestor.

Dec. 31. Barom^r in 1782 at S. Lambeth.

lowest April 1	28 : 5-10
highest Nov. 13	30 : 13-20
	or 6-$\frac{1}{2}$
Therm : lowest Feb. 12	23
highest June 18	81

CHAPTER XVI

"My Lord St. Albans said that nature did never put her precious jewels in a garret four stories high." BACON: Apothegm No. 17.

1783

Jan. 2.   Vast hoar frost.   Frost comes within doors.

Jan. 8.   Blowing with driving rain.   Walls sweat again.

Jan. 15.   Hailstorm in the night.

Jan. 25.   Snow gone.   The wryneck pipes.[1]

Jan. 30.   Lambs fall apace.   Ground full of water.

Feb. 7.   Much rain in the night.   Flood at Gracious Street.

Feb. 9.   Vast rain in the night, with some thunder, & hail.   Peter Wells's well runs over.

Feb. 10.   Sheep rot very much.   Ewes & lambs are much distressed by the continual wet.

Feb. 14.   Showers.   A perfect & lovely rain-bow.

Feb. 15.   Drank tea at Newton by day-light. Sheep on the down look deplorably!

Feb. 17.   Partridges are paired.   Foot-paths in many places very clean.

Feb. 18.   Cleaned-up the borders in the garden. Sowed radishes, & a few carrots under the fruit-wall.

Feb. 19.   Men busy in plowing for pease.   Timothy the tortoise awakes.

Feb. 20.   Men sow pease in their fields, & horse-beans.

Feb. 21.   Ashed the two meadows.

Feb. 26.   Venus begins to appear above the hanger.

Feb. 28.   Rain in Feb: 5 in. 54 hun.   Peter Wells's well ceases to run over.   It is only 36 feet in depth, & continues almost full all the winter in wet seasons. A lavant from the hanger fills it. The wells in this part of the street are 63 feet deep. [Cf. T.P., I.]

Mar. 1. [*Side note.*] Mercury was visible all the first half of March : but partly thro' bad weather, & partly for want of an horizon I never was able to get a sight of him.

Mar. 4. Snow on the ground 4 or 5 inc : deep. Snow melts on sunny roofs.

Mar. 5. Snow 7 inches deep : no drifting. Swift thaw.

Mar. 6. Flood at Gracious street. All the fields full of water. Snow much gone. The barometer strangely low ! [28°] A slight shock of an earthquake at this time at Paris.

Mar. 8. The crocus's make a gaudy appearance, & bees gather on yᵐ. The air is soft. Violets blow. Snow lies under hedges. Men plow.

Mar. 10. [Alton] Vast lavants at Chawton.

Mar. 11-*Apr.* 10. [South Lambeth] Note : Thomas kept the rain at Selborne.

Mar. 14. Daffodil blows.

Mar. 17. Full moon. Ice. Nect. & peach blow. Moon totally eclipsed.

Mar. 20. Water sinks in Thomas White's field.

Mar. 31. Cellars almost dry by pumping.

Apr. 1. Field almost dry.

Apr. 2. *Timothy*, my tortoise, came-out for the first time at Selborne. Aurora bor :

Apr. 8. *Swallow* appeared at Liss [Hants].

Apr. 9. *Red-start* at Selborne.

Apr. 10. Therm. 72 ! ! ! Prodigious heat : clouds of dust. Thermʳ at Selborne only 62 ! Nightingale at Bradshot.

Apr. 11. [Newton] Several bank-martins near the great lake on this side Cobham.² Two swallows at Ripley [Surrey].

Apr. 12. [Selborne] Wheat mends. Barley-grounds work well. My grass-walks begin to be mown. Grass-lamb six pence p^r pound: veal 5^d: fresh butter 9½^d.

Apr. 13. Three swallows at Goleigh. [Farm in Priors Dean.]

Apr. 16. Fine barley-season. Vast bloom of plums, & cherries.

Apr. 17. Tortoise weighs 6 ae 11¼ oun. He begins to eat.

Apr. 18. A nightingale sings in my fields. Young rooks.

Apr. 19. One *house-martin* about the stables. Sultry out of the wind. The garden is watered every day.

Apr. 20. Some whistling plovers in the meadows towards the forest.

Apr. 22. [Reading] Young goslings abound. No hirundines.

Apr. 23-25. [Oxford] No hirundines.

Apr. 26. Several swallows on the road. Fine clover in Oxfordshire, & Berks. Barley-fields work finely. Vivid Aurora.

Apr. 27. [Selborne] Many swallows. Strong Aurora ! ! !

Apr. 30. Gardens want rain. We water every day. Cucumbers come.

May 1. Peat is brought in from the forest. Cut some asparagus.

May 3. Honey-suckles against walls begin to blow. Early tulips blown-out: late begin to turn colour.

May 6. Some ponds, & ditches dry, & cleansed out.

May 8. Apple-trees in high bloom ; in danger from the frost.

May 14. Sowed a crop of kidney-beans. large white Dutch. Planted some basons in the field with China-asters, & China pinks. Pricked out many

China asters, on a mild hot-bed. Honey-suckles stocks, & wall-flowers smell sweetly. Tulips blow out ; but their cups will be much larger. The large Apricot-tree much infested with maggots, which twist & roll-up the leaves ; these we open, & destroy the maggots, which would devour most of the foliage. These maggots are the produce of small spotted phalaenae.[3]

May 17. Wells sink : Benham's is dry. Sprinkled & washed the foliage of the fruit-trees, that were honey-dewed, & began to be affected with aphides. Stocks blow finely. Tulips, thro' heat, will continue but a small time in bloom.

May 23. Stocks blow, & are very double & handsome !

May 26. The frost cut down all the early kidney-beans, injured the annuals ; & made the apple-trees cut much of their fruit.

May 27. Not one spring chafer this year.

May 29. Young redstarts.

May 31. Goose-berries, & currans are coddled on the trees by the frost. Planted the basons in the fields with annuals. Began to tack the vine-shoots : there will be a tollerable bloom. The potatoes in the meadow seem to be all killed. Aphides prevail on many fruit-trees. Medlar-tree blows. The sun at setting shines up my great walk.

June 1. The late frost cut-down the fern, & scorched many trees. Wheat spindles for ear.

June 3. Turned mould for future hot-beds. Showers about. Great rain at Farnham, Froil [Froyle] &c. Rain at London.

June 4. Cut the tall hedge down Baker's hill.

June 5. Hops are very lousy, & want a good shower. Washed the cherry-trees against the wall with a white-wash brush : they are full of aphides, but have a vast crop of fruit.

June 7. Tulips are faded. Honey-suckles still in beauty. My columbines are very beautiful : tyed

some of the stems with pieces of worsted, to mark
them for seed.   Planted-out pots of green cucumbers.
Dᵣ Derham says, that *all cold* summers are wet summers : & the reason he gives is that rain is the effect
not the cause of cold.   But with all due deference
to that great Philosopher, I think, he should rather
have said, that *most* cold summers are dry ; For it
is certain that sometimes cold summers are dry ; as
for example, this very summer *hitherto :* & in the
summer 1765 the weather was very dry, & very cool.
See Physico-theol : p : 22. Vast honey-dews this
week.   The reason of these seems to be, that in hot
days the effluvia of flowers are drawn-up by a brisk
evaporation ; and then in the night fall down with
the dews, with which they are entangled.⁴ This very
clammy substance is very grateful to bees, who gather
it with great assiduity, but it is injurious to the trees
on which it happens to fall, by stopping the pores
of the leaves.   The greatest quantity falls in still,
close weather ; because winds disperse it, & copious
dews dilute it, & prevent its ill effects.   It falls
mostly in hazey warm weather.

June 8.   The potatoes, killed-down by the frost,
shoot again.

June 10.   Wych-elm sheds it's seeds, which are
innumerable.

June 11.   Soft rain all days.   Snails come forth in
troops.   Mᵣ Beeke⁵ came from Oxford.

June 12.   *Ophrys nidus avis,* many in bloom on
the hanger, along the side of yᵉ Bostal.

June 13.   *Serapias latifolia* begins to bloom in the
hanger.   The *Serapias's* transplanted last summer
from the hanger to my garden, grow & thrive.   Mᵣ
Beeke returned.

June 17.   The potatoe-shoots, that were cut-down
by the frost, all spring again ; the kidney-beans do
not.   Lighted a fire in the parlor.

June 19.   Vast crops of cherlock among the spring-
corn.

June 21. The late ten dripping days have done infinite service to the grass, & spring-corn.

June 22. Corn-flags, fraxinella, martagons, pinks, & dark-leaved orange-lilies begin to blow. Bees swarm. Cherries look finely, but are not yet highly ripened.

June 23. Vaſt honey-dew; hot & hazey; miſty. The blades of wheat in several fields are turned yellow, & look as if scorched with the froſt. Wheat comes into ear. Red even: thro' the haze. Sheep are shorn.[6]

June 24. Vaſt dew, sun, sultry, miſty & hot. This is the weather that men think is injurious to hops. The sun " shorn of his beams " appears thro' the haze like the full moon.

June 25. Turned the swarths, but did not ted [scatter for drying] the hay. Much honey-dew on the honey-suckles, laurels, great oak.

June 26. Tedded the hay, & put it in small cock. Sun looks all day like the moon, & sheds a ruſty red light. M^r & Mrs Brown, & niece Anne Barker came from the county of Rutland.

June 27. Nose-flies, & ſtouts [either gad-flies or gnats] make the horses very troublesome.

June 28. Ricked the hay of the great meadow in lovely order: six jobbs. The little meadow is hardly made. The country people look with a kind of superſtitious awe at the red louring aspeċt of the sun thro' the fog . . . " Cum caput obscurâ nitidum ferrugine texit."[7]

June 29. My garden is in high beauty, glowing with a variety of solſtitial flowers. A person lately found a young cuckow in a small neſt built in a beechen shrub at the upper end of the boſtal. By watching in a morning, he soon saw the young bird fed by a pair of hedge-sparrows. The cuckow is but half-fledge; yet the neſt will hardly contain him: for his wings hang out, & his tail & body are much

compressed, & streightened. When looked at he
opens a very red, wide mouth, & heaves himself
up ; using contorsions with his neck by way of menace,
& picking at a person's finger, if he advances it
towards him.

July 1.   Thatched the hay-rick.   M^r & Mrs Brown
& Niece Anne Barker left me.

July 1 & 2.   Tremendous   thunder-storms   in
Oxford-shire & Cambridge-shire ![8]

July 2-3.   [Bramshot place] The foliage on most
trees this year is bad.   Vast damage this day by lightening in many counties ! !   Great thunder-shower at
Lymington, & in the New forest, & in Wilts, &
Dorset, & at Birmingham, & Edinburg.

July 3.   Mr Richardson's garden abounds with
fruit, which ripens a fortnight before mine.   His
kitchen-crops are good, tho' the soil is so light &
sandy.   Sandy soil much better for garden-crops
than chalky.

July 5.   [Selborne] Tim : Turner bought, &
carried-off my Saint-foin, the 16th crop.   It was over-
ripe, & not so large a burden as the last.   The S^t
foin was all run to seed.   The garden wants rain.

July 6.   Some young martins came out of the nest
over the garden-door.   This nest was built in 1777,
& has been used ever since.   As the summer has
been dry, & we have drawn much water for the garden,
I caused my well to be plumbed, & found we have
yet 13 feet of water.   When we were measuring I
was desirous of trying the depth of Benham's well,
which becomes dry every summer ; & was surprized
to find it 25 feet shallower than my own : the former
being only 38 feet deep, & the latter 63.

July 7.   The young cuckow sits upon the nest,
which will no longer contain him.

July 9.   Bees have thriven well this summer,
being assisted by the honey-dews, which have abounded
this year.[9]

July 10.   About 8 o'clock on the evening of the
10th a great tempest arose in the S.W. which steered-
off to the N.W. : another great storm went to the
N.E. with continued thunder, & lightening.   About
10 another still heavier tempest arose to the S.E. &
divided, some part going for Bramshot & Headley,
& Farnham ; & the rest for Alresford, Basingstoke,
&c.   The lightening towards Farnham was prodigious.   It sunk all away before midnight.   Vast
showers around us but none here.

July 11.   The heat overcomes the grass-mowers
& makes them sick.   There was not rain enough
in the village to lay the dust.   The water in my
well rises ! tho' we draw so much daily ! watered
much.   No dew, sun, & hase, rusty sunshine !¹⁰
The tempest on friday night did much damage at
West-meon, & burnt down three houses & a barn.
The tempests round on thursday & friday nights
were very aweful ! There was vast hail on friday
night in several places.   Some of the standard-
honey-suckles, which a month ago were so sweet &
lovely, are now loathsome objects, being covered
with aphides, & viscous honey-dews.   Gardens
sadly burnt.

July 13.   Five great white sea-gulls flew over the
village toward the forest.

July 14.   When the owl comes-out of an evening,
the swifts pursue her, but not with any vehemence.

July 15.   No rain since June 20th at this place ;
tho' vast showers have fallen round us, & near us.

July 17.   The jasmine, now covered with bloom,
is very beautiful.   The jasmine as so sweet that I
am obliged to quit my chamber.

July 19.   Men talk that some fields of wheat are
blighted : in general the crop looks well.   Barley
looks finely, & oats & pease are very well : Hops
grow worse, & worse.

July 21.   Lapwings flock.   Lark-spur figures.

July 23. Turnips (field) thrive, & are hoeing.

July 25. Trenched two more rows of celeri in the upper end of the plot by W. Dewey's : the ground mellow. We plant out the cabbage-kind some few at a time. The boys bring me a large wasp's nest full of maggots.

July 26. Some wheat reaped at Faringdon. Boys bring two more wasps nests.

July 27. My china-holly-hocks, after standing a year or two, lose all their fine variegated appearance, & turn to good common sorts, being double, & deeply coloured.

July 28. Wasps swarm so that we were obliged to gather-in all the cherries under the net.

July 29. Preserved a good quantity of currans, & some rasps, cherries, & apricots.

July 30. Few hazel-nuts. Men house field-pease. Ponds are dry. Grass-walks burn. Ripening weather.

July 31. The after-grass in the great meadow burns. The sheep-down burns & is rusty. Much water in the pond on the hill! This morning Will Tanner shot, off the tall meris-trees in the great mead, 17 young black-birds. The cherries of these trees amuse the birds & save the garden-fruit.

Aug. 1. Much smut in some fields of wheat. Goody Hampton left the garden to go gleaning. Barley cut about the forest-side . We shot in all about 30 black-birds. Vast shooting star from E. to N. My nephew Sam : Barker came from Rutland thro' London by the coaches.

Aug. 2. Burning sun. Workmen complain of the heat.

Aug. 3. My white pippins come-in for kitchen uses. The aphides, of various species, that make many trees & plants appear loathsome, have served their generation, & are gone, no more to be seen this year : perhaps are all dead. Thistle-down flies. [Cf. 23 Aug. 1774.]

" Wide o'er the thistly lawn, as swells the breeze,
A whitening shower of vegetable down
Amusive floats. The kind impartial care
Of Nature nought disdains ; thoughtful to feed
Her lowest sons, & clothe the coming year,
From field to field the feathered seed she wings."

Thomson's Summer.

Aug. 4. Wheat seems very good. Hops are quite gone. They have some weak side-shoots without any rudiments of bloom.

Aug. 6. Our fields & gardens are wonderfully dryed-up : yet after all this long drought Well-head sends forth a strong stream. The stream at the lower end of the village has long been dry. Mr Barker came from Rutland thro' Oxford on horse-back.

Aug. 7. Sarah Dewey came to assist in the family.

Aug. 9. Flies come in a door, & swarm in the windows ; especially that species called Conops calcitrans. Nep : John White came by the coach from London.

Aug. 13. Farmer Spencer of Grange finished wheat-harvest. Mr Pink of Faringdon finished Do Mr Yalden finished wheat-harvest. Farmer Bridger of Black-more finished harvest of all sorts.

Aug. 15. Took this morning by bird-lime on the tips of hazel-twigs several hundred wasps that were devouring the goose-berries. A little attention this way makes vast riddance, & havock among these plundering Invaders.

Aug. 16. Farmer Knight of Norton finishes wheat-harvest. Farmer Lassam of Priory Do Farmer Hewet of Temple finished Do.

Aug. 18. The *Colchicum*, or autumnal crocus blows. On the evening of this day, at about a quarter after nine o' the clock, a luminous meteor of extraordinary bulk, & shape was seen traversing the sky from N.W. to S.E. It was observed at Edinburg, & several other Ern parts of this Island. No accounts of it, that I have seen, have been published from any

of the western counties.   It was also taken notice of
at Ostend.   This meteor, I find since, was seen at
Coventry, & Chester.   4 swifts at Guildford ; 1
swift at Meroe [Merrow] ; 1 swift at Dorking.

Aug. 24.   Paid for four wasps-nests.   On this
day the Duke of Kingston India man, outward
bound. Captain Nutt, was burnt at sea off the island
of Ceylon.   Mr Charles Etty,[11] one of the mates, was
wonderfully saved, tho' he could not swim an inch,
by clinging to a yard-arm that had been flung over
board ; by which he was kept above water about
an hour & ¼, 'till he was taken up by a boat, & carryed,
naked as he was, aboard the Vansittart India-man
Captain Agnew, who treated him with great human-
ity, & landed him in a few days at Madras.   This
ship, cargoe, & more than 70 lives were lost by the
carelessness of a mate in drawing rum, who permitted the candle to catch the spirits ; so that the
whole vessel was in flames at once, without any
chance of extinguishing them.   She burnt about
four hours, & then blew up : so that nothing was
saved except what cloaths some had on their backs.
She had soldiers aboard, & some passengers, & a few
women, & children.   Potatoes very fine, tho' the
ground has scarce ever been moistened since they were
planted.   They were also very good last year, tho'
the summer was mostly wet & cold.   Fern-owl
glances, & darts about in my garden in pursuit of
phalænæ, with inconceivable swiftness.

Aug. 25.   Muscae domesticae swarm in the kitchen.
When the sun breaks-out, the roofs, & grass-walks
reek.   Men cut their field-beans.

Aug. 26.   Some fly-catchers, that haunt about the
church, take the flies off the sides of the towers with
much adroitness.   Swallows do the same in the
decline of the summer.

Aug. 27.   Sowed a plot of *spinage* with 3 ounces
of seed : the ground still too dry, & wanted much
pressing down with the garden-roller.

Aug. 30. Planted-out in a bed a great number of Seedling-polyanths : seed fᵐ Bramshot-place.

Aug. 31. Tremendous thunder-storm in London. The stream [bourne] which rises in James Knight's upper pond has failed all this summer, as it does all very dry summers ; so that the channel is dry down to the middle of the short Lithe ; from whence there is always water running 'till it joins the Well-head stream at little Dorton. This spring, which is at the bottom of the Church-litten-closes,¹² seems to rise out of the hill on which the Church is built. Timothy begins to frequent the border under the fruit-wall for the sake of warmth.

Sept. 1. Red sunshine. Sowed a bed of Coss-lettuce.

Sept. 2. Nectarine, one of the new trees, the fruit delicate. Two of the new peach, & Nect : trees this year are distempered, & one barren : one nect : has a crop of well-flavoured fruit.

Sept. 4. *Tremella nostoc* appears on the walks. Tho' the weather may have been ever so dry & burning, yet after two or three wet days this strange jelly-like substance abounds.

Sept. 6. Nep : John White left us, & went to London. *Sun-dew* blows. Planted 50 curled endive-plants, which we had from Daniel Wheeler. The heat of the summer prevented our last sowings from growing.

Sept. 8. Ponds are filled. Hirundines skulk about to avoid the cold wind. Mʳ Sam : Barker left us, & went to Fyfield.

Sept. 9. Mʳ Etty's well is still foul. Began to light fires in the parlor. Brother Thomas, & Molly White came.

Sept. 10. Gathered-in the white pippins, a great crop. Cleansed-out the zigzag. Tho : Holt White, & Henry Holt White came. Bessy White, Sam White, & Ben Woods [nephew] came from Fyfield. The *Virginian creeper* is grown up to the eaves ; but will probably shoot no farther, as the leaves at bottom begin to turn red. Total eclipse of the moon.

Sept. 11. Sam White, & Ben Woods returned to Fyfield. Fly-catcher. Harvest moon. Selborne hopping lasts only two days in Farmer Spencer's, & Master Hale's gardens; many gardens afford no pickings at all. Mr Hale will have only about 200 weight. The great garden at Hartley, late Sr Sim: Stuart's, consisting of 20 acres, produced only about two tons.

Sept. 12. Tyed-up endives: they are backward this year, & not well grown. One sowing never came up. The barley about Salisbury lies in a sad wet condition.

Sept. 13. Began to mend the dirty parts of the bostal with chalk.

Sept. 14. Mr Yalden's tank is full. Brought down by Brother Thomas White from South Lambeth & planted in my borders. Dog toothed violets. Persian Iris. Quercus cerris [Turkey oak]. Double ulmaria [*Spiraea*]. Double filipendula [Dropwort]. Double blue campanula. Large pansies. Hemerocallis [day lily]. White fox-glove. Iron fox-glove. Double wall-flower. Double scarlet lychnis.

Sept. 17. Planted from Mr Etty's garden a root of the *Arum dracunculus*, or *Dragons;* a species rarely to be seen; but has been in the vicarage-garden ever since the time of my Grandfather, who dyed in spring 172$\frac{7}{8}$.

Sept. 19. Ivy begins to blow on Nore-hill & is frequented by wasps. Pd for a wasps nest, full of young.

Sept. 20. Fungi on the hanger are *Clavaria,* several sorts: *Boleti,* several. Mr & Mrs Richardson came.

Sept. 21. Green wheat in the N. field. Stormy wind all night, which has blown down most of my apples & pears.

Sept. 22. Thunder: rather the guns at Portsmouth. Splendid rain-bow. After three weeks wet, this vivid rain-bow preceded (as I have often known before) a lovely fit of weather. M^r & Mrs Richardson left us.

Sept. 23. Black snails lie out, & copulate. Vast swagging clouds.

Sept. 25. My wall-nut tree near the stable, which is usually barren, produces this year 5, or 600 nuts: the sort is very fine. The vast tree at the bottom of the garden bears every year, but the nuts are bad. Charles White, & Harry Woods came from Fyfield.

Sept. 29. Gathered-in the apples, knobbed russets, & non-pariels. Royal russets none. All the baking pears were blown down. No dearlings.

Sept. 30. Lovely weather, red even. True Michaelmas summer.

Oct. 2. Erected an alcove[13] on the middle of the bostal. Charles Henry White, & his sister Bessey returned to Fyfield.

Oct. 3. The hanger is beautifully tinged. Leaves fall apace. Dug up carrots. Many flesh-flies: here & there a wasp. The cat frolicks, & plays with the falling leaves. Acorns innumerable.

Oct. 4. This day has been at Selborne the honey market: for a person from Chert [near Farnham] came over with a cart, to whom all the villagers round about brought their hives, & sold their contents. This year has proved a good one to the upland bee-gardens, but not to those near the forest. Combs were sold last year at about 3¾d per pound; this year from 3½-4^d. Women pick up acorns, & sell them for 1^s p^r bushel. A splendid meteor seen at half hour past six in the evening; but not so large as that on the 18th of August.

Oct. 5. In the High-wood, under the thick trees, & among the dead leaves, where there was no grass, we found a large circle of *Fungi* of the *Agaric* kind, which included many beeches within its ring. Such

circles are often seen on turf, but not usually in
covert. We found a species of Agaric in the high
wood of a very grotesque shape, with the laminae
turned outward,[14] & the cap within formed into a
funnel containing a good quantity of water.

Oct. 8. Neps. Th: H. & Hen: H: White went
to Fyfield.

Oct. 10. Full moon. Sweet moonshine.

Oct. 11. Mr John Mulso[15] came. Hunter's moon
rises soon after sunset. Muscae domesticae abound
in the kitchen & enjoy the warmth of the fire. Where
they lay their eggs does not appear. The business
of propagation continues among them.

Oct. 12. The crop of acorns is so prodigious that
the trees look quite white with them ; & the poor
make, as it were, a second harvest of them, by gathering them at one shilling pr. bushel. At the same
time not one beech-mast is to be seen. This plenty
of acorns has raised store-pigs to an extravagant
price.

Oct. 13. Mr John Mulso left me.

Oct. 14. The potatoes in the meadow small, &
the ground very stiff. Low creeping fogs.

Oct. 15. Nep: Harry Woods left me, & went to
Funtington.

Oct. 16. Rover [White's mongrel dog] finds
pheasants every day ; but no partridges. The air
is full of gossamer. There is fine grass in the meadows.

. . . " see, the *fading, many-coloured* woods,
Shade deepening over shade, the country round
Imbrown." Thomson.

Oct. 17. Mowed & burnt the dead grass in my
fields. Rooks on the hill attended by a numerous
flock of starlings. The tortoise gets under the laurel-
hedge, but does not bury himself. Neps. T: H:
& H: Holt White returned from Fyfield.

. . . " a crouded umbrage, *dusk* & *dun,*
Of ev'ry hue, from *wan,* declining green
To sooty dark." Thomson.

Oct. 19. [Long descriptions of remains found at the Priory; see *Antiq.* XXVI.] Brother Thomas White, & daughter, & sons left us.

Oct. 21. *Nasturtiums* in high bloom, & untouched by the frost!

Oct. 23. The poor make quite a second harvest by gathering of acorns. Timothy Turner has purchased upwards of 40 bushels. Two *truflers* came with their dogs to hunt our hangers, & beechen woods in search of *truffles*; several of which they found in the deep narrow part of the hill between coney-croft-hanger, & the high wood; & again on each side of the hollow road up the high-wood, known by the name of the coach-road.

Oct. 25. The firing of the great guns at Portsmouth on this day, the King's accession, shook the walls & windows of my house.

Oct. 26. If a masterly lands-cape painter was to take our hanging woods in their autumnal colours, persons unacquainted with the country, would object to the strength & deepness of the tints, & would pronounce, at an exhibition, that they were heightened & shaded beyond nature. Wonderful & lovely to the Imagination are the colourings of our wood-land scapes at this season of the year!

" The pale descending year, yet pleasing still,
    A *gentler mood* inspires ; for now the leaf
    Incessant *rustles* from the *mournful* grove,
    Oft startling such as, studious, walk below,
    And slowly *circles* thro' the *waving* air.
    But should a quicker breeze amid the boughs
    Sob, o'er the sky the *leafy deluge streams* ;
    Till *choak'd*, & *matted* with the *dreary* shower,
    The *forest-walks*, at every rising gale,
    Roll wide the *wither'd waste*, & *whistle bleak.*"
                                   Thomson's Autumn.

Oct. 27. A couple of wood-cocks were shot in the high wood.

Oct. 28.   Planted many slips of pinks.

Oct. 29.   Tortoise begins to bury himself in the laurel-hedge.

Nov. 3-13.   [Fyfield] The runs dusty, & the chaises run on the summer tracks, on the downs.   Lovely clouds, & sky!!   Turnips on the downs are bad. Wheat looks well.   Men chalk the downs in some parts.

Nov. 4.   The stream at Fyfield is dry.   My brother Henry's crops of trufles have failed for two or three years past.   He supposes they may have been devoured by large broods of turkies that have ranged much about his home-fields, & little groves.

Nov. 5.   Wild-geese appear.   On the downs, & Salisbury plain they feed much on the green wheat in the winter, & towards the spring damage it much, so that the farmers set up figures to scare them away.

Nov. 7.   A chaced hind ran thro' the parish, & was taken at Penton [near Weyhill].   She ran but two hours the ground being too hard for her feet.   She was carryed home in a cart to Grateley.

Nov. 8.   My niece of Alton (Clement) was brought to bed of a girl.   This child makes my 40th nephew & niece, all living; Mr Clement, & Mr Brown inclusive.

Nov. 11.   This country swarms with pigeons from dove-houses.   Millers complain for want of water.

Nov. 14.   [Winchester] Mr Mulso's grapes at his prebendal-house are in paper bags: but the daws descend from the Cathedral, break open the bags, & eat the fruit.   Looked sharply for house-martins along the chalk-cliff at Whorwel, but none appeared. On Novr 3rd 1782: I saw several at that place.

Nov. 15.   [Selborne] Wind all night.   At Selborne, a storm at 11 A: M.   Sea-gulls abound on the Alresford-stream: they frequent those waters for many months in the year.

Nov. 16.   Winter is established.

" Fled is the *blasted* verdure of the fields ;
    And, *shrunk* into their beds, the *flowery race*
    Their *sunny* robes resign.   E'en what remain'd
    Of *stronger* fruits falls from the *naked* tree :
    And woods, fields, gardens, orchards, all around
    The *desolated* prospect *thrills* the soul."

                          Thomson's Autumn.

Nov. 23.   The stream in Gracious-street runs,
after having been dry for many months.

Nov. 26.   The farmers have long since sown all
their wheat, & ploughed-up most of their wheat-
stubbles.

Dec. 1.   Some ivy-berries near full grown : others,
& often on the same twig, just out of bloom.   Farmer
Lassam has more than 20 young lambs : some fallen
some days, near a fortnight.

Dec. 4.   Mowed some of the grass-walks ! Farmer
Lassam cuts some of his lambs : they are near a
month old.

Dec. 5.   Fetched some mulleins, foxgloves, &
dwarf-laurels from the high-wood & hanger ; &
planted them in the garden.

Dec. 11.   Planted 4 small spruce firs, & 2 Lombardy poplars in Baker's hill ; & one spruce fir in farmer
Parsons's garden.

Dec. 18.   Hares make sad havock in the garden :
they have eat-up all the pinks ; & now devour
the winter cabbage-plants, the spinage, the parsley,
the celeri, &c.   As yet they do not touch the
lettuces.

Dec. 24.   A fine yellow wagtail appears every day.

Dec. 27.   Mr Churton[16] came from Oxford.

Dec. 28.   Ground so icy that people get frequent
falls.

Dec. 29.   Carryed some savoy-heads, endive, &
celeri into the cellar : the potatoes have been there

some days.   Red breasts die.   Wag-tail.   Some sleet
in the night.   Ground covered with ice & snow.

  Dec. 31.   Ice under people's beds.   Water bottles
burst in chambers.   Meat frozen.   The fierce weather
drove the snipes out of the moors of the forest up
the streams towards the spring-heads.   Many were
shot round the village.

CHAPTER XVII

"The noble letters of the dead." (*In Memoriam*, XCV.)
"Specimen Days." (WALT WHITMAN.)

1784

Jan. 3. Snow gone : flood at Gracious ſtreet.

Jan. 7. Hoar froſt lies all day. Froſt comes in a door.

Jan. 8. Some wild-ducks up the ſtream near the village. Much wild fowl on the lakes in the foreſt.

Jan. 9. A grey crow shot near the village. This is only the third that I ever saw in this parish. Some wild-geese in the village down the ſtream.

Jan. 10. Small snow on the ground. Mr Churton left us.

Jan. 18. Clouds put up their heads.

Jan. 20. Vaſt snows in Cornwal for two days paſt.

Jan. 21. [Abroad] Ice in chambers. Hares frequent the garden & do much damage.

Jan. 22. Snipes come up the ſtream.

Jan. 24. The Thermomr at Totnes, in the county of Devon abroad this evening was, I hear, at 6.

Jan. 25. The turnips, that are not ſtacked, are all frozen & spoiled.

Jan. 26. Cut my laſt year's hay-rick.

Jan. 29. [Abroad] The dung & litter freeze under the horses in the ſtable. The hares nibble off the buds of the espalier-pear-trees.

Jan. 30. [Abroad] A long-billed curlew has juſt been shot near the Priory. We see now & then one in the very long froſts. Two, I underſtand, were seen.

Feb. 3.   A near neighbour of mine shot at a brace of hares out of his window ; & at the same discharge killed one, & wounded another. So I hope our gardens will not be so much molested. Much mischief has been done by these animals.

Feb. 4.   Hard frosts. Paths thaw. Fleecy clouds. Sky muddled. Halo.

Feb. 6.   Sowed 48 bushels of peat-ashes on the great meadow, which covered more than half. 31 bush : were bought of my neighbours.

Feb. 10.   No hares have frequented the garden since the man shot, & killed one, & wounded an other.

Feb. 11.   Snow covers the ground. Hares again in the garden.

Feb. 13.   This evening the frost has lasted 28 days.

Feb. 14.   Sent Thomas as Pioneer to open the road to Faringdon : [1] but there was little obstruction, except at the gate into Faringdon Hirn.

Feb. 15.   Snow deep, & drifted thro' the hedges in curious, & romantic shapes.

Feb. 16.   [Abroad] No hares frequent the garden.

Feb. 23.   The tops of the blades of wheat are scorched with the frost.

Feb. 24.   [Alton] The laurels, & laurustines are not injured by the severe weather. Snow scarce passable in Newton-lane !

Feb. 25-*Apr.* 1   [S. Lambeth] Little snow on the road. Thomas Hoar kept an account of the rain at Selborne in my absence.

Mar. 1.   Brother Tho : found a grass-hopper lark dead in his out-let : it seemed to be starved. I was not aware that they were about in the winter.[2]

Apr. 2.   [Selborne] No snow 'till we came to Guild-down ; deep snow on that ridge ! Much snow at Selborne in the fields : the hill deep in snow ! The country looks most dismally, like the dead of winter ! A few days ago our lanes would scarce have been passable for a chaise.

Apr. 3.   The crocus's are full blown, & would make a fine show, if the sun would shine warm. The ever-green-trees are not injured, as about London. On this day *a nightingale* was heard at Bramshot ! !

Apr. 4.   The rooks at Faringdon have got young. Very little spring-corn sown yet. Snow as deep as the horses belly under the hedges in the North field.   A brace more of hares frequenting my grounds were killed in my absence : so that I hope now the garden will be safe for some time.

Apr. 5.   My crocus's are in full bloom, & make a most gaudy show. Those eaten-off by the hares last year were not injured.

Apr. 7.   Many lettuces, both Coss, & Dutch, have stood out the winter under the fruit wall. They were covered with straw in the hard weather, for many weeks.

Apr. 8.   Men open the hills, & cut their hops. [Here follows an account of the timber in Holt Forest : T.P., IX.]

Apr. 13.   Mutton per pound 5ᵈ, Veal 5ᵈ, Lamb 6ᵈ, Beef 4ᵈ.   At Selborne.

Apr. 15.   Dogs-toothed violets blow.

Apr. 16.   Nightingale heard in Maiden-dance. [See map showing place-names.] Ring-dove builds in my fields.   Black-cap sings.

Apr. 17.   The buds of the vines are not swelled yet at all.   In fine springs they have shot by this time two or three inches.

Apr. 19.   Timothy the tortoise begins to stir : he heaves-up the mould that lies over his back.   Redstart is heard at the verge of the highwood against the common.

Apr. 20.   No garden-crops sowed yet with me ; the ground is too wet.   Artichokes seem to be almost killed.

Apr. 22.   The spring backward to an unusual degree ! Some swallows are come, but I see no insects except bees, & some phalaenae in the evenings. Daffodils begin to blow.

Apr. 23.   Timothy the tortoise comes forth from his winter-retreat.

Apr. 24.   Planted ten rows of potatoes against the Wid : Dewye's garden.   Planted one in the best garden.   John Carpenter buys now & then of Mr Powlett of Rotherfield a chest-nut tree or two of the edible kind : they are large, & tall, & contain 60 or 70 feet of timber each.   The wood & bark of these trees resemble that of oak ; but the wood is softer

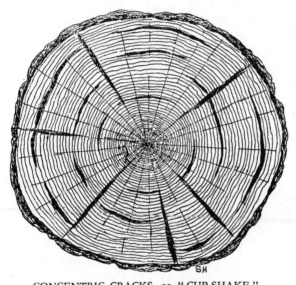

CONCENTRIC CRACKS, or " CUP-SHAKE,"
or RING-SHAKE IN TIMBER ;
due to poor soil, fungi, fire, frost, wind, &c.   The radial
cracks are called " star-shakes."

& the grain more open.   The use that the buyer turns them to is cooperage ; because he says the wood is light for buckets, jets[3] &c. & will not shrink.   The grand objection to these trees is their disposition to be *shaky* ; & what is much worse, *cup-shaky :* viz : the substance of these trees parts like the scales of an onion, & comes out in round plugs from the heart. This, I know, was also the case with those fine cheſt-nut-trees that were lately cut at Bramshot-place againſt the Portsmouth-road. Now as the soil at Rotherfield is chalk, & at Bramshot, sand; it seems as if this disposition to be shaky was not owing to soil alone, but to the nature of that tree. There are two groves of cheſt-nuts in Rotherfield-park, which are tall, & old, & have rather over ſtood their prime. J: Carpenter gives only 8ᵈ pr foot for this timber, on account of the defeᶜt above-mentioned.

Apr. 26. Sowed a crop of onions, & several sorts of cabbage: pronged the asparagus beds. Radishes grow.

Apr. 28. Grass-hopper lark whispers.

Apr. 29. The hoar-froſt was so great that Thomas could hardly mow. Bats out for the firſt time, I think, this spring: they hunt, & take the phalaenae along by the sides of the hedges. There has been this spring a pretty good flight of wood-cocks about Liss. If we have any of those birds of late years, it has been in the spring, in their return from the Weſt, I suppose, to the Eaſtern coaſt.

Apr. 30. Goose-berry bushes leaf: quicksets ſtill naked. Pile-wort in full bloom. Tulips shoot, & are ſtrong. Sowed a pint of scarlet kidney-beans.

May 1. Men pole hops; sow barley, & sow clover in wheat. Saw a cock white-throat.

May 2. No ring-ouzels this spring: the severity of the season probably disconcerted their proceedings.

May 3. Earthed the annual beds. Set up a copper-vane (arrow) on the brew-house. Goody hampton came to work in the garden for the summer. Timothy the tortoise weighs 6 ae 13 oun.; he weighed at firſt coming out laſt year only 6 ae 11¼ oun. He eat this morning the heart of a lettuce.

May 5. Cut the firſt cucumber, a large one. Golden weather. The polyanths blow finely, especially the young seedlings from Bramshot-place, many of which will be curious. Shot three green-finches, which pull-off the blossoms of the polyanths.

May 6. Pulled the first radishes. Crown-imperials & fritillaria's blown. Shot two more green-finches. There is a ring-dove's nest in the American Juniper in the shrubbery : but as that spot begins to be much frequented, the brood will scarcely come to good.

May 8. Auricula's blow finely in the natural ground. Owls have eggs. The hanger almost all green. Many trees in the Lythe in full leaf. Beeches on the common hardly budding.

May 10. The black-birds & thrushes are so reduced by the severe weather, that I have seen in my out-let only one of the former, & not one of the latter ; not one missle-thrush.

May 11. Sowed sweet alyssum in basons on the borders. Wheat improves very much : the women weed it.

May 12. There seem to be two, if not three nightingales singing in my out-let.

May 13. Cut the first bundle of asparagus.

May 14. Swallows build. They take up straws in their bills, & with them a mouthful of dirt. Fern-owl churs. The bark of felled oaks runs remarkably well ; so that the barkers earn great wages.

May 15. The tortoise is very earnest for the leaves of poppies, which he hunts about after, & seems to prefer to any other green thing. Such is the vicissitude of matters where weather is concerned, that the spring, which last year was unusually backward, is now forward.

May 16. Sultry. Left off fires in the parlor. So much sun hurries the flowers out of bloom. Flesh-flies begin to appear.

May 19. Flowers fade, & go-off very fast thro' heat. There has been only one moderate shower all this month. Bees thrive. Asparagus abounds.

May 21. Men bring up peat from the forest.

May 22.   Columbine & Monkshood blow.   The sycomores, & maples in bloom scent the air with a honeyed smell.   Lily of the valley blows.   Lapwings on the down.

May 23.   Field-crickets cry, & shrill in the short Lythe.

May 24.   A pair of swifts frequent the eaves of my stable.   The birds soon forsook the place, & did not build.

May 26.   Grasshopper lark in my outlet.

May 27.   My great single oak shows many catkins.

May 28.   Timothy the tortoise has been missing for more than a week.   He got out of the garden at the wicket, we suppose; & may be in the fields among the grass.   Timothy found in the little bean-field short of the pound-field.   The nightingale, fern-owl, cuckow, & grass-hopper lark may be heard at the same time in my outlet.   Gryllo-talpa [mole cricket] churs in moist meadows.

May 31.   Cinnamon rose blows.

June 1.   The single white thorn over the ash-house is one vast globe of blossoms down to the ground.   Laburnums, berberries, &c. covered with bloom.   Peonies in flower.

June 3.   Corn looks finely.   Pricked-out some good celeri-plants.   Turned the horses into Berriman's field.

June 4.   A pair of fern-owls haunt round the zig-zag.   Columbines make a fine show : this the third year of their blowing.

June 5.   Much damage done to the corn, grass & hops by the hail; & many windows broken! Vast flood at Gracious Street! vast flood at Kaker bridge! [S. of Alton.] Hail near Norton two feet deep. Nipped-off all the rose-buds on the tree in the yard opposite the parlor window in order to make a bloom in the autumn.   No bloom succeeded.

June 6.   [Long entry about the great hailstorm of the preceding day, partly reproduced in D.B., LXVI.

The following particulars are additional.] The flood at Gracious ſtreet ran over the goose-hatch, & mounted above the fourth bar of Grange-yard gate. . . The rushing & roaring of the hail as it approached was truely tremendous. The thunder at the village was little & diſtant. . . . At half hour after three on the same day a ſtill more deſtru&ctive one befell in Somersetshire. It extended seven miles in length, & about 2 in breadth, covering 7000 acres of fertile country; beginning in the S.W. and passing on slowly to the N.E. The centre of the ſtorm was seven miles W. of Taunton. The damage is very great!

June 10. Sold by Sᵗ foin again to Timothy Turner; it looks well, & is in bloom. The 17ᵗʰ crop. The buyer is to cut it when he pleases.

June 12. Men wash their sheep. Hoed carrots, parsneps, &c. Received 5 gallons & a quart of French brandy from Mʳ Edmᵈ Woods [of Godalming].

June 13. On this day arrived here from India Mʳ Charles Etty. In his passage out, the ship he belonged to was burnt off the Island of Ceylon. He came back from Madras to the Cape of good hope in the Exeter man of war; & from thence worked his passage in the Content transport, which brought him to Spit-head. The Exeter was so crazy, & worn-out, that they broke her up, & burnt her at the Cape. Mʳ Ch: Etty brought home two species of Humming-birds which he shot at the Cape of good hope; & two Oſtriches eggs from the same place: several fine shells from Joanna island [off Mozambique] & several turtle's eggs* from the Isle of Ascension. Also the *Gnaphalium squarrosum*, a curious Cudweed, from a Dutch-mans garden at the Cape. * Turtles-eggs are round, & white; a little variegated with fine ſtreaks of red, & as large as the eggs of a kite; perhaps larger.

June 16. *Phallus impudicus*, a ſtink-pot [fungus] comes up in Mʳ Burbey's asparagus-bed. Received a Hogsh: of port-wine, imported at Southampton.

June 20.   Narrow-leaved iris, cornflag, & purple martagons blow.   Butter-fly orchis in the hanger.

June 21.   Dark & chilly, rain.   Cold & comfortless.

June 25.   Towards the end of June they had snow in Austria, & the vines were frozen.

June 26.   Fire in the parlor.

June 29.   Mr & Mrs Richardson came.

July 2.   Began to cut my meadow grass; a good crop.   Mr & Mrs Richardson left us.   Low creeping mists.   Yellow even.

July 4.   On this day my Godson, Littleton Etty discovered a young Cuckow in one of the yew hedges of the vicarage garden, sitting in a small nest that would scarce contain the bird, tho' not half grown. By watching in a morning we found that the owners of the nest were hedge-sparrows, who were much busied in feeding their great booby.   The nest is in so secret a place that it is to be wondered how the parent Cuckow could discover it.   Tho' the bird is very young it is very fierce, gaping, & striking at peoples fingers, & heaving up by way of menace, & striving to intimidate those that approach it.   This is now only the fourth young cuckow that I have ever seen in a nest: three in those of h: sparrows, & one in that of a tit-lark.   As I rode up the N. field-hill lane I saw young partridges, that were about two or three days old, skulking in the cart-ruts; while the dams ran hovering & crying up the horse-track, as if wounded, to draw off my attention.

July 5.   Timothy Turner cuts Baker's hill, the crop of which he has bought.   It is St foin run to seed, the 17th crop.

July 7.   Vast damage done in various parts of the kingdom by thunder-storms & floods, from Yorkshire all across to Plymouth.

July 8.   Gloomy & heavy.   Much hay housed. Cool gale.   Pitch-darkness.

July 10.   The young cuckow gets fledge, & grows bigger than its nest.   It is very fierce, & pugnacious.

July 11. My horses, which lie at grass, have had no water now for about 8 weeks : nor do they seem to desire any when the pass by a pond, or stream. This method of management is particularly good for aged horses, especially if their wind is at all thick. My horses look remarkably well.

July 13. Finished ripping, furring,[4] & tiling the back part of my house ; a great jobb. Garden-beans come in.

July 14. *Papilio Machaon* in Mrs Etty's garden. They are very rare in these parts.

July 16. Made curran, & raspberry jam ! the fruit is hardly ripe ; but the small birds will steal it all away.

July 17. M^r Ch: Etty has taken the young Cuckow, & put it in a cage, where the hedge-sparrows feed it. No old Cuckow has been seen to come near it. Mr Charles Etty brought down with him from London in the coach his two finely-chequered tortoises, natives of the island of Madagascar, which appear to be the *Testudo geometrica*, Linn., and the *Testudo tessellata*, Raii. One of them was small, & probably a male, weighing about five pounds ; the other, which was undoubtedly a female, because it layed an egg the day after it's arrival, weighed ten pounds & a quarter. The egg was round, & white, & much resembling in size & shape the egg of an owl. Ray says of this species that the shell was " Ellipticae seu ovatae figurae solidae plus quam dimidia pars " : & again, " Ex omnibus quas unquam vidi maximè concava." Ray's quadrup : 260. The head, neck, & legs of these were yellow. These tortoises in the morning when put into the coach at Kensington were brisk, & well ; but the small one dyed the first night that they came to Selborne ; & the other, two nights after, having received, as it should seem, some Injury on their Journey. When the female was cleared of the contents of her body, a bunch of eggs of about 30 in number was found in her.

July 20. Bro : Henry & his son Sam came. Saw an old swift feed it's young in the air : a circumstance which I could never discover before.

July 22. The wind broke-off a great bough from Molly White's horse-chestnut tree.

July 24. Planted bore-cole, &c. Yellow horizon. Bro͉ Henry left us.

July 26. Bottled-out the hogshead of port-wine : my two thirds ran to 16 dozen & four of my bottles, some of which are Bristol bottles, & therefore large. M͉ & M͉͉ Mulso, & Miss Mulso, & Miss Hecky Mulso came.

July 29. Drew-out from the port-wine hogsh : for my share, eleven bottles more of wine : so that my proportion was 17 dozen, & three bottles. Thanksgiving for the peace.

Aug. 2. Wall-cherries, may dukes, lasted 'till this time, & were very fine.

Aug. 4. Skimmed my two pasture fields.

Aug. 7. Many hop-poles are blown down. Cool, autumnal feel. Days much shortened.

Aug. 10. M͉ & M͉͉ Mulso, &c : left us.

Aug. 12. Wheat husing at Heards.

Aug. 14. Plums show no tendency to ripeness. Scalded codlings[5] come in. The wheat that was smitten by the hail [June 5] does not come to maturity together : some ears are full ripe, & some quite green. Wheat within the verge of the hail-storm is much injured, & the pease are spoiled. A puff-ball, lycoperdon bovista, was gathered in a meadow near Alton, which weighed 7 pounds, & an half, & measured 1 Yard and One Inch in girth the longest way 3 feet two inches. There were more in the mead almost as bulky as this.

Aug. 15. Women bring cran-berries, but they are not ripe.

Aug. 17. Farmer Spencer, & farmer Knight are forced to stop their reapers, because their wheat ripens so unequally.

Aug. 18. Spinage very thick on the ground. Men hoe turnips, stir their fallows, & cart chalk.

Aug. 20. On this day my Niece Brown was delivered of her 4th child, a girl, which makes the 41st of my nephews & nieces now living. Boiled up some apricots with sugar to preserve them.

Aug. 24. White turnip-radishes mild, & good, & large.

Aug. 25. Sad harvest weather. This proves a very expensive, & troublesome harvest to the farmers. Pease suffer much & will be lost out of the pod.   My great apricot-tree appeared in the morning to have been robbed of some of it's ripe fruit by a dog that had stood on his hind legs, & eaten-off some of the lower apricots, several of which were gnawn, & left on the ground, with some shoots of the tree.   On the border were many fresh prints of a dogs feet.   I have known a dog eat ripe goose-berries, as they hung on the trees. Many wallnuts on the tree over the stable : the sort is good, but the tree seldom bears.

Aug. 29.   A Faringdon man shot a young fern-owl in his orchard.

Sept. 1.   Farmer Town began to pick his hops : the hops are many, but small.   They were not smitten by the hail, because they grew at the S.E. end of the village.   Hopping begins at Hartley.   The two hop-gardens, belonging to Farmer Spencer & John Hale, that were so much injured, as it was supposed, by the hail-storm on June 5th shew now a prodigious crop, & larger & fairer hops than any in the parish.   The owners seem now to be convinced that the hail, by beating off the tops of the binds, has encreased the side-shoots, & improved the crop.   *Que.*   Therefore should not the tops of hops be pinched-off when the binds are very gross, & strong ? [6] We find this practice to be of great service with melons, & cucumbers.   The scars, & wounds on the binds, made by the great hailstones are still very visible.

Sept. 4. My Nep: Edm^d White launched a balloon on our down, made of soft, thin paper; & measuring about two feet & a half in length, & 20 inches in diameter. The buoyant air was supplyed at bottom by a plug of wooll, wetted with spirits of wine, & set on fire by a candle. The air being cold & moist this machine did not succeed well abroad: but in Mr Yalden's stair-case it rose to the ceiling, & remained suspended as long as the spirits continued to flame, & then sunk gradually. These Gent: made the balloon themselves. This small exhibition explained the whole balloon affair very well: but the position of the flame wanted better regulation; because the least oscillation set the paper on fire.[7] Golden weather, red even.

Sept. 10. [Bramshot place] *Uncrested wrens* seem to be withdrawn. Mr Richardson's wall fruit at Bramshot-place is not good-flavoured, nor well-ripened: & his vines are so injured by the cold, black summer, as not to be able to produce any fruit, or good wood for next year. Mr Dennis's vines at Bramshot also are in a poor state.

Sept. 11. [Selborne] Mr Randolph the Rector of Faringdon came. I saw lately a small Ichneumon-fly attack a spider much larger than itself on a grass-walk. When the spider made any resistance, the Ich: applied her tail to him, & stung him with great vehemence, so that he soon became dead & motionless. The Ich: then running backward drew her prey very nimbly over the walk into the standing grass. The spider would be deposited in some hole where the Ich: would lay some eggs; & as soon as the eggs were hatch'd, the carcase would afford ready food for the maggots. Perhaps some eggs might be injected into the body of the spider, in the act of stinging. Some Ich: deposit their eggs in the aurelia of butterflies & moths.

Sept. 14. The heats are so great, & the nights so sultry, that we spoil joints of meat, in spite of all the care that can be taken.

Sept. 15. Mr Randolph left us. The autumn-sown spinage turns-out a fine crop : but it is much too thick. We draw it for use.

Sept. 16. Martins cling, & cluster in a very particular manner against the wall of my stable and brewhouse; also on the top of the may-pole. This clinging, at this time of the year only, seems to me to carry somewhat significant with it.

Sept. 17. Nep : Ben White left me : he stayed a few days.

Sept. 21. Gathered-in the early pippins, called *white apples :* a great crop.

Sept. 25. Sister Henry White, & her daughter Lucy came.

Sept. 26. Mr Taylor[8] took possession of Selborne vicarage.

Sept. 29. Took possession of Selborne curacy.[9]

Oct. 1. Gathered-in the Swan's egg, autumn-burgamot, Cresan-burgamot, Chaumontelle, & Virgoleuse pears : a great crop. The Swan-eggs are a vast crop. A wood-cock was killed in Blackmoor-woods ; an other was seen the same evening in Hartley-wood.

Oct. 3. Two young men killed a large *male otter*, weighing 21 pounds, on the bank of our rivulet, below Priory longmead, on the Hartley-wood side, where the two parishes are divided by the stream. This is the first of the kind ever remembered to have been found in this parish.

Oct. 6. A vast flock of ravens over the hanger : more than sixty !

Oct. 7. Mr Harry White, & Lucy left us.

Oct. 8. Mr Richardson came.

Oct. 9. M^r R : left us. It has been the received opinion that trees grow in height only by their annual upper shoot. But my neighbour over the way, Tanner, whose occupation confines him to one spot, assures me, that trees are expanded & raised in the lower parts also. The reason that he gives is this :

the point of one of my Firs in Baker's hill began for the first time to peep over an opposite roof at the beginning of summer ; but before the growing season was over, the whole shoot of the year, & three or four joints of the body beside became visible to him as he sits on his form in his shop. According to this supposition, a tree may advance in height considerably though the summer shoot should be destroyed every year.

Oct. 10. A person took a trout in the stream at Dorton, weighing 2 pounds, & an half ; a size to which they seldom arrive with us, because our brook is so perpetually harassed by poachers.

Oct. 11. Men draw & stack turnips.

Oct. 14. Finished gathering-in the apples. Apples are in such plenty, that they are sold for 8ᵈ per bushel.

Oct. 16. Mʳ Blanchard passed by us in full sight at about a quarter before three P : M : in an air-balloon ! ! ! He mounted at Chelsea about noon ; but came down at Sunbury to permit Mʳ Sheldon to get out ; his weight over-loading the machine. At a little before four P : M : Mʳ Bl : landed at the town of Romsey in the county of Hants. [*News-paper cutting pasted on inserted leaf.*] Extract of a Letter from a Gentleman [G. White]¹⁰ in a village fifty miles S.W. of London, dated Oct. 21. "From the fineness of the weather, and the steadiness of the wind to the N.E. I began to be possessed with a notion last Friday that we should see Mʳ Blanchard [in his balloon] the day following, and therefore I called upon many of my neighbours in the street, and told them my suspicions. The next day proving also bright and the wind continuing as before, I became more sanguine than ever ; and issuing forth, exhorted all those who had any curiosity to look sharp from about one to three o'clock [towards London] as they would stand a good chance of being entertained with a very extra-ordinary sight. That day I was not content to call

at the houses, but I went out to the plow-men and
labourers in the fields, and advised them to keep an
eye at times to the N. and N.E. But about one o'clock
there came up such a haze that I could not see the hill;
however, not long after the mist cleared away in some
degree, and people began to mount the hill. I was
busy in and out till a quarter after two and observed
a cloud of *London smoke*, hanging to the N. and N.N.E.
This appearance increased my expectation. At twenty
minutes before three there was a cry that the balloon
was come. We ran into the orchard, where we found
twenty or thirty neighbours assembled, and from the
green bank at the end of my house, saw a dark blue
speck at a most prodigious height dropping as it
were out of the sky, and hanging amidst the regions of
the air, between the weather-cock of the Tower and
the Maypole: at first, it did not seem to make any
way, but we soon discovered that its velocity was very
considerable, for in a few minutes it was over the
Maypole; and then over my chimney; and in ten
minutes more behind the wallnut tree. The machine
looked mostly of a dark blue colour, but some times
reflected the rays of the sun. With a telescope I
could discern the boat and the ropes that supported
it. To my eye the balloon appeared no bigger than a
large tea-urn. When we saw it first, it was north of
Farnham over Farnham heath; and never came on
this (east) side the Farnham road; but continued
to pass on the N.W. side of Bentley, Froil, Alton, &c.
and so for Medstead, Lord Northington's at the
Grange, and to the right of Alresford and Winchester.
I was wonderfully struck with the phaenomenon,
and, like Milton's "Belated Peasant," felt my heart
rebound with joy and fear at the same time. After
a while I surveyed the machine with more composure,
without that concern for two of my fellow creatures;
for two we then supposed there were embarked
in that aerial voyage. At last seeing how securely
they moved, I considered them as a group of

cranes or storks intent on the business of emigration, who had

> '. . . Set forth
> Their airy caravan, high over seas
> Flying, and over lands, with mutual wing
> Easing their flight . . . ' "
> [*Par. Lost*, VII, ll. 427-30.]

Oct. 21. This day at 4 o'clock P : M : Edm^d White launched an air-balloon from Selborne-down, measuring about 8 feet & ½ in length, & 16 feet in circumference. It went off in a steady, & grand manner to the E, & settled in about 15 minutes near Todmoor on the verge of the forest.

Oct. 24. I have seen no ants for some time, except the *Jet-ants*, which frequent gate-posts. These continue still to run forwards, & backwards on the rails of gates, & up the posts, without seeming to have anything to do. Nor do they appear all the summer to carry any sticks or insects to their nests like other ants.

Oct. 25. Hard frost, thick ice. In my way to Newton I was covered with snow ! Snow covers the ground, & trees ! !

Oct. 26. Horses begin to lie within.

Oct. 27. Dunged, trenched, & earthed the asparagus-beds, & filled the trenches with leaves, flower-stalks, &c.

Oct. 28. Mr John Mulso came.

Oct. 29. Foliage turns very dusky : the colour of the woods & hangers appears very strange, & what men, not acquainted with the country, would call very unnatural.

Oct. 31. Many people are tyed-up about the head on account of tooth-aches, & face-aches.

Nov. 1. Mr John Mulso was shot in the legs.[11]

Nov. 4. Timothy out. Great meteor.

Nov. 5. The deep, golden colour of the larches amidst the dark evergreens makes a lovely contrast.

Nov. 8. The hanger almost naked : some parts of my tall hedges still finely variegated : the fading foliage of the elm is beautifully contrasted to the beeches.

Nov. 14. No acorns, & very few beech-mast. No beech-mast last year, but acorns innumerable.

Nov. 23. Brother Thomas, & his daughter, & two sons came. The chaise that brought some of them passed along the king's high road into the village by Newton lane, & down the N. field hill ; both of which have had much labour bestowed on them, & are now very safe. This is the first carriage that ever came this way. Planted tulips again in the borders ; & the small off-sets in a nursery-bed.

Nov. 26. Haws in such quantities that they weigh down the white-thorns.

Dec. 2. Timothy is buried we know not where in the laurel hedge.

Dec. 6. Dismally dark : no wind with this very sinking glass.

Dec. 9. Much snow in the night. Vast snow. Snow 16 inches deep on my grass-plot : about 12 inches at an average. Farmer Hoar had 41 sheep buried in snow. No such snow since Jan. 1776. In some places much drifted.

Dec. 10. Extreme frost ! ! ! yet still bright sun. At 11 one degree below zero [*Added note*]. On the 9th and 10th of Decr when my Thermr was down at o, or zero ; & 1 degree below zero :—Mr Yalden's Thermr at Newton was at 19, & 22. On Dec. 24, when my Thermr was at 10½ that at Newton was at 22, & 19. At Newton, when hung side by side, these two instruments accorded exactly. [cf. D.B., LXIII.] Thomas Hoar shook the snow carefully off from the evergreens. The snow fell for 24 hours, without ceasing. The ice in one night in Gracious street full four inches ! Bread, cheese, meat, potatoes, apples all frozen where not secured in cellars under ground.

Dec. 11. My apples, pears, & potatoes secured in the cellar, & kitchen-closet; my meat in the cellar. Severe frost, & deep snow. Several men, that were much abroad, made sick by the cold; their hands, & feet were frozen. We hung-out two thermometers, one made by Dollond, & one by B: Martin: the latter was graduated only to 4 below ten, or 6 degrees short of zero: so that when the cold became intense, & our remarks interesting, the mercury went all into the ball, & the instrument was of no service.

Dec. 13. Shoveled out the bostal. Snow very deep still.

Dec. 14. Finished shoveling the path to Newton. Dame Loe came to help.

Dec. 15. Deep snow still. Snow drifts on the down, & fills up the path which we shoveled.

Dec. 16. Titmice pull the moss off from trees in search for insects.

Dec. 20. My laurel-hedge is scorched, & looks very brown!

Dec. 22. Farmer Lassam's Dorsetshire ewes begin to lamb. His turnips are frozen as hard as stones.

Dec. 23. Many labourers are employed in shoveling the snow, & opening the hollow, stony lane, [Honey Lane] that leads to the forest. Snow frozen so as almost to bear.

Dec. 25. Stagg, the keeper, who inhabits the house at the end of Wolmer Pond, tells me that he has seen no wild-fowl on that lake during the whole frost; & that the whole expanse is entirely frozen up to such a thickness that the ice would bear a waggon. 500 ducks are seen some times together on that pond.

Dec. 31. Much snow on the ground. My laurel-hedge, & laurustines, quite discoloured, & burnt as it were with the frost. [Here follow columns showing the monthly rainfall for 1784, at Fyfield, Selborne, & S. Lambeth respectively. Selborne has much the greatest total.]

CHAPTER XVIII

"A poor life it is if, full of care,
We have no time to stand and stare."
                                    W. H. Davies : *Leisure.*

1785

Jan. 1. Much snow on the ground. Ponds frozen-up & almost dry. Moles work: cocks crow. Ground soft under the snow. No field-fares seen; no wag-tails. Ever-greens miserably scorched; even ivy, in warm aspects.

Jan. 3. Began the new rick: the hay is very fine. Tho' my ever-greens are almost destroyed; Mr Yalden's bays, & laurels, & laurustines seem untouched. Berberries, & haws frozen on the trees. No birds eat the former.

Jan. 5. Brother Thomas left us.

Jan. 6. No snipes in the moors of the forest, or on the streams. No wood-cocks to be found this winter.

Jan. 7. Shook the snow from the ever-greens, & shovelled the walks. Snow-scenes very beautiful! On this day Mr Blanchard, & Dr Jeffries rose in a balloon from Dover-cliff, & passing over the channel towards France, landed in the forest De Felmores, just 12 miles up into the country. These are the first aëronauts that have dared to take a flight over the Sea ! ! !

Jan. 8. Received five gallons, & seven pints of French brandy from Mr Edm^d Woods.

Jan. 11. Men begin to plough again.

Jan. 21. Made a seedling-cucumber-bed. The glazier mended the light of the seedling-frame broken by the hail.

Jan. 23. Boys play on the Plestor at marbles, & peg-top. Thrushes sing in the coppices. Thrushes & blackbirds are much reduced.

Jan. 26. Planted two rows of garden-beans.

Jan. 31. The wind blowed-off the fox's tail.[1]

Feb. 1. On this cold day about noon a bat was flying round Gracious street-pond, & dipping down & sipping the water, like swallows, as it flew : all the while the wind was very sharp, & the boys were standing on the ice !

Feb. 2. The scorched laurels cast their leaves, & are almost naked.

Feb. 4. Arbutus's, Cypresses, Ilex's seem to be dead : even Portugal-laurels are injured, & Cedars of Libanus, American, & Swedish Junipers, & firs, Scotch & Spruce untouched.

Feb. 6. Young sheep suffer much by the weather, & look poorly.

Feb. 16. Men sow peat-ashes on their Grasses. Winter aconites make a gay show.

Feb. 18. Carried the apples, pears, & roots into the cellars.

Feb. 19. Thick ice. Ice in warm chambers. Boys slide. Ground as hard as iron. Snow on the ground.

Feb. 23. Snow-scenes very beautiful. Venus makes a most beautiful appearance.

Mar. 1. Carted in six loads of hot dung for the cucumber-bed.

Mar. 4. New worked up, & mended the garden-lights broken by the hail last summer.

Mar. 7. Glazier's bill . . . £2 - 5 - 10 for garden-lights, & hand-glasses.

Mar. 8. Sowed radishes under the hot-bed screen.

Mar. 9. On this day Mr Charles Etty sailed in the Duke of Montrose India-man, Captain Gray, for Madeira, & Bombay.

Mar. 10. Much beech-woods, & faggots carted home.

Mar. 17.   Made the four-light bearing cucumber-bed with five dung-carts, & ½ of dung.

Mar. 19.   Sowed a bed of spinage : the winter-spinage killed. Tulips, & crown-imperials, & hyacinths sprout. Planted eight larches in Baker's hill. Cucumbers thrive. Ice still in water-tubs. Men plough : the frost pretty much out of the ground wch is mellow.

Mar. 21.   Mr Charles Etty sailed from the mother bank, near the Isle of Wight, where they stopped to take in passengers.

Mar. 22.   Wheat-fields look naked like fallows. The surface of the ground is all dust.

Mar. 25.   Shoveled the alleys, & threw the mould on the borders, & quarters.

Mar. 26.   Sowed the great mead with ashes.

Mar. 29.   My niece Clement was brought to bed of a boy. This child makes my 42 nephew, & niece now living.

Mar. 30.   Thick ice. White frost. Winter-aconites out of bloom : snow drops make still a fine [show]. Violets, & coltsfoot blow.

Apr. 1.   Snow hangs in the trees, & makes a perfect winter scene !

Mar. 8.   Many lettuces under fruit-wall have stood the winter : they were covered with snow. The bloom of the wall-fruit seems to be killed in the bud.

Apr. 11.   Farmers wish much for rain.

Apr. 15.   Hot sun. Muddy sky. Goose-berries & honey-suckles begin to bud, & look green. Pronged the asparagus beds. My fine jasmine is dead. *Timothy* the *tortoise* roused himself from his winter-slumbers & came forth. He was hidden in the laurel-hedge under the wall-nut tree, among the dead leaves.

Apr. 18.   [Alton] The large *shivering* willow-wren. The Cuckow is heard this day.

Apr. 19-*May* 12.   [S. Lambeth.]

Apr. 21.   My brother Tho : planted his potatoes. He sowed purple broccoli. My brother cut four cucumbers. His plants, & Benjamin's[2] are strong, & in good order.

Apr. 25. Radishes dry, & hot.

Apr. 26. My brother Tho: made his melon-bed. *Red-start* sings.

Apr. 27. Quick-set hedges look green. Roads are choaked with dust. Swallows frequent houses: some sit & dress themselves on trees, as if wet, & dirty.

May 1. The dust on the roads insufferable! Saw one *swift*. Two *house-martins* in Fleet Street.

May 3. Blanchard & Miss Simonet ascended.

May 7. Pastures yellow with dandelions. Meadow-foxtail-grass, alopecurus pratensis, in bloom.

May 8. There is a great want of rain in France as well as in England. A cuckow haunts my brother's fields [S. Lambeth]; so that probably there will be a young cuckow hatched in the quickset-hedge. Millions of *empedes*, or *tipulae*, come forth at the close of day, & swarm to such a degree as to fill the air. At this juncture they sport & copulate: as it grows more dark they retire. All day they hide in the hedges. As they rise in a cloud they appear like smoke: I do not ever remember to have seen such swarms except in the fens of the Island of Ely. They appear most over grass-grounds.

May 9. The grass in my Brother's fields burns, & does not look so well as it did when I came.

May 11. Severe drying exhausting drought. Cloudless days. The country all dust. Timothy the tortoise weighs 6 ae 11¾ oun.[3] He spoils the lettuce under the fruit-wall: but will not touch the Dutch, while he can get at any coss.

May 12. Dragon-flies come out of their aurelia-state. Great bloom of apples round S. Lambeth.

May 13. [Alton] The country strangely burnt-up.

May 14. [Selborne] My fields have more grass than my brother's at S. Lambeth, which burn. My

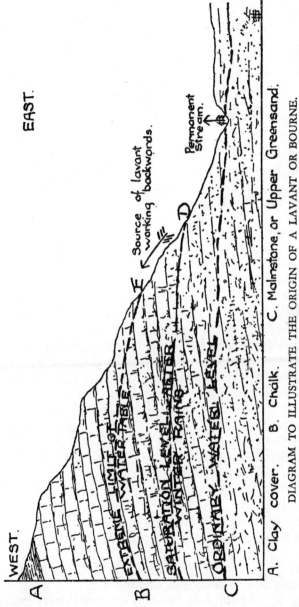

EAST.

WEST.

A

Source of lavant working backwards.

F

D

E

Permanent Stream.

EXTREME LIMIT OF WATER TABLE

B

SATURATION LEVEL AFTER WINTER RAINS

ORDINARY WATER LEVEL

C

A. Clay cover.   B. Chalk.   C. Malmstone, or Upper Greensand.

DIAGRAM TO ILLUSTRATE THE ORIGIN OF A LAVANT OR BOURNE.

After very wet seasons the water-level rises and a temporary stream (lavant) issues at D. Should the level still
rise, the source moves successively backward to E.

S^t foin looks well, & is grown. Ponds in bottoms are dry. Our down burnt brown.

May 18. My wall-nut trees seem much injured by the frost. The laurels shoot at the bottom of the boughs. Sycomores are injured. Chafers swarm about Oakhanger, & on the chalky soils ; but not with us on the clays.

May 19. Planted some red cabbages from S. Lambeth. Planted some green cucumber-plants to fill the void spaces in the early-frames.

May 21. The kidney-beans are cut down. Potted several balsoms, a fine sort saved last year.

May 22. Field-crickets cry round the forest.

May 24. Swifts copulate in the air, as they flie.

May 25. Wood-ruff blows.

May 26. *Rose-fly*, a green scarab. Tho' the stream has been dry for some time at *Gracious street* quite down to Kimber's mead ; yet, when it meets *Well-head* stream at *Dorton*, it is little inferior to that. This shows that there are several springs along the foot of the short *Lithe*, as well as a constant one at Kimber's.[4]

May 28. Planted some basons of green early cucumbers from S. Lambeth under the melon-screen. [Long note concerning the vine coccus, D.B., LIII, with the following additional particulars.] Thomas has been employed in killing them [cocci] for many hours. . . . I have observed the mischievous effects of these animals four years on this vine ; before which I never saw any of this kind. . . . The females stick motionless to the boughs, & look like large flat wood-lice. . . . The ants resort much to the *Cocci*, but do not destroy them, as I could wish. . . . I am sorry to find that the insects on my vine were not suppressed by the severity of last winter : if that could not check them, we must not expect any assistance from cold.

May 31. Thomas persists in picking the *cocci* off the vine, & has destroyed hundreds.

June 2. Abram Loe came. My well is very low.

June 4.   Several halo's & mock-suns this morning.
Wheat looks black, & gross.   Crickets sing much
on the hearth this evening : they feel the influence
of moist air, & sing against [at the approach of] rain.
As the great wall-nut tree has no foliage this year, we
have hung the meat-safe on Miss White's Sycomore,
which she planted a nut ; where it will be much in
the air, & be well sheltered from the sun by leaves.

June 5.   Dame's violets blow, & are very double.

June 8.   Planted the bank in the garden, & the
opposite border with China asters all the whole
length.

June 10.   The late severe winter, & spring seem to
have destroyed most of the black snails.   Planted-
out all the *annuals* in general down Baker's hill, & in
the garden.   The plants are strong, & vigorous, &
the season very favourable ; the earth is well moistened, & the weather warm, still, shady, & dripping.

June 11.   My potatoes do but just sprout above
ground.   Sweet Williams blow.   When the hen fly-
catcher sits on her eggs, the cock feeds her with great
assiduity, even on 'till past nine in the evening.

June 13.   Established summer.   Corn-flag blows.

June 14.   Fly-catchers have young. Standard honey-
suckles beautiful, & very sweet.

June 18.   The yew-hedges at the vicarage half-
killed by the winter.   My tall hedges are much
injured by the severity of last winter : many boughs
are killed, & the foliage in general is thin.

June 19.   Most of our oaks are naked of leaves, &
even the *Holt* in general, having been ravaged by the
caterpillars of a small *phalaena*, which is of a pale,
yellow colour.   These Insects tho' a feeble race,
yet from their infinite numbers are of wonderful
effect, being able to destroy the foliage of whole
forests, & districts.   At this season they leave their
aurelia, & issue forth in their *fly*-state, swarming &
covering the trees, & hedges.   In a field at Greatham
I saw a flight of *Swifts* busied in catching their prey

near the ground; & found they were hawking after these *phalenae*. The *aurelia* of this moth is shining, & as black as jet; & lies wrapped-up in a leaf of the tree, which is rolled round it, & secured at the ends by a web, to prevent the maggot from falling-out.[5]

June 22.   Turbid sunset : the disk of the sun looked like three suns.   Full moon.

June 25.   Fallows dusty, & in mellow order. Young fawns in the Holt.   My walnut-trees are almost naked, & half-killed by the winter ; while those at *Rood* are in full foliage, & shew fruit.

June 26.   Annuals die thro' heat.   Hops run their poles.   M[r] Powlett of Rotherfield has no water for his cattle in the park, but what he fetches from Alton ![6] He has a well for the house.   Many years ago M[r] Powlett's grandfather fetched water from Alton for all his cattle, deer & all, for three months together. My well is low ; but affords plenty of fine clear water. We draw great quantities for the garden.   A constant spring runs into it.

June 27.   The *Flycatchers*, five in number, leave their nest in the vine over the parlor-window.   *Hemero-callis*, day-lily, blows.   Chaffers fall dead from the hedges ; they have served their generation, & will be seen but little longer.

June 29.   Distant thunder.   The storm arose in the S. & parted ; so that we had only the skirts. When thunder arises in the S. we hardly ever receive the storm over us, because the clouds part to the right, & left before they reach us, influenced, I suppose, by the hills that lie to that quarter.   The walnut-trees throw-out shoots two or three feet below the extremities of the boughs : all above is dead.

June 30.   Mossed the white cucumber-bed.

July 1.   Timothy Turner cuts the S[t] foin on Baker's hill : this is the 18th crop ; & not a bad one, the severity of the drought considered.   My *balsams* are fine tall plants, & well-variegated, except a few, which blow white.

July 2. The heat at noon yesterday was so great that it scorched the white cucumbers under the hand-glasses, & injured them much. Annuals die with the heat. Took away the moss from the white cucumbers, because it seemed to scald [burn, scorch] them.

July 4. Gathered several pounds of cherries to preserve : they are very fine.

July 5. Young *cocci* abound again on the vines. Began to cut the meadow-grass : it is very scanty, not half a crop. Men sow turnips ; but the seeds lie on the ground without vegetating. Those that sprout are soon eaten by the fly.

July 6. Some young *Swifts* seem to be out : they settle on, & cling to the walls of houses, & seem to be at a loss where to go ; are perhaps looking for their nest.

July 8. *Ricked* my *hay*, which makes but a very small cob [heap, stack]. All the produce of the great mead was carried at two loads ; & all that grew on the slip was brought up by the woman & boy on their backs. My quantity this year seems to be about one third of a good crop. In a plentiful year I get about seven good Jobbs. Thatched the rick.

July 9. Ants swarm on the stairs : their male-flies leave them, & fill the windows : their females do not yet appear.

July 10. The spring in *Kimber's mead* is dry ; & also that in *Conduit-wood* ; from whence in old times the Priory was supplyed with water by means of leaden pipes. The pond on the common is also empty. All the while *Well-head* is not much abated, nor the spring at the bottom of the *church-litten* closes, where you pass over the foot-bridge to the Lithe. [Cf. 26 May, 1785.] Preserved cherries, & currans ; & made curran-jelly. Not one mess of wood-straw-berries brought this year.

July 11. The down is so burnt, that it looks dismally.

July 12-15. [Bramshot-place] My vines are nicely trimmed: not a superfluous shoot left. Cleared the cherry-trees, & took-in the nets. Mr Richardson's garden was not so much burnt-up as might be expected. There was plenty of pease, & kidney-beans; & much fruit, such as currans, gooseberries, melons, & cherries. The wheat at Bramshot looks well; but the spring-crops are injured by the drought. Turnips come-up pretty well. The pair of *Fly-catchers* in the vine are preparing for a second brood, & have got one egg. This is the first instance that I remember of their breeding twice.

July 14. Vast shower in the evening towards Odiham [Hants]. Wheat on the strong lands looks finely. The crop in the Ewel looked so thin, as if there would [be] nothing all the spring: but now there is fine even wheat. Fine rain at London.

July 15. [Selborne] Boys brought the fourth wasp's-nest.

July 17. Newton great pond is almost dry; only two or three dirty puddles remain, which afford miserable water for the village. My nephew Edm^d White of Newton turns his sheep into five acres of barley, which is spoiled by the drought. Mr Pink of Faringdon does the same by a field of oats.

July 18. Savoys & artichokes over-run with aphides. The *Fly-catcher* in the vine sits on her eggs, & the cock feeds her. She has four eggs.

July 21. Heavy showers. Ponds fill.

July 22. Made black curran-jelly, & rasp: jam.

July 23. Some water in the pond on the down. Mr Edm^d White's tank has four feet of water.

July 25. Boys bring the sixth & seventh wasps nest. My Nep: Edm^d White sends me some fine wall-nuts for pickling. The trees at Newton were not at all touched by the severity of last winter; while mine were so damaged that all the bearing twigs were destroyed. My wall-nut trees have this summer pushed out shoots thro' the old bark, several feet from the extremities of the boughs. While the hen-fly-catcher sits, the cock feeds her all day long : he also pays attention to the former brood, which he feeds at times.

July 26. By frequent picking we have much reduced the *Cocci* on the vines.   Vast storm of thunder, & rain at Thursley, which damaged the crops. Thursley is in Surrey, to the N.E. of us.

July 30. Boys bring the 8th & 9th wasps nest. Pyramidal campanula blows.

July 31. Hops begin to form on their poles : but the gardens in general, fall off, & look lousy, since the rains.

Aug. 1-5. [Meon-stoke] All the way as we drove along, we saw wheat harvest beginning.   The ponds at Privet, where they have been much distressed for water, are nearly full.   The down-wheat, about Meonstoke a poor crop.   Many turnips fail.   The *fly-catchers* hover over their young to preserve them from the heat of the sun.

Aug. 3. Harvest-bugs   are   troublesome. Fly-catchers in Mr Mulso's garden,[7] that seem to have a nest of young. *Tremella nostoch* abounds in Mr Mulso's grass-walks.

Aug. 6. [Selborne] My young fly-catchers near fledge.

Aug. 7. Wasps begin on the goose-berries.

Aug. 8. Pease lie in a sad state, & shatter-out. Gleaning begins : wheat is heavy. *Agaricus pratensis* champignion, comes-up in the fairey-ring on my grass-plot.

Aug. 9. Mushrooms come in.   Fire gleams. Fly-catchers, second brood, forsake their nest.

Aug. 10. Men bind their wheat as fast as they reap it.   Hops look black.

Aug. 12. *Black-caps* eat the berries of the honey-suckle, now ripe. Pheasant-cocks crow.

Aug. 13. My Nephew Edmd White's tank at Newton runs over. On the first of August, about half an hour after three in the afternoon the people of Selborne were surprized by a shower of *Aphides* which fell in these parts. I was not at home ; but those who were walking the streets at that juncture found themselves covered with these insects, which settled also on the trees, & gardens, & blackened all the vegetables where they alighted. My annuals were covered with them ; & some onions were quite coated over with them when I returned on Aug. 6th. These armies, no doubt, were then in a state of emigration, & shifting their quarters ; & might come, as far as we know, from the great hop-plantations of Kent or Sussex, the wind being that day at E. They were observed at the same time at Farnham, & all along the vale to Alton. Of the conveyance of Insects from place to place, see Derham's Physico-theology : p. 367.

Aug. 15. Sam, & Charles come from Fyfield. The harvest-scenes are very beautiful ! Farmer Spencer makes a wheat-rick.[8] Wheat very fine & heavy.

Aug. 16. My goose-berries are still very fine, but are much eaten by the dogs.

Aug. 17. Few mushrooms to be found. Sowed second crop of white turnip-radishes. Abram Loe came the second time.

Aug. 18. *Colchicum, autumnal crocus*, emerges, & blows.

Aug. 19. Sam & Charles leave us. Gleaners get much wheat.

Aug. 20. Men house, & rick wheat in cold, damp condition.

Aug. 23. Martin's & swallows congregate by hundreds on the church & tower. These birds never cluster in this manner, but on sunny days. They are chiefly the first broods, rejected by their dams, who are busyed with a second family.

Aug. 25. The dripping season has, this day, lasted six weeks ; it has done some harm to the wheat,

& retarded wheat-harveſt; but has been of infinite
service to the grass, & turnips, &c.

Aug. 28.   Boys bring the 22nd, 23rd, & 24th wasps
neſt.   Many wasps at the plum-trees.

Aug. 29.   John Hale, & Farmer Spencer begin to
pick hops.

Aug. 30.   The kings field is open to the down.[9]
No mushrooms to be found with us: the case was
the same laſt year.

Sept. 1.   Dogs eat the goose-berries when they
become ripe; & now they devour the plums as they
fall; laſt year they tore the apricots off the trees.

Sept. 4.   Boys bring the 25th wasp's neſt.

Sept. 6.   Stormy wind, which broke-down great
part of my Orleans plum-tree, & blew-down Molly
White's horse-cheſt-nut, & did vaſt damage to the
hop-gardens, which are torn, & shattered in a sad
manner!   This ſtorm was very extensive, being very
violent at the same time at Lyndon, in Rutland.
Much mischief was done at London, & at Portsmouth,
& in Kent; at Brighthelmſtone also; & in Devonshire.

Sept. 8.   Mr S. Barker came.   Planted a *Parnassia*,
[grass of Parnassus] which he brought out of Rutland
in full bloom, in a bog at the bottom of Sparrow's
hanger.

Sept. 10.   Boys bring the 26th wasp's neſt.   Mens
second crop of clover cut, & spoiled by the rains.
A bad prospeƈt with respeƈt to winter fodder!
Farmer Spencer sows some wheat-ſtubbles with rye
for spring-feed.

Sept. 12.   Wasps much subdued.

Sept. 14.   Turned the horses into the great
meadow: there is a vaſt after grass, more than when
the meadow was mowed in the summer.

Sept. 15.   The dripping weather has laſted *this
day nine weeks*, all thro' haying, & harveſt: much hay
is also spoiled of the second cutting: so that men,
having loſt both crops, will in many parts be very

short of fodder, especially, as turnips have missed in many places.

Sept. 18.   A ring-ouzel shot on Hindhead.

Sept. 19.   No mushrooms : plenty in Rutland.

Sept. 21.   Bro^r Henry came.

Sept. 22.   Charles & Bessy White came.

Sept. 24.   Bro^r Henry left us.

Sept. 25.   Vast rain.   Violent current in the street.

Sept. 27.   My well, notwithstanding the rains is very low still, so that we let out all the rope to draw a bucket of water.[10]

Sept. 28.   Several   ring-ouzels   on   Nore   hill. Farmer Tull mows mill-mead, a second crop, which it is expected will produce near 3 tuns on an acre. Men mow also clover, hoping to get some hay at last.   Timothy the tortoise spends all the summer in the quarters of the kitchen-garden among the asparagus, &c. but as soon as the first frosty mornings begin, he comes forth to the laurel-hedge, by the side of which he spends the day, & retires under it at night ; 'till urged by the encreasing cold he buries himself in Novr amidst the laurel-hedge.

Sept. 30.   Will Tanner thinks he saw in the high wood marks where a wood-cock [Qy. woodpecker ?] had been boring.   Mr Barker, who rode this day to Rake, Rogate, & Furley-hill, saw much grass, & clover cut, & cutting.   Some barley out.

Oct. 4.   Bro^r Henry comes.

Oct. 6.   Gathered-in the swans-egg pears, a bushel ; more to be gathered.

Oct. 8.   Brother Henry, Bet, & Charles left us. Finished turning the mould in the mead.   Received from Mr Ed^d Woods 5 gallons, & 1 pint of French brandy.

Oct. 10.   Mr S. Barker left us.

Oct. 12.   The grass cut the last week in Septr all lies rotting.   My well begins to rise.   It has been so low all this autumn as not to afford water sufficient for the occasions of the family.   Had it not been for the frequent rains, we should have been at a loss, when we wanted to wash or brew.

Oct. 13.   [Alton] Barley abroad at Faringdon.

Oct. 14.   [Reading] Grey & mild.

Oct. 15-19   [Oxford] Hay lies about in Berkshire & Oxfordshire.

Oct. 17.   Timothy Turner finished the mowing of Bakers-hill.

Oct. 20.   [Alton] Much hay making all the way. Hay housing at Alton.

Oct. 21.   [Selborne] Timothy the tortoise lies in the laurel-hedge, but is not buried.

Oct. 22.   My well is risen six or seven yards.

Oct. 24.   Dug up my potatoes, a poor crop: many of them are rotten.

Oct. 27.   Water in the well very deep.

Oct. 28.   Saw seven ring-ouzels on the old hawthorns at Clay's barn.   Part of the hay in Baker's-hill was cocked & housed: it smells well, & is not so much damaged as might have been expected.

Oct. 29.   Snow lies on the hay-cocks in Baker's hill !

Oct. 31.   21 cocks of hay lying still in Baker's hill.

Nov. 1.   Bror Tho : Mr & Mrs Ben White, & Nep : Thos. Holt White came from Fyfield.

Nov. 5.   Wild *wood-pigeons* appear in a large flock in the coppices above Coombwood pond.   Timothy Turner housed the remainder of the hay in Baker's hill.   Dame Loe came.

Nov. 8.   A Gent : writes word from St Mary's, Scilly, that in the night between the 10th & 11th of this month, the wind being W. there fell such a flight of Woodcocks within the walls of the Garrison, that he himself shot & carryed home 26 couple, besides 3 couple which he wounded, but did not give himself the trouble to retrieve.   On the following day, the 12th the wind continuing W. he found but few.   This person further observes, that *easterly* & *Northerly* winds only have usually been remarked as propitious in bringing Woodcocks to those islands, viz. Scilly. So that he is totally at a loss to account for this *western* flight, unless they came from Ireland. As they took their departure in the night between the 11th & 12th, the wind still continuing West, he supposes they were gone to make a visit to the Counties of Cornwall & Devonshire.[11] From circumstances in the letter, it appears that the ground within the lines of the Garrison abounds with furze. Some Woodcocks settled in the street of St Mary's, & ran into the houses & out-houses.

Nov. 9. The great holly at Burhant-house is now beginning to blow.[12] Farmer Lasham finishes haymaking ! !

Nov. 11. Began to use celeri : it is very large, & somewhat piped [coarse and hollow]. Ring-ousel on the common.

Nov. 12. The ring-ouzel is killed by a hawk.

Nov. 13. Mr Ben White left us, & went to London.

Nov. 15. We find several pheasants in our walks. The hills thro' the fog appeared like vast mountains.

Nov. 16. Found some rasp-bushes on the down among the furze : & some low yew-trees, gnawn down by the sheep, among the bushes.

Nov. 17. Found the feathers of a ring-ouzel on the down, that had been killed by a hawk. Mrs Ben White left us & went to London.

Nov. 19. Harry Holt White left us, & went to Fyfield.

Nov. 21. Partridges associate in vast coveys.

Nov. 25. Mosses begin to grow, & look vivid ; & will begin to blow in a few weeks.

Nov. 28. We have had this week in Hartley-wood, & those parts a considerable flight of woodcocks : while in the upland coverts few or none were found.

Nov. 29. There was about this time, as the newspapers say, a vast flight of wood-cocks in Cornwall.

Nov. 30. Grapes are at an end.

Dec. 2.   Mem : to send Thomas on this day to Mr Collis colleƈtor of the excise. Bror Thomas White, & Tho : Holt White left us. Mrs & Miss Etty came home.

Dec. 5.   Some sportsmen beat the bogs of Wolmer-foreƨt carefully : saw but three brace of snipes.

Dec. 8.   Some few flights of wild fowl come to Wolmer-pond ; but do not ƨtay.

Dec. 9.   Swans egg pears continue good.

Dec. 12.   Young crickets of all sizes in my kitchen-chimney.

Dec. 17.   *Antirrhinum cymbalaria* thrives ƨtill, & is in full bloom, & will so continue 'till severe froƨts take place. Planted several firs from S. Lambeth : & several seedlings of the *Helleborus foetidus* [from the Hanger].

Dec. 18.   Sweet weather.

Dec. 19.   Cut down the artichokes, & covered them ; firƨt with earth, & then with long dung. Covered the asparagus with long dung.

Dec. 20.   Dug up carrots, second crop.

Dec. 21.   Planted 20 Scotch-firs round Benham's orchard.

Dec. 23.   Mr Churton came from Oxford.

Dec. 26.   Many wild-fowls, ducks & widgeons, at Wolmer-pond 'till the hard weather came : since which they have disappeared.

Dec. 27.   Tapped my new rick of hay this day, which, tho' made without rain, is vapid, & without much scent, & consiƨts more of weeds than grass. The summer was so dry, that little good grass grew, 'till after the firƨt crop was cut. The rick also is very small.

Dec. 31.   Snow covers the ground.

CHAPTER XIX

"The spirit walks of every day deceased."
(YOUNG: *Night Thoughts*, II, 180.)

1786

Jan. 3. Fierce frost. On this day at 8 o'the clock in the evening Captain Lindsey's hands were frozen, as he & Mr Powlett were returning from Captain Dumeresque's to Rotherfield. The Gent: suffered great pain all night, & found his nails turned black in the morning. When he got to Rotherfield, he bathed his hands in cold water. Snow on the ground six inches deep at an average.

Jan. 4. One of the most severe days that I ever remember with a S. wind. *Footnote.* The snow on wednesday [Jan. 4] proved fatal to two or three people who were frozen to death on the open downs about Salisbury. Much damage happened at sea about that time. In particular the Halsewell outer-bound India-man was wrecked, & lost on the shore of Purbeck.[1]

Jan. 5. The fierce drifting of wednesday proved very injurious to houses, forcing the snow into roofs, & flooding the ceilings. The roads also are so blocked up with drifted snow that the coaches cannot pass. The Wintõn coach was overturned yesterday near Alresford.

Jan. 9. Mr Churton left us.

Jan. 14. Sowed 36 bushels of peat-ashes on part of my farthest field, which has never been ashed since it was laid-down to grass. Qu: if it be right to sow ashes amidst so much rain & snow? So much moisture must probably dilute the ashes too much, & render them of no effect. Much snow on the ground. These ashes did no manner of good.

Jan. 18. Covered the spinage-bed with straw: the celeri & winter-lettuces are also covered.

Jan. 21. Snow gone.

Jan. 28. Mr Richardson, & son William came.

Jan. 31. Mr Richardson left us.

Feb. 1. The hazels are finely illuminated with male bloom. Female bloom of hazels appears, & the male-bloom sheds it's farina.

Feb. 3. The marsh-titmouse begins his two harsh, sharp notes.

Feb. 4. Sowed a good coat of ashes on Baker's hill, & also on the great meadow. Bought 40 bushels of ashes of Mrs Etty, & 36 bushels of sundry others. Sowed my own also.

Feb. 7. Driving rain. Strong flaws, & gusts with rain, hail, & thunder.

Feb. 14. Bullfinches eat the buds of honey-suckles.

Feb. 18. Pleasant season : paths dry. Men plough & sow. Large titmouse sings his three notes.

Feb. 20. Bror Henry & his son Gil : came.

Feb. 22. Sowed a crop of radishes, under the melon-screen : & a crop of onions.

Feb. 28. Bror Henry & his son left us.

Feb. 27. Snow shoe-deep. Wrote to Dr Chandler at Nismes.

Feb. 28. The snow is at an average about seven inches deep. As it fell without any wind, it is lodged much on the trees, so that the prospects are very grotesque, & picturesque.

Mar. 2. Bull-finches injure the fruit-trees by eating the buds.

Mar. 3. Netted the wall-cherry-trees, to preserve the buds from the finches.

Mar. 5. Vast icicles on eaves.

Mar. 6. The birds are so distressed, that ring-doves resort to my garden to crop the leaves of the bore-cole ! blackbirds come down to the scullery door. Snow little abated.

Mar. 7. Snow drifted over hedges, & gates! Ring-doves, driven by hunger, come into John Hale's garden, which is surrounded by houses! Black-birds, & thrushes die. A starving wigeon settled yesterday in the village, & was taken. Mention is made in the newspapers of several people that have perished in the snow. As Mr Ventris came from Faringdon, the drifted snow, being hard-frozen, bore his weight up to the tops of the stiles. The net hung over the cherry-trees is curiously coated over with ice.

Mar. 11. Snow wastes very fast. Roofs clear of snow. The ground appears. About this time my niece Brown was brought to bed of her fifth child, a girl, who encreases the number of my living nephews & nieces to 43.

Mar. 14. Took away the netting from the wall cherries.

Mar. 20. Sowed six-rows of garden-beans in the meadow; & two in the garden. *Chif-chaf* is heard : his notes are loud, & piercing.

Mar. 22. Some patches of snow still on the hanger : much snow in Newton hollow lane.

Mar. 26. Viper comes out. Two swallows were seen at Nismes in Languedoc : & on the 28th several, tho' the air was sharp, & some flakes of snow fell.[2]

Mar. 28. On this day the streets of Lyons were covered with snow.

Mar. 30. Mr Taylor[3] & his Bride came to Selborne.

Apr. 3. Earthed the cucumber-bed : plastered some fresh cow-dung under the hills. Sowed two ounces of carrot seed in the garden-plot in the meadow.

Apr. 4. Planted 1 doz : of white currans, & six goose-berry-trees, with many rasp-plants on the orchard-side of the bank. Turned-out the cucumber-plants into the hills of the bearing-bed; they are large & strong, & began to be too big for the pots. Sowed onions, & parsnips : the ground is dusty, & works well. 10 pots of Cucumber-plants remain. Sowed radishes, & lettuce. Planted one Roman, & one Newington Nectarine-tree against the fruit-wall.

Apr. 10. Planted 12 goose-berry-trees, & three monthly roses, & three Provence roses. M^r & Mrs Taylor left Selborne.

Apr. 13. Daws are building in the church. *Nightingale* sings in French-mere [at the north of Selborne parish].

Apr. 14. *Timothy* heaves up the mould, & comes out of his hibernacula under the wallnut-tree.

Apr. 16. Timothy the tortoise, after a fast of more than five months, weighs 6 ae. 12 oz. 11 dr. Some snow in Shalden lanes. Crown Imperial blows.

Apr. 17. Sowed a box with polyanth-seeds, our own saving.

Apr. 18. Men sow clover in their wheat.

Apr. 19. Sowed holly-hocks, columbines, & sweet Williams.

Apr. 20. Slipped out & planted many doz. of good polyanths. Young Geo: Tanner shot a *water-ouzel, merula aquatica,*[4] near Ja: Knight's ponds. This is the first bird of the sort that was ever observed in this parish. This bird, being only pinioned, was caught alive, & put into a cage, to which it soon became reconciled ; & is fed with woodlice, & small snails. W : ouzels are very common in the mountainous parts of the N. of England, & in N. Wales. They haunt rocky streams, & water-falls ; & tho' not web-footed often dive into currents in pursuit of insects.

Apr. 21. The voice of the cuckow as heard in the hanger.

Apr. 23. Grass lamb. Timothy, if you offer him some poppy leaves, will eat a little ; but does not seek for food.

Apr. 26. My hay is out [used up]. Many cock-pheasants are heard to crow on Wick-hill farm. We have a large stock of partridges left to breed-round the parish.

Apr. 27. Farmer Knight brought me ½ a ton of good meadow-hay.

Apr. 29. Red-breasts have young.

May 1. Bombylius minor [one of the bee-flies] appears.

May 2. [Alton] White frost, sun, cold air.

May 3. [Selborne] Made the annual-bed for a large three-light frame with 3 loads of dung.

May 4. Cut two fine cucumbers ; & began to eat the brown lettuces under the fruit-wall, where they stood the winter. Lettuce well-loaved, & very fine.

May 6. Great showers, & hail all round. Showers of hail at a distance look of a silvery colour. Rain-bow. The hanger is bursting into leaf every hour. A progress in the foliage may be discerned every morning, & again every evening.

May 8. Polyanths make a fine show. Pastures yellow with bloom of dandelion, & with cowslips.

May 9. Timothy, contrary to his usual practice lies out all day in the rain.

May 10. My grass is long enough to wave before the wind. Wheat turns some what yellow.

May 12. The water-ouzel is living, & recovered of its wound.

May 13. The wind beats the buds off the trees, & blows the cabbages out of the ground. The planet Venus appears. On this day my niece Clement was brought to bed of her fifth child, a boy, who makes my 44th nephew, & niece, all now living. [Two long notes on fossil-wood and the stone-curlew reproduced in D.B., LIX.]

May 15. Timothy began to march about at 5 in the morning.

May 17. Timothy Turner's Bantham⁵ sow brings 20 pigs, some of which she trod-on, & overlaid ;

so that they were soon reduced to 13. She has but
12 teats. Before she farrowed her belly swept on
the ground.

May 18. Dandelions are going out of bloom;
& now the pastures look yellow with the *Ranunculus
bulbosus*, butter-cups.

May 19. Mrs Yalden came. Many pairs of daws
build in the church: but they have placed their
nests so high up between the shingles, & the ceiling,
that ye boys cannot come at them. These birds go
forth to feed at ½ hour after four in the morning.

May 23. Slipped-out the artichokes, & earthed
them up. Mrs Yalden left us.

May 25. The prospect from my great parlor-
windows to the hanger now beautiful: the apple-
trees in bloom add to the richness of the scenery!
The grass-hopper lark whispers in my hedges. That
bird, the fern owl, & the nightingale of an evening
may be heard at the same time: & often the wood-
lark, hovering & taking circuits round in the air at a
vast distance from the ground.

    While high in air, & pois'd upon its wings,
    Unseen the soft, enamour'd wood-lark sings.[6]
Wood-larks in summer sing all night in the air.

May 26. Much gossamer. The air is full of
floating cotton from the willows. There are young
lapwings in the forest. Female wasps about: they
rasp particles of wood from sound posts & rails,
which being mixed-up with a glutinous matter form
their nests. Hornets collect beech-wood.

May 27. Mr Richardson came.

May 30. Honey-suckles begin to blow. Columbines very fine. Mr Richardson left us.

May 31. Swifts are very gay, & alert. Tulips
are gone off. Chafers abound: they are quite a pest
this year at, & about Fyfield.

June 1. Potted nine tall *balsams*, & put the potts
in a sunk bed. Dragon-flies have been out some
days. The oaks in many places are infested with

caterpillars of the *Phalaena quercus* [*Tortrix viridana*] to such a degree as to be quite naked of leaves. These palmer-worms hang down from the trees by long threads. The apple-trees at Faringdon are annoyed by an other set of caterpillars that strip them of all their foliage. My hedges are also damaged by caterpillars.

June 3. Daws from the church take the chafers on my trees, & hedges. Thomas picks the caterpillars that damage the foliage of the apricot-trees, & roll up their leaves.

June 6. Began to tack the vines; they are again infested with the cotton-like appearance which surrounds the eggs of the *Coccus vitis viniferae*. . . . For some account of this insect, see my Journal for summer last 1785.

June 9. [Alton] Captain Dumaresque cuts his S^t foin.

June 10. [Selborne] Men have a fine season for their turnip-fields, which work very well, & are well pulverized.

June 11. In Richd Butler's garden there is a Flycatcher's nest built in a very peculiar manner, being placed on a shelf that is fixed against the wall of an out-house, not five feet from the ground; & behind the head of an old rake lying on the shelf. On the same spot a pair of the same birds built last year; but as soon as there were young the nest was torn down by a cat.

June 13. Grey, sprinkling, gleams with thunder. Wavy, curdled clouds,[7] like the remains of thunder.

June 14-*July* 6. [South Lambeth] About Newton men were cutting their S^t foin: & all the way towards London their upland meadows, many of which, notwithstanding the drought, produce decent crops. We had a dusty, fatiguing journey. Bro: Thos has made his hay; & his fields are much burnt-up.

June 19. My brother's gardeners plant-out annuals. The ground is well moistened. They prick-out young cabbages, celeri, &c.

June 20. On this day Miss Anne Blunt, by being married to Mr Edm^d White, encreased the number of my nephews & nieces to forty & five.

June 22. [London] Jasmine in warm aspects begins to blow.

June 24. Wheat is in bloom, & has had a fine, still, dry, warm season for blowing. Nights miserably hot, & sultry.

June 25. Cauliflowers, Coss-lettuce, marrow-fat pease, carrots, summer-cabbage, & small beans in great profusion, & perfection. Cherries begin to come in : artichokes for supper. Bro^r Ben's outlet swarms with the *Scarabaeus solstitialis*, which appears at Midsummer. My two brothers gardens abound with all sorts of kitchen-crops.

June 27. Many of Bro^r Thomas's young fowls pine, & die ; & so they did last summer.

June 28. Bro : Thomas's gardener stops his vines, & tacks them. Bro : Ben's vines (those that came from Selborne) have very weak, scanty shoots. Bro^r Thos vines have good wood, & show for much fruit.

June 30. Bro^r Ben : cuts his Lucern a second time : the second crop is very tall.

July 3. The fruit of Dr Wesdale's great St. Germain pear swells, & grows large. Dwarf kidney-beans begin to pod. A cloud of swifts over Clapham : they probably have brought out their young. On this day Thomas got up all my hay in good order, & finished my rick, which contains eight good jobbs or loads ; at least six tuns. Thatched & secured my hay-rick. Two jobbs of the hay were from Baker's hill, the other six from the meadow, & slip. Baker's hill cut the 19th year : the Saint foin is got very thin, but other grasses prevail.

July 7. [Alton] Many swifts near Kingston. Vast rain at Bagshot. Hops are healthy round Alton, & Selborne.

July 8. [Selborne] The rick sweats, & fumes, & is in fine order. The pond at Faringdon is dry ; my well is very low, having been much exhausted by long waterings. Received five gallons, & a pint of brandy from Mr Edm^d Woods.

July 9. Roses, sweet-williams, pinks, white & orange lilies make a gaudy show in my garden. Annuals are stunted for want of rain. Mr White's tank at Newton measures three feet in water.

July 12. Gathered the wall-cherries, & preserved them with sugar : they are very fine.

July 15. Made jellies, & jams of red currans. Gathered broad beans. Mushrooms begin to come in Mr Edm^d White's avenue, under the Scotch firs. The cat gets upon the roof of the house, & catches young bats as they come forth from behind the sheet of lead at the bottom of the chimney.

July 17. Rye, & pea-harvest begins. Several nightingales appear all day long in the broad walk of Baker's hill.

July 18. Gathered & preserved some Rasps.

July 19. Oaks put-out their midsummer shoots, some of which are red, & some yellow ; & those oaks that were stripped by caterpillars begin to be cloathed with verdure. Many beeches are loaded with mast, so that their boughs become very pendulous, & look brown, I see no acorns. Selborne down is very rusty : the pond still is one part in three in water.

July 24-26. [Bramshot place] Mr Richardson's garden abounds with all sorts of crops, & with many sorts of fruits. His sandy soil produces an abundance of every thing ; & does not burn in droughts like the clays, which are now bound-up so as to injure the growth of all garden matters. The watered [irrigated] meadows at Bramshot flourish & ook green, the upland grass is much scorched. M^r R: has a pretty good show of Nectarines.

July 25. Pease are hacked : rye is reaping : turnips thrive & are hoing.

July 27. [Selborne] Saw a nightingale. Stifling dust.

July 29. Plums fail in all gardens. The sharp wind soon dries the surface of the ground. The wind damages the flowers, & blows down the apples, & pears.

July 30. Some hop-gardens injured by the wind of yesterday. Artichokes so dried-up that they do not head well.

Aug. 1. The poor begin to glean wheat. The country looks very rich, being finely diversifyed with crops of corn of various sorts, & colours.

Aug. 3. The fallows of good husbandmen are in a fine crumbling state, & very clean. Sowed a crop of prickly-seeded spinage to stand the winter : the ground was very hard & cloddy, & would not rake ; so we levelled it down as well as we could with a garden-roller, & sprinkled it over with fine, dusty mould to cover the seeds.

Aug. 6. Mrs Ben White, by being delivered of a boy this morning, has encreased my nephews, & nieces to the number of 46.

Aug. 15. Planted cuttings of dames violets, & slips of pinks under hand-glasses : planted also more sweet williams, & polyanths.

Aug. 16. Colchicum [meadow saffron] blows.
    Say what retards, amidst the summer's blaze,
    Th' autumnal bulb, 'till pale, declining days ?[8]

Aug. 19. Mushrooms come in Mr White's avenue at Newton.

Aug. 21. Kidney-beans bear by heaps ; & cucumbers abound. Coveys of partridges are said to be very large. Butchers meat keeps badly.

Aug. 22. Mushrooms are brought me from Hartley. I do not meet with one wasp. Young fern-owls are found, a second brood.

Aug. 23. We kept a young fern-owl for several days in a cage, & fed it with bread, & milk. It was moping, & mute by day; but, being a night bird, began to be alert as soon as it was dusk, often repeating a little piping note. [*Later entry*.] Sent it back to the brakes among which it was first found.

Aug. 26. Earwigs damage the wall-fruit before it gets ripe, warm & moist. Young fowls die at Newton. Mushrooms are brought in great plenty.

Aug. 27. Made five bottles, & a pint of catsup.

Aug 29. Tyed-up the unmoved endives.

Aug. 30. Hop picking-becomes general. The women earn good wages this year : some of them pick 24 bush: in a day, at 3 half-pence pr. bushel.

Sept. 2. [Account of producing echoes by swivel-guns; D.B., LX. The "friend" is Mr White of Newton.]

Sept. 4. Cut my new rick : the hay is good.

Sept. 6. Hirundines cluster on Hartley-house, & on the thatch of the Grange barn.

Sept. 8. Made a pint of catsup. Heavy rain.

Sept. 15. Golden-crowned wren, & the creeper, certhia, seen in my fields.

Sept. 17. Much damage has been done at sea & land by the late strong winds ; in particular about London. The vines were very forward in June : but now the grapes are quite backward, having made no progress in ripening for some weeks, on account of the blowing, black, wet weather. The bunches are of a good size, & the grapes large, & much want hot sunshine to bring them to perfection. My potted balsoms, which stand within, are still in beauty, tho' they have been blowing now more than three months. One in particular is more showy now than ever, & has such double flowers that they produce no seed. The blossoms are as large as a crown-piece.

Sept. 22. [Alton] Great dew, cold air, cloudless.

Sept. 23.   Gathered berberries.   Bro: Thomas & sons came.

Sept. 24.   Dame Loe [Low] came.

Sept. 25.   Niece Betsey came from Fyfield.

Sept. 26.   Saw a nest full of young swallows, nearly fledged, in their nest under Captain Dumaresq' gate way at Pilham-place.   Saw the same day many martins over Selborne village.   I have often seen young house-martins in their nests in the Mich: week;   but never swallows before.

Oct. 1.   About Octob.ʳ 1, the weather was cold & wet at Vevey, in Switzerland; when the Hirundines flew so near the ground as to be a prey to cats, which watched for them; & some entered mens windows so tame & hungry as to sit on a finger, & take flies when offered to them, or which they saw on the glass or walls.

Oct. 3.   Gathered-in the apples called *dearlings*, which keep well, & are valuable kitchen apples. My only tree of the sort stands in the meadow, & produced ten bushels of fruit.   Apples this year have sold at 8ˢ per bushel: so had the price continued the produce would be worth four pounds.   Next year probably there will be no crop; because I do not remember to have seen this tree bear two years following.

Oct. 4.   On this day an woodcock was seen in a coppice at Froyle.   Gathered-in the Royal-russets, & knobbed russets; the former are fine shewy apples. There is a good crop of each sort.

Oct. 7.   The great rains do not influence our wells in the least.   Niece Betsey returned to Fyfield.   On this day Miss Mary Haggitt of Rushton, Northamptonshire, by being married to my Nephew Sam Barker [R.H.W., II., pp. 158, 160] encreased the number of my nephews & nieces to 47.

Oct. 8.   "We saw a great number [of swallows] on the wing at Rolle: & about that time their departure seems to have been general."⁹

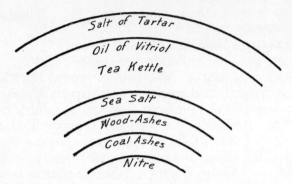

With the above mentioned articles Broʳ Thoˢ has attempted to make a fairey-ring,[10] circle within circle : & we are to take notice in the spring which circle, & whether any, will produce grass of a deeper green than before. The tea kettle which has occasioned the dots was set-out, time after time, full of boiling water. The circles made with oil of vitriol, with sea salt, & with salt petre have discoloured the grass : those with Sal. Tartar, wood, & coal ashes have no visible effect at present. [*Later note.*] The grass seems killed where the tea-kettle stood.

Oct. 9. Nep: Tom, & Harry White went to Fyfield.

Oct. 11. The news papers mention vast floods about the country ; & that much damage has been done by high tides, & tempestuous winds.

Oct. 14. Men sow wheat in good order at Temple, & Wick-hill [Selborne parish]. The hop-planters of this parish returned from Wey-hill fair with chearful faces, & full purses ; having sold a large crop of hops for a good price. The hops of Kent were blown away by the storms, after the crop of this country was gathered in.

Oct. 15. Prodigious damage appears to be done by the late tempests, all over the kingdom ; & in many places abroad.

Oct. 16. Broʳ & Sister Benjⁿ came.

Oct. 19. Men pull up turnips, & stack them. My balsoms in pots, that have been in bloom four months, now begin to fade.

Oct. 20. Rover [White's dog] springs several pheasants in Harteley-wood. We find many large coveys of partridges.

Oct. 22. Bro^r & Sister Benjn left us.

Oct. 23. *Red-wings* are late this autumn. Perhaps the vintage was late this year in Germany ; so that these birds were detained by the grapes, which they did not wish to exchange for our hips, & haws. Red-wings do much damage in vineyards, when the grapes are ripe. My tall hedges, & the hanging woods, do not shew their usual beautiful tints, & colours : the reason is because the foliage was so much torn & shattered by the rains & tempests.

Oct. 26. Several wet, floated fields are now sown, that must have missed their wheat-crop, & have lain 'till spring, had not this fine dry season drained them, & rendered them fit for sowing.

Oct. 30. Rover springs several pheasants, & some coveys of partridges.

Nov. 3. The oaks in Comb-wood & below the Temple are in full leaf, & many of them in good verdure. The beeches in general have lost their foliage.

Nov. 5. Here & there a wood-cock is found. Sweet moonshine. [Evening behaviour of the rooks, D.B., LIX.] The beautiful planets Jupiter, & Venus appear now of an evening soon after sunset almost opposite, the former in the N.E. & the latter in the S.W.

Nov. 8. Mrs Etty returned to Selborne.

Nov. 9. This day compleats the 28th of this dry fit, which has done infinite service to the low districts, that were floated with water by the heavy rains in the beginning of last month.

Nov. 12. The hogs gave been turned for some weeks into the high-wood, & hanger, where they have availed themselves much of the large crop of beech-mast. The hogs find, no doubt, many trufles in the high-wood, where they are said to abound. Last week Wolmer-pond was sewed [drained off], & fished after an interval of more than 20 years. And yet there was no quantity of fish; for the carps did not amount to one hundred; nor was there any young stock: tench there were none; many young perch; a few large, lank pikes; & a few large eels. It is said that the pond is to lie a-sew [as drained] all next summer. The pond being an area of more than 60 acres, was several days in running dry. If this pond continues dry next spring, more Roman coins[11] may be found, in windy weather, on the surface of the sand. Many hundreds were found when it last lay dry, about the year 1741.

Nov. 14.   Boys slide on the Ice! Flocks of *hen*-chaffinches are seen.

Nov. 15.   Covered the rows of the celeri with straw. This day compleats the 34 of the dry weather. Horses begin to lie within.

Nov. 16.   I have often observed many titmice in beechen woods: by a heap of beech-mast now lying in my orchard I see that t: mice feed on the kernels of the fruit of that tree; & that *marsh-titmice* are employed all day in carrying them away.

Nov. 21.   Bought 61 bushels of peat-ashes, & laid them up in the ash-house.

Nov. 22.   I sent a woman up the hill with a peck of beech-mast which she tells me she has scattered all round the down amidst the bushes & brakes, where there were no beeches before. I also ordered Thomas to sow beech-mast in the hedges all round Baker's-hill.

Nov. 26.   Mr Cane[12] saw in one flock some hundreds of whistling plovers on the downs.

Nov. 27-30.   [Fyfield] Grey crows on the downs.

Nov. 28.   Mr Talbot turned-out a stag, which after wounding some hounds, & an horse, was taken alive.

Dec. 1.   [Winton] The downs are very heavy.

Dec. 2.   [Selborne] Several white gulls, as usual, wading about in the stream beyond Alresford.

Dec. 13.   Peter Wells's well is 36 feet deep, my own 63.   Peter Wells's well runs over: when this is the case, the springs are very high. [*Later.*] This overflowing lasted only two or three days.

Dec. 15.   A cellar in the back-street Faringdon is full of water.

Dec. 16.   The walks in my fields are strewed with the berries of misseltoe, blown from the hedges.

Dec. 22.   Ice in my chamber.

Dec. 23.   Snow all day.

Dec. 26.   Mr Churton came.

Dec. 31.   Deep fog, grey, & mild, & moist.

[*Fly-leaf.*]

Revd Mr Randolph[13]

Dr.				Cr.			
1783	ae	s	d	1783	ae	s	d
To p^d Beagley, brick-layer at Faringdon	2 :	5 :	2	By bank-bill sent by post	10 :	0 :	0
To p^d J. Finden, carpenter at Do. ...	4 :	2 :	0				
To p^d Shawyer, brick burner ...	1 :	6 :	9				
	7 :	13 :	11				
Xmass: 1783							
To p^d to sundry poor	2 :	2 :	0				
	9 :	15 :	11				
Receipted 1784							
To balance received Septr 13 ...	0 :	4 :	1				
	10 :	0 :	0				
Herb^t Randolph							

CHAPTER XX

"Line upon line, here a little, and there a little."

(Isa. xxviii, v. 10.)

1787[1]

Jan. 1.   Slept at Newton.

Jan. 3.   On this evening there was a total eclipse of the moon : but the sky was so cloudy, that we saw nothing of the progress of it.

Jan. 6.   Paths dry : boys play at taw on the Plestor.

Jan. 8.   Wheeled dung into the garden, & to the basons in Baker's hill.   Mr Churton left us.

Jan. 11.   This afternoon I saw at the house of my neighbour Mr Burbey, [village shopkeeper] 54 young girls, which he entertained with tea, & cakes : they were, except a few, natives of this village.

Jan. 17.   Strong aurora.

Jan. 19.   Mice eat the crocus roots.

Jan. 31.   Small frost, sun, still, & pleasant. Beautiful dappled sky.

Feb. 2.   Storm-cock sings.   Brown wood-owls come down from the hanger in the dusk of the evening, & sit hooting all night on my wall-nut trees. Their note is like a fine vox humana, & very tuneable. The owls probably watch for mice about the buildings. White owls haunt my barn, but do not seem to perch often on trees.

Feb. 10.   Took Mrs Etty's ashes, 28 bushels ; paid her.

Feb. 17.   Cucumber-plants thrive, & are pinched to make them throw-out side shoots.

Feb. 19.   Sowed Baker's hill, & the great mead with ashes.   Crocus's make a glorious show : bees much out.   The air-full of insects, & gossamer. *Bat* appears.

Feb. 21.   Male yew-trees shed clouds of farina.

Feb. 22.   Planted five rows of long-pod beans in the meadow-garden.   The air somewhat sharp.

Feb. 23.   On Feb. 23 the cuckow was heard at Rolle in Switzerland.   Rooks build at Faringdon parsonage.

Feb. 27.   On this day my niece [wife of] Edm^d White was delivered of a daughter, who encreases my Nephews, & nieces to the number of 48.

Mar. 8.   Frost, sun, pleasant, spring-like. Pile-wort [lesser celandine] blows.

Mar. 12.   Dogs-toothed violets begin to blow.

Mar. 14.   The male-bloom of the cucumbers opens : the bed is warm, & the plants thrive.   Planted more roses from South Lambeth.

Mar. 16.   The cats brought-in a dead house-martin from the stable.   I was in hopes at first sight that it might have been in a torpid state ; but it was decayed, & dry.   Polyanths blow.   Jet-ants appear.

Mar. 18.   Timothy the tortoise heaves up the earth : he lies under the wall-nut tree.

Mar. 19.   Women hoe wheat.   Gossamer abounds. Sowed a bed of Celeri under a hand-glass.

Mar. 20.   Sent me from South Lambeth, two Nectarine-trees ; several sorts of curious pinks ; some mulberry-rasps ; some scarlet lichnis's : a root of Monk's rhubarb [*Rumex alpinus*].

Mar. 22.   The tortoise comes forth from his hole. Men open their hop-hills & cut their hops.

Mar. 23.   Timothy hides his head under the earth.

Mar. 26.   Transplanted some of the best, blowing seedling polyanths from the orchard to the bank in the garden.   Planted some scorpion-sennas [*Coronilla Emerus*] from S. Lambeth.

Mar. 27.   Swallows were first seen this year at Messina in Sicily.

Mar. 28.   Timothy continues to lie very close.

Mar. 29. Some swallows were seen over the lake of Geneva, & at Rolle. On March 30 several were seen at the same place.

Mar 30. Chaffinches pull off the blossoms of the polyanths, which are beautifully variegated.

Mar. 31. Turner shoots the chaffinches. Mackarals come. Bantam-hen lays. Black & grey snails without shells.

Apr. 1. Crown imperials, double hyacinths, cherries against walls, blow.

Apr. 2. Lined the back of the Cucumber-bed with hot dung.

Apr. 3. Cowslips blow under hedges.

Apr. 6. *Stone-curlews* pass along over my house of an evening with a short quick note after dark. *Wry-neck* pipes in the orchard. Nightingale sings at Citraro in the nearer Calabria.[2]

Apr. 8. Mrs Clement's daughter, born this day, makes my nephews & nieces 49.

Apr. 9. [Alton] Sun, sharp wind.

Apr. 10. [Reading] Cloudless. Goslings. No swallows.

Apr. 11-15. [Oxford.]

Apr. 13. Sam White[3] elected fellow of Oriel College in Oxford.

Apr. 16. [Caversham] Men are busy in their barley season.

Apr. 17. [Alton] Pears, cherries, & plums in fine bloom along the road. Some hundreds of martins were seen to pass over Rolle towards Geneva, & two swifts : the day was wet & cold.

Apr. 18. [Selborne] Cut a *brace* of fine *cucumbers*.

Apr. 21. Mowed the grass-walks in part : they were crisp with hoar frost. Cut some grass in the orchard for the horses. Swallow on a chimney.

Apr. 23. *Cuckow* sings on the hill. Nightingale sings in my outlet.

Apr. 25. Sent 9 bantam's eggs to be put under a sitting-hen at Newton.

Apr. 28.   Set Gunnory, the Bantam hen, on nine of her own eggs.

Apr. 30.   April 30 was cold & sharp at Rolle; when a number of martins formed two thick clusters on a ledge projecting in front of an house in one of the streets of that town.   They descended gently as they arrived one on another.

May 1.   Young brown owls.

May 2.   The foliage of peach, & nectarine-trees scorched by the winds : the leaves are shrivelled, & blotched.

May 4.   Sowed a plot of red beet.

May 5.   Sowed ten weeks stocks, & radishes, & lettuces.

May 6.   Timothy, the tortoise, who has just begun to eat, weighs 6 ae, 12½ oz.   Agues are much about ; at Hawkley, & Emshot, & Newton ; & in Selborne street.

May 7.   The large white pippin-tree full of bloom. No house-martin seen yet !

May 10.   Farmer Spencer's orchard in fine bloom.

Mav 12.   House martin appears : only one.

May 13.   Ice at Nore-hill.   Tulips make a show.

May 14.   [Marelands   (near   Farnham)]   Dew, bright, showers : thunder, gleam of sun.

May 15-*June* 14.   [South Lambeth] The showers did not extend to the east beyond Cobham.

May 16.   Agues abound around S. Lambeth. Cucumbers not plenty.

May 18.   Leaf-cabbages very fine.   Spinage good.

May 20.   The red-start sits, & sings on the fane [vane] in Bro : Ben's garden upon the top of an high elm.

May 21.   Mr *Charles Etty* returns from Canton. He left England in March 1785, & sailed first for Bombay.   White-thorn bloom fragrant.

May 22.   Medlars blow. Mushrooms in a bed under a shed in Brother Thomas's garden.

May 23. A pair of red-backed *Butcher-birds, lanius collurio,* have got a nest in Bro : Tho : outlet. They have built in a quickset-hedge. We took one of the eggs out of the nest : it was white ; but surrounded at the big end by a circle of brown spots, coronae instar.

May 24. Bro : Ben cuts three rows of Lucern daily for his three horses : by the time that he has gone thro' the plot the first rows are fit to be cut again.

May 27. [Note on nests of chaffinch and wren : see D.B., LVI.]

May 30. *Lactuca virosa* [stinking lettuce] spindles for bloom : the milky juice of this plant is very bitter, & acrid.

June 1. Some fly-catchers : but they do not yet begin to build. Carrots drawn.

June 2. Hay is making at Vaux-hall.

June 3. Bro^r Thomas cuts cauliflowers. The foliage on the Lombardy-poplars is very poor.

June 4. Bro^r Ben cuts his hay. Pease are cryed about at 1s. 6d. per peck. Kidney-beans & potatoes are injured by the frost of saturday night.

June 5. [London] The tortoise took his usual ramble, & could not be confined within the limits of the garden. His pursuits, which seem to be of the amorous kind, transport him beyond the bounds of his usual gravity at this season. He was missing for some days, but found at last near the upper malt-house.

June 7. Ice thick as a crown piece. Potatoes much injured, & whole rows of kidney-beans killed : nasturtiums killed.

June 10. The gale rises, & falls with the sun. Levant weather. Some house-martins at Stockwell-chappel.

June 11. Straw-berries, scarlet, cryed about. Straw-berries dry, & tasteless. Quail calls in the field next to the garden.

June 12. A poor gardener in this parish [S. Lambeth] who had three acres of kidney-beans, has lost them all by the frost of last week! Hay finely made, & making. The rudiments of the vine-bloom does not seem to be injured by the late frost.

June 13. The late frost, I find, has done much damage in Hants.

June 15. [Alton] Field-pease in fine bloom. Many swifts at Wansworth, Kingston, Cobham, &c. Hay-making general about London; some meadow hay cut at Farnham.

June 16. [Selborne.]

June 18. A pair of fly-catchers build in my vines. The late frost did much damage at Fyfield, but little or none at Selborne. My potatoes, kidney-beans, & nasturtiums were not injured: some balsoms, that touched the glasses, were scorched.

June 22. Netted the wall-cherries. Boys bring wood-strawberries; not ripe.

June 23. Brood of nightingales frequents the walks. The number of swifts are few, because they are stopped-out from the eaves of the church, which were repaired last autumn. The nest of a *Flusher*, or red-backed *Butcher*-bird was found near Alton. Pease, barley, & oats look well, especially the first, which show fine bloom: wheat looks but poorly. Wheat at market rises. Sheep are washed.

June 25. Nep. and niece Ben White brought little Ben.

June 27. A brood of little partridges was seen in Baker's hill among the Sainfoin.

June 29. Gracious street pond dry, & cleaned-out. Much water in the pond on the hill. The pond at Faringdon dry.

June 30. [Note on the mode in which the squirrel, the field-mouse, and the nuthatch open nuts. See D.B., LVI.] The space behind my alcove is covered with the shells of nuts which the bird [nuthatch] had bored after he had fixed them in the corners of the cornice of that edifice.

July 4. Timothy Turner cuts Baker's hill, the 20th crop : over ripe.

July 5. Flowers hurried, & injured by the heat. Curious pinks.

July 7. Preserved some Duke cherries, very fine fruit. The pupils of the eyes of animals[4] are diversifyed : in all the birds & fishes that I have seen they are round, as in men : but those of horses, & cows, & sheep & goats & I think deer & camels, are oblong from corner to corner of the eye. The pupils of the domestic cat differ from those of all other quadrupeds ; for they are long & narrow, yet capable of great dilatation, & standing near at right angles with the opening of the eye-lids. The eyes of wasps are said to be lunated in shape of a crescent.

July 11. Planted a line of kidney-beans.

July 13. The apricots drop off in a surprizing manner. Planted a bed of Savoys.

July 14. Hops are dioecious plants : hence perhaps it might be proper, tho' not practised, to leave purposely some male plants in every garden, that their farina might impregnate the blossoms.[5] The female plants without their male attendants are not in their natural state : hence we may suppose the frequent failure of crop so incident to hop grounds. No other growth, cultivated by man, has such frequent & general failures as hops.

Daniel Wheeler's boy found a young fledge cuckow in the nest of an hedge-sparrow. Under the nest lay an egg of the hedge-sparrow, which looked as if it had been sucked. In the late hot weather the cock bird has been crying much in the neighbourhood of the nest, but not since last week.

July 15. Mr White of Newton finds mushrooms in his fir-avenue. *Tremella* abounds on my grass-walks.

July 16. The hedge-sparrow feeds the young cuckow in it's cage.

July 20. Planted-out African Marigold's, & sunflowers. Made a fire in the parlor. Kept the horses within because of the cold, & wet. Sam White entered on his year of probation at Oriel College in Oxford.

July 21. Vast crop of goose-berries : white currans very fine.

July 22. Mushrooms appear on the short Lythe.

July 23. Young red-breasts, a second brood. Notwithstanding the showery season, the aphides encrease on the hops.

July 26. The farmers talk much that wheat is blighted. Kidney-beans do not thrive.

July 27. Rooks in vast flocks return to the deep woods at half past 8 o'clock in the evening.

July 30. Wheat-harvest will be backward. Mr White's tank at Newton runs over; but Captain Dumaresque's, which is much larger, is not full.

July 31. Vast rain, an inch & quarter in 8 hours.

Aug. 1. Several golden-crowned wrens appear in the tall fir-tree at the upper end of Baker's hill : they were probably bred in that tree.

Aug. 10. When the *redbreasts* have finished the currans, they begin with the berries of the honey-suckles, of which they are very fond.

Aug. 11. The children in strawberry time found & destroyed several pheasant's nests in Goleigh wood. Nep. and Niece Ben White came from London.

Aug. 12. Bull-finches feed on the berries of honey-suckles. B. Hall came.

Aug. 13. M^r & Mrs Richardson & son came.

Aug. 14. Gleaning begins.

Aug. 16. M^r & Mrs R. left us. Farmer Parsons harvests wheat. Gleaners carry home large loads.

Aug. 19. Showers about : Rain-bows. Vivid Aurora.

Aug. 20. Nep. T. H. Wh : came from Fyfield.

Aug. 23.   Much wheat carried.   The Ewel, & Pound-field thrown open [to gleaners]. Cool autumnal feel.   Nightingales seen in Honey-lane : they were the laſt that I observed.   Cut at one time 191 fine cucumbers.

Aug. 24.   Nep. Ben White left us.

Aug. 26.   Timothy the Tortoise, who has spent the two laſt months amidſt the umbrageous foreſts of the asparagus-beds, begins now to be sensible of the chilly autumnal mornings ; & therefore suns himself under the laurel-hedge, into which he retires at night.   He is become sluggish, & does not seem to take any food.

Aug. 27.   Molly White & Nep. Tom rode to Fyfield.

Aug. 31.   Young hirundines cluſter on the dead boughs of the walnut tree.

Sept. 1.   The shooters find many coveys, but not large ones.   Not one wasp.

Sept. 3.   Broʳ Thoˢ sons & daughter came from Fyfield.

Sept. 4.   Vaſt numbers of partridges.   A young fern-owl shot at Newton.

Sept. 5.   Stone-curlews pass over followed by their young, which make a piping, wailing noise.

Sept. 8.   Mʳˢ Brown [Mary Barker] brought to bed of a boy, who added to 49 before, encreased my nephews & nieces to the round number of fifty.

Sept. 9.   [Note about Chinese dogs : D.B., LVIII. The following notes are additional.] Mʳ Charles Etty has brought home with him from Canton. . . . They might probably therefore become springing-spaniels ; but have more the appearance of vermin-hunters. . . . Mʳ Etty's Chinese dogs bark in a thick short manner like foxes.   Mʳ E. says that Quiloh is Chinese for a dog.

Sept. 10.   Hops so small that a notable [clever, capable] woman & her girl can pick but nine bushels in a day, where laſt year they could pick 20.

Sept. 11.   Cow-grass housed.   Gathered heaps of Cucumbers.

Sept. 12.   Lapwings leave the low grounds, & come to the uplands in flocks.   A pair of honey-buzzards, & a pair of wind-hovers appear to have young in the hanger.   The honey-buzzard is a fine hawk, & skims about in a majestic manner.

Sept. 13.   Gathered-in my early white pippins.

Sept. 15.   Women make poor wages in their hop-picking.   Housed all my potatoes, & tyed up many endives.

Sept. 18.   Mr Churton came from Cheshire.

Sept. 19.   Nep. Ben, & wife left us, & went to London.

Sept. 20.   Saw pheasants at long coppice.   My well sinks much, & is very low.

Sept. 21.   Vast halo round the moon.   Began fires.

Sept. 22.   Guns are heard much from Portsmouth.

Sept. 23.   Began to use the spinage sown the first week in August : very fine & abundant.

Sept. 24.   Many *swallows*, & some *bank-martins* at Oakhanger-ponds.   A multitude of *swallows* at Benes-pond [Wolmer Forest] ; & some few *house-martins*, which probably roost on the willows at the tail of that pond.   The swallows washed much ; a sure sign that rain was at hand.

Sept. 26.   Many ravens on the hill, & a flight of starlings.

Sept. 29.   Vast flock of ravens on the down.

Oct. 1.   Wheat not so good as last year : 50 sheaves do not yield more than forty did this time twelve month.

Oct. 3.   Men sow wheat ; but wish for more rain to moisten their fallows.   The quantity of potatoes planted in this parish was very great, & the produce, on ground unused to that root, prodigious.[6]   David Long had two hundred bushels on half an acre. Red or hog potatoes are sold for six pence p^r bushel. Mr Churton left us & went to Waverly [near Farnham] : Nep. Tom & Harry left us, & went to Fyfield.

Oct. 5. Bro^r Ben & wife came. Put my fine hyacinths into a bed, that were taken-up in the summer. Put also some good tulips, & striped crocus's from Bro : Tho^{s's} garden into beds.

Oct. 6. My well is very low ; & the stream from Gracious street almost dry.

Oct. 7. My tall, streight Beech at the E. corner of Sparrow's hanger, from a measurement taken by Rich: Becher & son, proves to be exactly 74 feet & ½ in height. The shaft is about 50 feet without a bough.

Oct. 8. One waggon carries this year all the Selborne hops to Weyhill : last year there were many loads. Jack Burbey's brown owl washes often when a pan of water is set in it's way. *Wood-cock* killed at Bramshot.

Oct. 9. Timothy sets his shell an edge against the sun. The best Selborne hops were sold for 15 pounds, & 15 Guineas per hund :

Oct. 12. Partridges, & pheasants are very shy, & wild. Bro^r Ben & wife left us & went to Newton.

Oct. 13. We saw several *Red-wings* among the bushes on the N. side of the common. There were swallows about the village at the same time : so that summer & winter birds of passage were seen on the same day.[7] The aurora was very red & aweful.

Oct. 14. [Note on the barometer at Newton: D.B., LX., with the information that it was fixed by " Bro^r Thos & Nep. Edm^d White ".]

Oct. 17. Gathered-in the last apples, in all about 8 bushels. Planted 100 cabbages to stand the winter.

Oct. 21. William Dewye Senr who is now living, has been a certificate man[8] at Selborne since the year 1729, some time in the month of April. He is a parishioner at the town of Wimborn-Minster in the County of Dorset.

Oct. 23. The number of partridges remain very great. Pheasants do not abound.

Oct. 28. Sam White saw three swallows at Oxford near Folly bridge.

Oct. 29.   About four o'clock this afternoon a flight of *house-martins* appeared suddenly over my house, & continued feeding for half an hour & then withdrew. Some thought that there were swallows among them.

Oct. 30.   Bro^r Thos. left us, & went to London.

Nov. 1.   Split-out the great Monk's-rhubarb plant into 7, or 8 heads, & planted them in a bed that they may produce stalks for tarts in the spring.   The N. Aurora made a particular appearance forming itself into a broad, red fiery belt, which extended from E. to W. across the welkin : but the moon rising, at about 10 o' the clock, in unclouded majesty on the E. put an end to this grand, but aweful, meteorous phenomenon.

Nov. 2.   Farmer Hoar saw one cock ring-ouzel at Nore hill.

Nov. 6.   Several wood-cocks in Harteley wood : they are in poor condition.

Nov. 7-15.   [Fyfield] Saw several grey crows on the downs between Winchester, & Andover ; & four pheasants feeding at the corner of Whorwel-wood. Green wheat beautiful on the downs, but not forward sown.   The Fyfield Comedians performed *Much ado about nothing, with the Romp*.[9]

Nov. 8.   Rain, blowing, & showers.   Stormy, hail.   Red, turbid N. Aurora.   Bro^r Henry's grapes did not ripen well.

Nov. 13.   The Fyfield players performed *Richard the third*.

Nov. 14.   The late hard winters killed the extreamities of my wall-nut trees, so that they have born no fruit since : but the same severe seasons killed many of the fyfield wall-nut trees quite down to the ground.

Nov. 16.   [Wintŏn] The stream at Fyfield encreases very fast.   Spent three hours of this day, viz. from one o' the clock till four, in the midst of the downs between Andover & Wintŏn, where we should have suffered greatly from cold & hunger, had not the day proved very fine, & had not we been opposite to the

house of Mr Treadgold's down farm, where we were
hospitably entertained by the labourer's wife with
cold sparerib, & good bread, & cheese, & ale, while
the driver went back to Andover to fetch a better
horse. The case was, the saddle-horse being new to
his business, became jaded and restiff, & would not
stir an inch; but was soon kept in countenance by
the shaft-horse, who followed his example : so we were
quite set-up 'till four o' the clock, when an other
driver arrived with an other lean jaded horse, & with
much difficulty assisted in dragging us to Wintŏn,
which we did not reach till six in the evening. We
set out from Fyfield at eleven; so we were seven
hours in getting 19 miles. During our long conversation with the dame, we found that this lone farm-
house & it's buildings, tho' so sequestered from all
neighbourhood, & so far removed from all streams,
& water, are much annoyed with Norway rats :[10]
the carter also told us that about 12 years ago he had
seen a flock of 18 bustards at one time on that farm,
& once since only two. This is the only habitation
to be met with on these downs between Whorwel
& Winchester.

Nov. 17. [Selborne] Sam White was chosen by the
favour of the Provost & fellows of Oriel Coll :
Bishop Robinson's Exhibitioner. This advantage
will last him three years, 'till he takes his Mrs degree.

Nov. 24. Housed all the billet-wood in dry, good
order. Covered the lettuce under the fruit-wall with
straw.

Nov. 26. Monthly roses now in bloom.

Nov. 28. Children slide on the ponds. Rake up,
& burn the leaves of the hedges.

Nov. 30. Frost comes within door : ice in the
pantry, & chambers.

Dec. 3. The yellow Bantham-pullet begins to lay.

Dec. 6. Five or six bats were flying round my
chimnies at the dawn of the day. Bats come forth
at all times of the year when ye Thermr is at 50,

because at such a temperament of the air.   Phalaenae
are stirring, on which they feed.

Dec. 11.   Bought a strong, stout white Galloway
mare, that walks well, & seems to be gentle.   She was
lately the property of Mr Leech, Surgeon at Alton,
deceased.

Dec. 15.   Began to cut my new hay-rick.

Dec. 21.   Shortest day.   Pleasant weather.   A
hunted hind came down Galley-hill into the street;
where being headed by the village dogs, it turned
back to Well-head, & was taken in Kircher's farm
yard, [See map] & put into the barn, being quite
run down.   One of the Gent: pursuers let it blood,
& hired a man to watch it all night.   In the morning
by seven o' the clock a deer-cart came, & took it
away.   There were several Gent: in with the dogs,
when they took the deer.   The dogs & hind were
said to belong to Mr Delmee, who lives near Fareham.
The deer was turned-out in the morning on Stevens
Castle down near Bishop's Waltham, which is at
least 18 miles from this place.   The dogs were short
& thick, but had shrill notes like fox-hounds, &
when they ran hard opened[11] but seldom, so that
they made but little cry.

Dec. 24.   Deep snow.   The Bantham fowls, when
first let out, were so astonished at the snow that they
flew over the house.

Dec. 25.   The snow, where level, about one foot
in depth: in some places much drifted.

Dec. 27.   A *musca domestica*, by the warmth of my
parlor has lengthened out his life, & existence to this
time: he usually basks on the jams of the chimney
within the influence of the fire after dinner, & settles
on the table, where he sips the wine & tastes the
sugar & baked apples.   If there comes a very severe
day he withdraws & is not seen.

Dec. 30.   Some of our hollow lanes are not
passable.

CHAPTER XXI

"So word by word, and line by line,
The dead man touch'd me from the past."

In Memoriam, XCV.

1788

Jan. 1. Contracted my great parlor chimney by placing stone-jams on the top of the grate on each side, & building brick-work on the jams as high as the work-man could reach. This expedient has entirely cured the smoking, & given the chimney a draught equal to that in the old parlor.

Jan. 3. A flood at Gracious street.

Jan. 7. The woodmen begin to fell beeches in the hanger for the second time : they now enter where they left off last year on the S.E. side of Shop-slidder.[1]

Jan. 8. Old John Carpenter planted a Sycamore tree on the Plestor near the pound. Furze blows. The amenta [catkins] of hasels open & shed their farina. Ivy-berries swell.

Jan. 9. Mr Churton left us & went to Waverly.

Jan. 10. [Newton] The dry summer killed the new planted Sycamore on the Plestor.[2]

Jan. 16. Some thaws. Load of straw brought in.

Jan. 19. Received from London a quarter of an hundred of Salt fish.

Jan. 20. Bro^r Thomas came from Fyfield.

Jan. 23. Dan. Wheeler plants his field with beans.

Jan. 26. Salt-fish proves good. The Creeper, certhia familiaris, appears in my orchard, & runs up the trees like a mouse. Golden-crowned wren is also seen.

Jan. 27. Snow lies still a yard deep at Stair's hill.

Jan. 28.   Several snow-worms found under the bottom of an old hay-rick in a torpid state, but not without some motion.

Jan. 29.   Rover sprung two brace of pheasants in long coppice.

Jan. 31.   Tubbed half an hog, weighing 8 score : put half a bush. of salt, & two ounces of salt petre. The pork was well trod into the tub, & nicely stowed.

Feb. 1.   Received a brace of pheasants from Wood-house farm.

Feb. 2.   Second Bantham pullet lays.   Cucumbers sprout.

Feb. 4.   Bro^r Thomas left us, & went to London.

Feb. 5.   A couple of woodcocks were given me.

Feb. 14.   Bro. Thomas & Molly came from London.

Feb. 15.   Taw & hop-scotch come in fashion among the boys.

Feb. 17.   M^r Ch : Etty sailed for Bombay, & Canton aboard the Montrose India man in the capacity of third mate.   [*Later.*] He was delayed, & did not sail till the last week in March.

Feb. 18.   Turnip-tops come into eating. The ground dries at a wonderous rate.

Feb. 19.   Bro^r Thos & Molly White went to London.   M^r & Mrs Clements & two children came, with *Zebra*,[3] the Nurse-maid.

Feb. 21.   Full moon. Barom. at Newton 28! no wind.

Feb. 22.   Mrs Edm^d White was brought to bed of a boy, who makes my nephews & nieces 51 in number.

Feb. 24.   Partridges, & missel-thrushes are paired. M^r & Mrs Clement left us.

Feb. 27.   Dug up the suckers of rose-trees, & planted them in a nursery.

Feb. 28.   Sowed the great mead, & Baker's Hill with a good dressing of ashes : of my own 31 bushels ; bought 54.

Feb. 29.   Spread Mrs Etty's coal-ashes on the bank, & other borders.

Mar. 2.   A strong smell of London smoke.

Mar. 3.   A squirrel in my hedges.

Mar. 8.   Earthed the bearing cucumber-bed. The bed comes gently to its heat.

Mar. 12.   The sun mounts and looks down on the hanger. The air is milder.

Mar. 15.   The hot-bed streams very much; but the plants thrive, & put out their roots down the sides of the hills; & the weeds spring in the bed. Yet a little neglect, should there come hot sun-shine, would burn the plants. Holes are bored on every side in the dung.

Mar. 17.   Sowed the border opposite the great parlor-windows with dwarf upright larkspurs; a fine sort.

Mar. 18.   The wheat-ear, a bird so called, returns & appears on Selborne down.

Mar. 20.   Violent hail-storm, which filled the gutter, & came in & flooded the stair-case; & came down the chimnies & wetted the floors.

Mar. 21.   Young squab red-breasts were found this day in a nest built in a hollow tree.

Mar. 22.   On the 27th of February 1788, *Stone-curlews* were heard to pipe; & on March 1st, after it was dark some were passing over the village, as might be perceived by their quick, short note, which they use in their nocturnal excursions by way of watch-word, that they may not stray, & lose their companions. Thus we see, that retire whithersoever they may in the winter, they return again early in the spring, & are, as it now appears, the first summer birds that come back. The smallest uncrested wren has been deemed the earliest migrater, but it is never heard 'till about the 20th of March. Perhaps the mildness of the season may have quickened the emigration of the curlews this year. They spend the day in high elevated fields & sheep-walks; but seem to descend in the night to streams & meadows, perhaps for water which their upland haunts do not afford them.

Mar. 23. Mʳ Churton, who was this week on a visit at Waverley, took the opportunity of examining some of the holes in the sand-banks with which that diſtrict abounds. As these are undoubtedly bored by bank-martins, & are the places where they avowedly breed, he was in hopes they might have slept there also, & that he might have surprised them juſt as they were awakening from their winter slumbers. When he had dug for some time he found the holes were horizontal & serpentine, as I had observed before ; & that the neſts were deposited at the inner end, & had been occupied by broods in former summers : but no torpid birds were to be found. He opened & examined about a dozen holes. Mʳ Peter Collinson made the same search many years ago, with as little success. These holes were in depth about two feet.

Mar. 24. [Alton] Bright, grey, wet.

Mar. 25-30. [Oxford.]

Mar. 26. Large Mackarel.

Mar. 31. [Reading] Mʳ Loveday's tortoise is come-out. Young goslins.

Apr. 1. [Selborne] Daffodils in bloom. Mʳ Churton came.

Apr. 2. Sowed a bed of Onions. Mʳ Churton left us.

Apr. 3. Dogs-tooth violets blow.

Apr. 5. The firſt radishes failed. After all Mʳ Charles Etty did not sail fᵐ Sᵗ Hellens [I. of Wight] 'till this morning.

Apr. 6. NIGHTINGALE heard in the church-litten coppice : qu.

Apr. 7. Cucumbers begin to set. Put some honey in the frames to tempt the bees.

Apr. 8. Timothy heaves up the earth.

Apr. 10. Crown Imperials blow, & ſtink. Much gossamer. Bat.

Apr. 12. Mowed the grass of the fairey-ring on the grass-plot. Sent Mr White of Newton some male cucumber-blossoms in a box, to set some fruit in bloom in his frames. Fritillaria blows.

Apr. 13. Bees frequent the cucumber-frames. *Nightingales* heard below Temple [farm].

Apr. 15. Pronged the asparagus-beds. Sowed Mrs Eveleigh's curious asters in a hot-bed; & several perennials in the cold ground.

Apr. 19. Mended the fences this week all round my outlet. Insects abound; yet no swallows to be seen. The voice of the Cuckoo is heard in the land.[4]

Apr. 21. Timothy begins to eat: he crops the daisies, & walks down to the fruit-wall to browse on the lettuces. Mr Ventris observed at Faringdon a little whirl-wind, which originated in the road before his house, taking up the dust & straws that came in it's way. After mounting up thro' one of the elms before the Yard, & carrying away two of the rooks nests in which were young squabs; it then went off, leaving the court-yard strewed with dust & straws, & scraps of twigs, & the little naked rooks sprawling on the ground. A pair of rooks belonging to one of these nests built again & had a late brood.

Apr. 23. Gave away 24 eggs of the Bantham kind among my neighbours.

Apr. 24. Grass-hopper lark whispers. Cowslips blow.

Apr. 25. Wall-cherries loaded with bloom. The wild meris, cherry, blossoms.

Apr. 26. Harsh, windy unpleasing, weather for many days.

Apr. 30. Began to mow the orchard for the horses. Timothy weighs 6 ae 13 oz. 10d. Mole-cricket churs.

May 1. Mrs Ben White was brought to bed of a boy, who encreases my nephews, & nieces to the number of 52. Bro^r Thos Wh. tulips from South Lambeth make a gaudy appearance. The bloom of cherries, pears, & apples is great; of plums, bullace, sloes, little. Currans promise for much fruit, gooseberries for little. Peaches & nectarines are set: cherries begin to set.

May 3. Men cart peat & chalk. The deepest roads are quite dry.

May 4. Shade the best tulips from the vehemence of the sun. Polyanths are hurried out of bloom. Vine-shoots are forward. Sowed the great annual-frame with flower-seeds: sowed two hand-glasses with cucumbers, green & white. Timothy wanders round the garden, & strives to get out: he is shut-up in the brew-house to prevent an escape.

May 5. The great oak in the mead abounds with male bloom.

May 6. The wood-lark sings in the air at three in the morning: stone curlews pass over the village at that hour.

May 11. In some districts chafers swarm: I see none at Selborne. Cotton blows from the willows, & fills the air: with this substance some birds line their nests. Mr Burbey's brown owl, which was a great washer, was drowned at last in a tub where there was too much water.

May 12. *Fern-owl* chatters: it comes early this year.

May 15. Sheared my mongrel dog Rover, & made use of his white hair in plaster for ceilings. His coat weighed four ounces. The N.E. wind makes Rover shrink. A black bird has made a nest in my barn on some poles that lie on a scaffold.

May 18. A thunder-storm at London that damaged houses.

May 20. Fly-catcher begins to make a nest in my vine.

May 22. Saint-foin & fiery lilly begin to blow.

May 25. My winter lettuces all run-off to seed. The Culture of Virgil's vines corresponds very exactly with the modern management of hops. I might inftance in the perpetual diggings, & hoeings, in the tying to the ftakes & poles, in pruning of the superfluous shoots &c : but lately I have observed a new circumftance, which was Farmer Spencer harrowing the alleys between the rows of hops with a small triangular harrow, drawn by one horse, & guided by two handles. This occurrence brought to my mind the following passage :

> ". . . . ipsa
> Flectere luctantis inter vineta juvencos."
>   Second Georgic. [II, 356-7.]

May 27.   Mr White of Newton fetches water from Newton pond to put into his tank.

May 28.   The Flycatcher, which was not seen 'till the 18th, has got a neft, & four eggs.

May 29.   On this day there was a tempeft of thunder & lightening at Lyndon in the County of Rutland, which was followed by a rain that lafted 24 hours. The rain that fell was 1 in 40 h.

June 2.   [Alton] Mr Edmd White, & Captain Dumaresque cut their Saint foin.

June 3-26.   [S. Lambeth] June 3. Blue mift. Hay-making is general about Clapham & South Lambeth : Bror Benjn has eight acres of hay down, & making.

June 4.   Dingy. Saw some red-backed butcher birds about Farnham.

June 6.   Scarlet ftrawberries at 2s. per pottle. Red-backed butcher-bird, or Flusher, in Bro. Ben's outlet.

June 7.   Bro. Ben ricked ye hay of eleven acres of ground in delicate order.

June 8.   The black-clufter vines from Selborne are in bloom, & small delicately !

June 9.   Mazagon beans[5] come in.

June 11.   Some good oats about S. Lambeth.

June 12.   My Brother's gardener cut his firft melon, a Romagna.

June 13.   The bloom of the vines fills the chambers with an agreeable scent some what like that mignonette.

June 14.   The *scarabaei solstitiales* begin to swarm in my Brother's outlet.   My Bro[r] this spring turned one of his grass-fields into a kitchen-garden, & sowed it with crops : but the ground so abounded with the

PURSE GALL
(*Pemphigus bursarius*),
on Lombardy Poplar.

maggots of these chafers, that few things escaped their ravages.   The lettuces, beans & cabbages were mostly devoured : & yet in trenching this enclosure his people had destroyed multitudes of these noxious grubs.   The stalks & ribs of the leaves of the Lombardy poplar are embossed with large tumours of an oblong shape, which by incurious observers have been taken for the fruit of the tree.   These Galls[6] are full of small insects, some of which are winged, & some not. The parent insect is of the Genus of *Cynips*. Some poplars of the garden are quite loaded with these excrescencies.

June 15. A double scarlet Pomegranade buds for bloom. A bunting appears about the walks : this is a very rare bird at Selborne. The solstitial chafers swarm by thousands in my Brother's grounds. They begin to flie about sun-set, but withdraw soon after nine, & probably settle on the trees, to feed & to engender. My chamber at S. Lambeth is much annoyed with gnats.

June 17. Cherries turn colour, & begin to be eatable ; but are small for want of moisture : are netted. A cat gets down the pots of a neighbour's chimney after the Swallows nests.

June 18. Neither the pease or beans have the same flavour, & sweetness as in moist summers.

June 19. *Muscae domesticae* swarm in every room. I have often heard my Brothers complain how much they were annoyed with flies at this place. They are destroyed by a poisonous water called *fly-water*, set in basons, & by bird-lime twigs laid across pans of water.

June 21. Bro. B. has in his grounds 77 rows of Lucerne, which are each 48 yards in length. This plot furnishes his three horses with green meat the summer thro', & is cut at an average four times in the year. His gardener cuts-up three rows at a time several evenings in the week, & observes that one row fodders one horse for 24 hours. The crop is kept clean at considerable expense ; & would soon be over-run with weeds, was not care & attention bestowed. As soon as the whole rows are gone thro', those that were cut at first are ready to be cut again. He has 15 lights of melons, & 16 lights for cucumbers ; & 40 hand glasses for ridge-Cucumbers, & other purposes.

June 22. My Fly-catchers left their nest this day.

June 24. Four women gather my Bror's gooseberries for sale.

June 27.   Met a cart of whortel-berries on the road [to Selborne].

June 28.   [Selborne] June 25ᵗʰ 1788.   Mʳ Reeve, a master Carpenter in the town of Lambeth, is employed in building a Conservatory for the Queen of Naples, the dimensions of which are 117 feet in length, 40 feet in breadth, 20 feet to the angle of the roof, & 10 feet to the eaves.   This noble greenhouse (the largest that has been constructed yet in this kingdom) is to be roofed with sash-work on both sides, the upper sashes of every other one of which are so contrived as to slide down with pullies : the sides also are to be lined with sashes which pull up & down : the South end also is to be sashed, but the N. end is to be close, thro' which there is an entrance by a pair of large doors.   That there might be no beams across to obstruct the view, the roof is supported by 16 pillars of cast iron, weighing 500 weight each, which are so ramifyed at the top as to give the roof something of a Gothic air, & to add to it's strength.   The area is to contain two beds or borders of earth, of 100 feet by 11 feet, around & between which there are roomy alleys or paths. This whole deal frame-work, when finished, is to be taken to pieces, & so sent by sea to Naples.   The whole area of this house will contain 4680 square feet, & the two beds, or borders 2200 square feet. As the soft & southern climate of Naples produces oranges, lemons, pomgranades, citrons, & many trees & plants with which we croud our greenhouses ; we are to suppose that this royal Conservatory will be furnished only with the most fragrant, choice, & rare vegetable productions of the Tropics. The Gardener of the King of the two Sicilies is a Scotchman, who went over two years ago ; & had been partner, or assistant to Mʳ Gordon.   If there is any defect in this edifice, it seems to lie in the sash-roofing, which appears to be rather too slight & delicate for the length of the bearings ; tho' each sash

is ftiffened by a small iron-rod; however, heavy
snows, we may suppose, are seldom or never seen in
Lat. 40-50 so as weigh-in the glass.   But with regard
to the brittle healing, large hail-ftorms from off the
Apennine may some times be dreaded, which would
occasion almoft as great an havock among that fragil
tiling, as a shower of cinders from the neighbouring
Vesuvius.   The clatter and jingle on such an occasion
would put a man in mind of that beautiful, & expressive line in Virgil,
       " Tam multa in tectis crepitans salit horrida
          grando."   [Georg : I, 449.]
June 30.   Crop of apples general.   The parsonage-
orchard at Faringdon, that has failed for many years,
has now a full burthen.
July 3.   Red-backed butcher-bird, or flusher at
Little comb.   Gathered a good mess of Rasps for
jam.
July 4.   Gathered cherries for preserving.   Cut
a doz. of artichokes.   Broad beans come in.   Sowed
endive.
July 5.   The fly-catchers build again in the vines
with a view to a second brood.   Timothy grazes on
the grass-plot.   Some dishes of wood ftraw-berries
are brought to the door.
July 6.   The late burning season has proved fatal
to many deer in elevated situations, where the turf
being quite scorched up, the ftock in part perished
for want.   This is said in particular to have been the
case at Up-park in Sussex.   A want of water might
probably gave been one occasion of this mortality.
Some fallow deer have dyed in the Holt.
July 7.   Mrs White^7 made much Rasp, & curran
jams.
July 8.   Made cherry-jam.   H.W. & Anne Woods
came.
July 9.   Bunches of snakes eggs are found under
some ftraw near the hot-beds.   Several snakes
haunted my out-let this summer, & caft their sloughs

in the garden, & elsewhere. Cran-berries are offered at the door. D.L. [Dame Loe].

July 10. There are now some fallow-deer, & a red deer in Hartley wood.

July 12. Codlins come in for stewing. Wasps encrease & gnaw the cherries. Hung bottles to take the wasps.

" Contemplator item, cum se Nux plurima silvis
   Induet in florem, & ramos curvabit olentis :
   Si superant foetus, pariter frumenta sequenter[ur];
   Magnaque cum magno veniet tritura calore."
                                        [Georg. I., 187-90.]

If by *Nux* in this passage Virgil meant the *Wall-nut*,[8] then it must follow, that he must also mean that a good wall-nut year usually proves a good year for wheat. This remark is verifyed in a remarkable manner this summer with us ; for the wallnut trees are loaded with myriads of nuts, which hang in vast clusters ; & the crop of wheat is such as has not been known for many seasons. The last line seems also to imply, that this coincident, even in Italy, does not befall but only in a dry, sultry summer. Tho' wall-nut-trees in England blow long before wheat ; yet it is probable that in Italy, where wheat is more early than with us, they may blossom together. And indeed unless these vegetables had accorded in the time of their bloom, the Poet would scarce have introduced them together as an instance of con-comitant fertility.

July 14. Piped[9] many shoots of elegant pinks. There are some buntings in the N. field : a very rare bird at Selborne. They love open fields, without enclosures. Jennetings,[10] apples so called, come in to be eaten. Potatoes come in.

July 16. Bull-finch eats the berries of the honey-suckle. Bro[r] Tho[s] came.

July 18. Fly-catcher feeds his sitting hen, M[rs] H. W., Bessy, & Lucy came.

July 19. Poultry begin to moult.

July 21. Began to cut my meadow-grass. Farmer Parsons begins wheat-harvest in the Ewel; farmer Hewet at the forest-side. A young man brings a large wasps nest, found in my meadow.

July 23. An other wasps nest. Wheat blited at Oakhanger. Oakhanger-ponds empty: they were sewed in the spring.

July 26. The fields are now finely diversifyed with ripe corn, hay & harvest scenes, & hops. The whole country round is a charming land-scape, & puts me in mind of the following lovely lines in the first book of the *Cyder* of *John Philips.*[11]

"Nor are the hills unamiable, whose tops
    To heaven aspire, affording prospect sweet
    To human ken; nor at their feet the vales
    Descending gently, where the lowing herd
    Chews verdurous pasture; nor the yellow fields
    Gaily interchang'd, with rich variety
    Pleasing; as when an Emerald green enchas'd
    In flamy gold, from the bright mass (foil) [*sic*]
        acquires
    A nobler hue, more delicate to sight."
                                    [Bk. I.]

July 27. We have had a few chilly mornings & evenings, which have sent off the swifts. I have remarked before, many times, how early they are in their retreat. Surely they must be influenced by the failure of some particular insect, which ceases to fly thus early, being checked by the first cool autumnal sensations; since their congeners will not depart yet these eight or nine weeks.

July 28. S. Loe. Two swifts sipping the surface of Bin's pond. The bed of Oakhanger pond covered with large muscle shells. The *flint*, or summer *snipe*. Large flock of lapwings in the Forest.

July 30. Some workmen, reapers, are made sick by the heat. Much wheat bound. Some housed by John Carpenter.

Aug. 2. Many bats breed under the tiles of my house. Five gallons, & one pint of brandy from London.

Aug. 5. Farmer Spencer's rick slipped down as it was building.

Aug. 6. Flight of lapwings comes up into the malm [Upper Greensand] fallows.

Aug. 7. Two or three beeches below Bradshot are quite loaded with maſt. The King's field[12] is cleared, & thrown open.

Aug. 9. Wheat-harveſt will moſtly be finished by monday ; viz. in old July.[13]

Aug. 14. H. W. & Miss W. left us & went to Newton. Bro. Henry, & B. White, & wife came with little Tom, & Nurse Johnson.

Aug. 19. Farmer Lasham has much wheat out, which was not ripe when other people cut, & housed.

Aug. 20. Nep. Ben returned to London.

Aug. 22. The swallows are very busy skimming & hovering over a fallow that has been penned [for grazing by sheep] ; probably the dung of the sheep attraɕts many inseɕts, particularly scarabs.

Aug. 23. Some mushrooms spring on my hot-bcds. Mʳ Sam Barker, from a measurement taken, adjudged Wolmer pond to contain 66 acres, & an half, exclusive of the arm at the E. end : the pond-keeper at Frinsham [Frensham] avers that his pond measures 80 acres. *Ziᶎania aquatica,* Linn : called by the English setlers-wild Rice ; & by the Canadian French—*Folle Avoin.*[14] In consequence of an application to a Gentleman at *Quebec,* my Bro. Thomas White received a cask of the seed of this plant, part of which was sent down to Selborne. His desire was to have received it in the ear, as it then would have been much more likely to have retain'd its vegeta-tive faculty : but this part of his requeſt was not attended to ; for the seed arrived ſtript even of it's husk. It has a pleasant taſte, & makes a pudding equal to rice, or millet. This kind of corn, growing naturally in the water, is of great service to the wild natives of the south west part of N. America : for as *Carver*[15] in his travels says, they have no farther care & trouble with it than only to tye it up in bunches when it first comes into ear, & when ripe to gather it into their boats ; every person or family knowing their own by some distinction in the bandage. *Carver* observes, that it would be very advantageous to new settlers in that country, as it furnishes at once a store of corn the first year ; & by that means removes the distress & difficulty incident to new colonies till their first crop begins to ripen. *Linnaeus* has given this plant the name of *Zizania* : but what could induce the celebrated Botanist to degrade this very beneficial grain with the title of that pernicious weed which the enemy in the parable served among the good-corn while men slept, does not so easily appear. (Math. 13 chapter.)

Aug. 24. A stag, which has haunted Hartley wood the summer thro' was roused by a man that was mowing oats just at the back of the village. Several young persons pursued him with guns, & happening to rouse him again on the side of Nore hill, shot at him ; & then collecting some hounds from Emshot, & Hawkley, they drove him to a large wood in the parish of Westmeon, where they lost him, & called-off their dogs.

Aug. 26. Mr Hale & Tim. Turner begin to pick hops in the Foredown. Hale picked 350 bushels : his hops are large & fine.

Aug. 28. A bat comes out many times in a day, even in sunshine to catch flies : it is probably a female that has young, & is hungry from giving suck : the swallows strike at the bat.

Sept. 1. A sand-piper shot at Hawkley.

Sept. 2. J. Hale's crop of hops under the S. corner of the hanger is prodigious : many hills together produce a bushel each, some two, & some three ! Mr White of Newton cuts some Saint foin a

second time.   Nep. Ben came from London.   Barley,
& seed-clover are housed.

Sept. 4.   Vaſt showers about.   Were all wet thro'
in our return from Faringdon.   Under the eaves of
an house at Faringdon are 22 martins neſts, 12 of
which contain second broods now nearly fledge:
they put out their heads, & seem to long to be on the
wing.

Sept. 9.   On the brow of the cliff that looks down
on Candover's farm-house [Hartley] my Brother
found a lime-tree which had been cut down to a
ſtool, when the coppice was cut formerly.   Was
this a wild tree[16] or planted?

Sept. 11.   Nep. Ben & wife, & nurse & baby
left us, & went to Newton.

Sept. 13.   Gathered in my golden pippins, a small
quantity.   Mr Churton came from Cheshire.

Sept. 14.   The gale snapped-off a large bough from
my Cadillac pear tree, which is heavily laden with
fruit.

Sept. 15.   Gathered many of the baking pears to
disburthen the boughs, & keep them from breaking.

Sept. 22.   The swallows seem to be diſtressed for
food this cold wet weather, & to hawk up & down the
ſtreet among the houses for flies with great earneſt-
ness.   Some of my rasps bears twice in the year,
& have now ripe fruit: these berries the partridges
have found out, & have eaten moſt of them.   Thomas
sprung two brace & an half among the bushes this
morning.   These birds were hatched in Baker's
hill.   A flood laſt week at Hedleigh mill.   The miller
at Hawkley has long been diſtressed for want of water.
Spinage very fine.   Herrings are brought to the door.

Sept. 29.   Mr Churton left us.   T. H. White
came from Fyfield.

Sept. 30.   Gathered such of the Cadillac pears, as
could readily be reached by ladders.   Thomas says
there are 13 bushels on my only tree.

Oct. 1.   H. H. White came from Fyfield.

Oct. 2.   Gathered six bush : & half of dearlings from the meadow-tree : four or five bush : remain on the tree.   The foliage of the Virginian creeper of a fine blood-colour.

Oct. 4.   Fyfield, the spaniel, rejects the bones of a wood-cock with horror.   Gathered in the non-pareils.   The prodigious crop of apples this year verified in some measure the words of Virgil made use of in the description of the Corycian garden ;

" Quotq' in flore novo pomis se fertilis arbos
    Induerat, totidem in autumno matura tenebat."
[Georg. IV, 143-4.]

Oct. 6.   Gathered-in some royal russets, very fine.

Oct. 7.   Many gulls, & wild fowls on Wolmer pond. Whitings brought.

Oct. 8.   Bought of bright hops—21 pounds ; of brown—49.

Oct. 9.   D.L. Virginian creeper sheds it's leaves. It's leaves have a silky appearance.   Tho. H. White, & H.H.W. went to Fyfield.

Oct. 10.   Nailed-up a Greek, & an Italian inscription on the front of the alcove on ye hanger.   Boys took a large round wasps nest in the Ewel, nearly as large as a gallon measure.   Several martins round the church.   Many flies on the tower, which come out from the belfry to sun themselves.

Oct. 14.   Women & children go a-*acorning*, & sell their acorns at one shilling p^r bushel.

Oct. 15.   Vast quantities of gossamer : the fields are covered with it :
" slow thro' the air
The *gossamer*-floats ; or stretch'd from blade to blade
The wavy net-work whitens all the fields."
[Cf. 3 Aug. 1783.]
Celeri comes in, very good.

Oct. 18.   Bro^r T. White planted two Lombardy poplars in the corners of the pound : & a Sycomore on the Plestor near the pound.

Oɓ. 20. Leaves fall. The pound field is sown with American wheat.

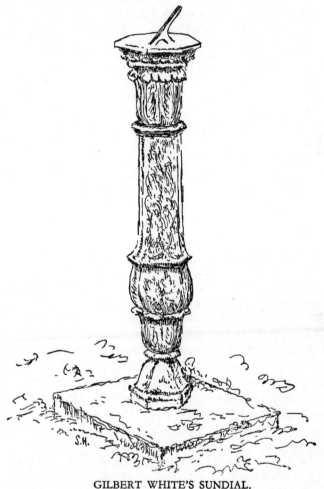

GILBERT WHITE'S SUNDIAL.

Oɓ. 22. Much wheat is sown. The fallows are very dry ; & the roads clean as in summer.

Oɓ. 23. Much peat carted thro' the village.

Oɓ. 24. Gave away many ſtone-less berberries : the tree every year bears vaſt burdens.

Oct. 26. Some woodcocks shot on the Barnet lately.

Oct. 27. Set up again my stone dial, blown down many years ago, on a thick Portland-slab in the angle of the terrass. The column is very old, came from Sarson house near Amport [near Andover], & was hewn from the quarries of Chilmarke.[17] The dial was regulated by my meridian line.

Oct. 29. Meridian line & dial accord well.

Oct. 30. Larches turn yellow; ash leaves fall; the hanger gets thin: my tall hedges finely diversifyed.

Nov. 1. Planted on the bank in the garden several dames violets raised from cuttings under hand-glasses. Sowed some seeds of the *Zizania aquatica* in Comb-wood pond. The King's stag-hounds came down to Alton, attended by a Huntsman & six yeoman prickers with horns, to try for the Stag that has haunted Hartley wood, & it's environs for so long a time. Many hundreds of people, horse & foot, attended the dogs to see the deer unharboured: but tho' the huntsmen drew Hartley wood, & Temple hangers; yet no Stag could be found. Lord Hinchinbroke, the master of the hounds, & some other Nobleman attended. The royal pack, accustomed to have the deer turned-out before them, never drew the coverts with any address or spirit, as many people that were present observed; & this remark the event has proved to be a just one. For as Harry Bright was lately pursuing a pheasant that was wing-broken in Hartley wood, he stumbled upon the stag by accident, & ran in upon him as he lay concealed amidst a thick brake of brambles, & bushes.

Nov. 3. Bro. Tho. sowed many acorns, & some seeds of an ash in a plot dug in Baker's hill. The King's hounds tryed our coverts for the stag, but with no success.

Nov. 5. Swarms of sporting gnats come streaming out from the tops of the hedges, just as at Midsum.

On this soft summer-like day some h. martins might have been expected along the hanger; but none appeared.

Nov. 6.   Bro. Tho. & T.H.W. left us, & went to London.

Nov. 9.   Many people are sick.

Nov. 10.   Planted 10 roses, 2 cypresses, 6 violae tricolores, 3 sent from S. Lambeth by Bro. Ben.

Nov. 11.   Men have taken advantage of this dry season, & have chalked their hop-gardens, & fields. The chalk at the foot of the hill is called marl,[18] but it is only a hard grey chalk.   This chalk is of service on the malms.

Nov. 18.   Farmer Lasham's Dorsetshire ewes have produced several lambs. Insects abound. Wheat comes up well.

Nov. 19.   M^r. Hale continues to chalk the black malm[19] field opposite to J. Berriman's house, called Hasteds.   He has laid on about 120 loads on less than 3 acres.

Nov. 22.   The smoke of the new lighted lime-kilns this evening crept along the ground in long trails: a token of a dry, heavy atmosphere.

Nov. 23.   The downy seeds of travellers joy [clematis] fill the air, & driving before a gale appear-like insects on the wing.   Mrs *Clement* brought to bed of a boy.   My nephews & nieces now 53.

Nov. 24.   Liss [nr. Selborne] hounds are hunting on the common.   My well very low: some wells are dry.   We have taken away much of the old wood from the vines.   Wheeled dung.

Nov. 26.   Finished shovelling the zigzag, & bostal. Wildfowl on Wolmer-pond.

Nov. 27.   Some light snow.   Boys slide on lakes. Turned up much fine rotten earth from among the rubbish carryed out of the garden.

Nov. 28.   [Newton] Mr White's tank at Newton has been empty some days.

Nov. 29. A vast flock of *hen* chaffinches are to be seen in the fields along by the sides of Newton-lane, interspersed, I think, with a few *bramblings*, which being rare birds in these parts, probably attended the finches on their emigration. They feed in the stubbles on the seeds of knot-grass, the great support of small, hard-billed birds in the winter.

Nov. 30. Many wild fowls haunt Wolmer pond : in the evenings they come forth & feed in the barley-stubbles.

Dec. 1. Several wells in the village are dry, & some ponds in the neighbourhood. Well-head runs much as usual. There is a fine perennial spring at the bottom of Hasteds. Men cart earth, & marl from Clays pond.[20]

Dec. 3. The grey chalk carried-out upon Hasteds falls to pieces. Good mackarel brought to the door.

Dec. 4. The plows have been stopped by frost some days. Men cart earth & dung for their hop-grounds. Covered the lettuces, artichokes, spinage, & celeri with straw. Took in the urns.

Dec. 6. The millers around complain that their streams fail, & they have no water for grinding.

Dec. 7. The wind & frost cut down the wheat, which seems to want a mantle of snow.

Dec. 8. Great want of water upon the downs about Andover. The ponds, wells, & brooks fail.

Dec. 9. J. Hale clears out the ponds at Little comb.

Dec. 10. Great complaint for want of rain, & water, round Dublin in Ireland.

Dec. 13. The Stag seen again about Oakhanger. He some times haunts about Hartley wood, & some times about the Holt.

Dec. 14. The navigation of the Thames is much interrupted thro' the want of water occasioned by the long dry season.

Dec. 15. Therm^r 20, 23, 17. Many have been disordered with bad colds & fevers at Oxford. The water in the *apparatus*[21] for making mineral water froze in the red room. The wind is so piercing that the labourers cannot stand to their work. Ice in all the chambers. The perforated stopple belonging to the apparatus broke in two by the frost. Apples preserved with Potatoes & carrots in the cellar. Shallow snow covers the ground, enough to shelter the wheat.

Dec. 18. Most of the wells in the street are dry ! Among the rest my own is so shallow as not to admit the bucket to dip ! Moved some apples & pears into the kitchen-closet. The horse-roads are dusty.

Dec. 20. The frost has lasted now five weeks.

Dec. 22. A considerable flight of wood-cocks around the Barnet.

Dec. 23. Moles work, & heave up their hillocks.

Dec. 29. Many wild geese in the moors of the forest.

Dec. 30. Ice within doors. Rime. Snow on the Ground.

CHAPTER XXII

"Not a tree,
A plant, a leaf, a blossom, but contains
A folio volume. We may read, and read,
And read again, and still find something new."

HURDIS: *The Village Curate.*

1789

Jan. 1. Snow thick on the ground. Timothy begins to sink his well at the malt-house.

Jan. 3. Rime hangs on the trees all day. Turner's well-diggers have sunk his well about six feet. It is now about on a level with mine, viz. 63 feet deep. They came to-day to a hard blue rag,[1] & a little water.

Jan. 4. Began the new hay-rick. Snow on the ground; but the quantity little in comparison with what has fallen in most parts. As one of my neighbours was traversing Wolmer-forest from Bramshot across the moors, he found a large uncommon bird fluttering in the heath, but not wounded, which he brought home alive. On examination it proved to be *Colymbus glacialis, Linn.* the great speckled Diver, or Loon, which is most excellently described in *Willughby's Ornithology.* Every part & proportion of this bird is so incomparably adapted to it's mode of life, that in no instance do we see the wisdom of God in the Creation to more advantage.[2] The head is sharp, & smaller than the part of the neck adjoining, in order that it may pierce the water; the wings are placed forward & out of the center of gravity, for a purpose which shall be noticed hereafter; the thighs quite at the podex, in order to facilitate diving; & the legs are flat, & as sharp backwards almost as the edge of a knife, that in striking they may easily cut the water; while the feet are palmated, & broad for swimming, yet so folded up when advanced forward to take a fresh stroke, as to be full as narrow as the shank. The two exterior toes of the feet are longest; the nails flat & broad, resembling the human, which give strength & increase the power of swimming. The foot, when expanded, is not at right angles to the leg or body of the bird; but the exterior part, inclining towards the head, forms an acute angle with the body; the intention being not to give motion in the line of the legs themselves, but by the combined impulse of both in an intermediate line, the line of the body. Most people know, that have observed at all, that the swimming of birds is nothing more than a walking in the water, where one foot succeeds the other as on the land; yet no one, as far as I am aware, has remarked that diving fowls, while under water, impell & row themselves forward by a motion of their wings, as well as by the impulse of their feet: but such is really the case, as any person may easily be convinced who will observe ducks when hunted by dogs in a clear pond. Nor do I know that any one has given a reason why the wings of diving fowls are placed so forward. Doubtless not for the purpose of promoting their speed in flying, since that position certainly impedes it; but probably for the encrease of their motion under water by the use of four oars instead of two; yet were the wings & feet nearer together, as in land birds, they would, when in action, rather hinder than assist one another. The *Colymbus* was of considerable bulk, weighing only three drachms short of three pounds averdupoise. It measured in length from the bill to the tail (which was very short) two feet; & to the extremities of the toes, four inches more; & the breadth of the wings expanded was 42 inches. A person attempted to eat the body, but found it very strong & rancid, as is the flesh of all birds living on fish. *Divers* or *loons*, though bred in the most northerly parts of Europe, yet are seen with us in very severe winters; & on the Thames are called *Sprat loons*, because they prey much on that sort of fish. The legs of the *Colymbi* & *mergi* are placed so very backward, & so out of all center of gravity, that these birds cannot walk at all. They are called by Linnaeus *Compedes*, because they move on the ground as if shackled or fettered.

Jan. 5.   Turner's well-diggers advance slowly thro' a blue rag.   Mr Churton left us, & went to Waverly.

Jan. 6.   Therm^r 25 ; 18.   Fierce frost, sun, cutting wind.   Severe day.

Jan. 7.   Salted-up a small hog in the pickling tub, —weight 8 scores, & 8 pounds : the meat was young, & delicate.   The people at Froxfield fetch their water [some six miles] from Petersfield up Stoner hill.

Jan. 8.   A severe frost prevails all over the Continent.

Jan. 9.   The farmers are in pain about their turnips, both those on the ground, & those that are stacked under hedges.   The people at Forestside drive all their cattle to be watered at a spring issuing out at Temple grounds at the foot of Temple hanger.[3] Oakhanger ponds, & Cranmer ponds are dry.   The frost has lasted now just seven weeks : it began Novr 23.   T. Turner has sunk his well 9 feet without coming to water.   He now desists on account of the expence.   My well, I now find, has more than three feet of water ; but the rope is too short to reach it.

Jan. 12.   Therm^r 8 ; at South Lambeth 2½.   This frost, as frosts usually do, went-off soon after the Therm^r was at the lowest.

Jan. 13.   Deep snow : snow drifted through every crevice.   Swift thaw.   Snow that had been driven in now melts & drips thro' the garret-ceiling.

Jan. 14.   The snow drifted in thro' the tiling now melts, & floats the ceiling.   A *Gooseander* & a Dun diver, a drake and a duck of the same species, *Mergus Merganser*, were brought me this morning.   They are beautiful birds, never to be seen in the South but in hard frosts : were shot on the stream at Hedleigh.

Jan. 15. Snow melts very fast. The frost, where a grave was dug, appeared to have ent'red the ground about 12 inches.[4]

Jan. 16. Now the rope is lengthened my well furnishes me with water.

Jan. 17. Fine thaw, snow decreased.

Jan. 18. A swan came flying up the Lythe, &, without regarding objects before it, dashed itself against Dorton-house, & fell down stunned. It recovered, & was sold to the miller at Hawkley.

Jan. 22. Now the ice is melted on Hartley-park pond, many dead fish come floating ashore, which were stifled under the ice for want of air.

Jan. 29. [Newton] Bantam-hens make a pleasant little note, expressive of a propensity towards laying. Fog so deep that we could not see the alcove in the garden.

Jan. 31. [Selborne] Farmer Knight's wheat of a beautiful colour. Children play at hop-scotch. Rain in Jan. 4 inc. 48h. I now see, that after the greatest droughts have exhausted the wells, & streams, & ponds, four or five inches of rain will completely replenish them.

Feb. 1. Boys play at taw on the plestor. Two of the Bantam (*sic*) hens lay each an egg.

Feb. 4. Green rye has a delicate soft tinge in its colour, distinguishable from that of wheat at a considerable distance.

Feb. 5. As one of farmer Spencer's cows was gamboling, & frisking about last summer on the edge of the short Lythe, she fell, & rolled over to the bottom. Yet so far was she from receiving any injury by this dangerous tumble, that she fattened very kindly, & being killed this spring proved fine beef.

Feb. 8. The open catkins illuminate the hazels; these are the male blossoms: the female are so minute as to be scarce discernible.

Feb. 12. About this time Miss Chase, & Miss Rebecca Chase sailed for Madras in the Nottingham India-man.

Feb. 13. Lined the hot-bed screen with reeds. Cucumbers come up well : bed works well.

Feb. 19. A large bank in Burrant⁵ garden covered with winter-aconites, which have been there more than 40 years. Missel thrush sings on one of the firs.

Feb. 20. Dug a plot of ground for beans.

Feb. 21. Yesterday I fixed some nuts in the chinks of some gate-posts in a part of my outlet where *Nut-hatches* used to haunt : & to day I found that several of them were drilled, & the kernels gone. [Cf. D.B., LVI.]

Feb. 26. Our butcher begins to kill grass-lamb.

Mar. 2. Sowed the great meadow with ashes ; with 49 bushels bought of neighbours, & with 28 bushels of our own : total 77.

Mar. 5. Male yew trees shed their farina in clouds.

Mar. 6. Mʳ Richardson came.

Mar. 7. Mʳ Richardson left us.

Mar. 9. Loud thunder at Hinckley in Leicester-shire, & lightening that did some damage : it happened in the midst of snow.

Mar. 10. Mʳ & Mʳˢ Clement, & 3 children came.

Mar. 13. Snow in the night : snow five inches deep. Snow melts on the roofs very fast : & runs thro' the ceiling of the garret.

Mar. 15. Snow on the ground. Raw & cold. Mʳˢ Clement left us.

Mar. 16. Mended the cucumber frames.

Mar. 17. Icicles hang in eaves all day. Snow melts in the sun.

Mar. 19. Snow lies on the hill. Made the bearing cucumber-bed : the dung is full wet, but warm.

Mar. 24. About this time sailed for Antigua Ned White [nephew], aboard the Lady Jane Halliday, Captain Martin.

Mar. 26. Icicles hang all day. Hot-bed smokes.

Mar. 28. Snow did not lie. Apricots begin to blow. Earthed the nearing cucumber-bed. The plants in the seedling-bed grow, & want room. N. Aurora.

Mar. 30. Sowed dwarf lark-spurs.

Mar. 31. Sowed a crop of onions, lettuce, & radishes.

Apr. 1. Rain in the night, spring-like. Crocus's make a gaudy show. Some little snow under the hedges.

Apr. 3. Some wood-cocks are now found in Hartley-wood: as soon as the weather grows a little warm, they will pair, & leave us.

Apr. 5. *Wry-neck* pipes. The smallest *uncrested wren* chirps loudly, & sharply in the hanger.

Apr. 6. Timothy the tortoise heaves up the sod under which he is buried. Daffodil blows.

Apr. 9. Brimstone butter-fly. The tortoise comes out. Dog violets blow. Summer like.

Apr. 11. White frost, sun. Timothy the tortoise weighs 6 ae. 14 oz. Dug several plots of garden ground & ground digs well. Sweet even.

Apr. 14. Pulled down the old forsaken martin's nests in some of which we found dead young. They grow fetid, & foul from long use. Redstart appears in my tall hedges.

Apr. 17. Five gallons of french brandy from London. Cucumbers show fruit in bloom. Cuculus cuculat: the voice of the cuckoo is heard in Black-moor woods. Sowed hollyhocks, columbines, snap-dragons, stocks, mignonette, all from S. Lambeth, in a bed in the garden: also sweet williams, & Canterbury bells.

Apr. 19. The vines of John Stevens, which were trimmed late, not till March, bleed much; & will continue to do so untill the leaf is fully expanded. It is remarkable, that tho' this is the case while the trees are leafless, yet lop them as much as you please when the foliage is out, they will not shed one drop. Dr Hales was not acquainted with this circumstance when he cut-off a large bough of his vine at Teddington late in the spring; & it was lucky for science that he was not. For his sollicitude for his vine, & his various attempts to stop the effusion of the sap, led him step by step to many expedients, which by degrees brought on abundance of curious experiments, & ended in that learned publication known by the name of *Vegetable Statics*, a work which has done much honour to the Author, & has been translated into many modern languages.

Apr. 20. Apricots set very fast. The willows in bloom are beautiful. Men pole their hops : barley is sowing at the forest side. Several swallows, h. martins, & *bank-martins* play over Oakhanger ponds. The horses wade belly deep over those ponds, to crop the grass⁶ floating on the surface of the water.

Apr. 22. Young broods of goslings. Woodsorrel, & anemony blow. The cuckoo cries along the hanger. Wheat thrives.

Apr. 23. Swallows & martins do not yet frequent houses. Women hoe wheat.

Apr. 26. This morning I saw a certificate from the town of Wymburn Minster in the country [sic] of Dorset to the parish of Selborne, acknowledging William Dewye to be parishioner of the said town. This paper is dated Apr. 20, 1729 : so that Will: Dewye, & wife, both still living, have been certificate people here exactly 60 years.⁷

Apr. 27 [Alton] Showers, windy. One beech in the hanger shows some foliage.

Apr. 28. Timothy the tortoise begins to eat dandelion.

Apr. 29. [Selborne] Scarce an hirundo has been seen about this village.

Apr. 30. Brother Thomas White, & daughter, & little Tom came.

May 2. The long frost of last winter has proved very destructive to pond-fish the kingdom over, except in those pools & lakes thro' which passed a constant current of water : nor did the expedient of

breaking holes in the ice avail. M^r. Barker, who has
been writing an account of the late frost, thinks that it
did mischief. A current of water introduces a constant
current of fresh air, which refreshes continually the air
of the pools & ponds, & renders it fit for respiration.

May 4. Beat the grass-banks in the garden. Put
up the urns. Martins come into the old nests. Bat
out. Nightingale in my out-let. Snails come out.

The *Fern-owl,* or *Goat-sucker* chatters in the hanger.
This curious bird is never heard till warm weather comes:
it is the latest summer bird except the fly-catcher.

May 8. Cut the first mesh of asparagus. The
bloom of plums is very great. Peat-carting begins.

May 10. Nep. Ben came. The beeches on the
hanger, now in full leaf, when shone down on by the
sun about noon, exhibit most lovely lights & shades,
not to be expressed by the most masterly pencil. The
hops are infested by the *Chrysomela oleracea,* called by
the country people the turnip-fly, or black dolphin,
which eats holes in their leaves. This species is—
" saltatoria, femoribus posticis crassissimis " :—
" *chrysomelae saltatoriae* plantarum cotyledonibus, &
tenellis foliis infestae sunt." Linn: [Cf. 22 Aug. 1772].

May 13. Nep. Ben & wife left us. Great tempest
at Winchester.

May 15. Caught a mouse in the hot-bed: cut
several cucumbers, but they are ill-shapen.

May 17. The mice have infested my garden much
by nestling in my hot-beds, devouring my balsoms, &
burroughing under my cucumber-basons : so that I
may say with Martial . . .
" Fines Mus populatur, & colono
Tanquam Sus Calydonius timetur."
Epigramm : XIX. lib : XI.

May 18. Very blowing all day.

May 19. *Stellaria holostea* greater stichwort, blows :
a regular, periodical plant.

May 20. Martins build briskly at the Priory, & in
the street. Oaks show prodigious bloom.

May 22. Hirundines keep out in the rain : when the rain is considerable. Swifts skim with their wings inclining, to shoot off the wet.

May 23. *White thorn* blows. The air is filled with floating willow-down. Martins begin to build against the end of my brew-house. Columbines blow. N. Aurora. Timothy the tortoise begins to travel about, & be restless.

May 24. Dr Chandler by letter dated Rolle en Suisse April 4th, 1789. " The *Swallows* disappeared here about the end of September, 1787, the weather being cold : but Octr 17th I saw a pair as we passed among the mountains towards Fort le Cluse on the road to Lyons ; & my servant saw a pair on the 19th when we had got thro' the mountains into Bresse. Passing an islet of the Rhone Octr 23 near Pont St Esprit, again I saw a *swallow*, which dipped to drink. As we approached nearer Marseilles, we saw wasps, dragon-flies, butter-flies, & other summer-insects. I was ashore Novr 10 at Porto Longona, in the isle of Sr Elbe, off the coast of Italy, towards the evening. Philip declared that a *swallow* had passed over his head, of which I doubted ; but presently after saw three crossing the Port towards us. They flew almost strait, very swiftly ; & I should have supposed were going to Italy, if the distance had been less, or the Sun not so near setting. Wasps were in full vigour, & numerous there. I was assured by a friend at Rome, March 16, 1788, that he had seen *swallows* at Naples six weeks before. Mr Morris informed me that *martins* had been busy under the eaves of the house, where he lodged, about a week. I saw there, two days after, four nests which they had begun to repair, & on the 26th a couple of the birds : but Mr M. declared that he had heard them twitter at least as early as the first of March. The first *swift* I observed was over the river Liris on my return from Naples April 27th: *Nightingales* sung there. On the 20th of last March Philip saw two *martins* about the lake of Geneva ; & was assured by a

man that he had seen them on the 18th.   On the 25th
he saw several swallows ; & supposes the *martins* to
have perished with the cold, as they have not been seen
since, & the weather has been bad.   They seem to have
disappeared again, as I have not yet seen one.   I
remarked *bees*, & a *brimstone-butterfly*, March 15th ;
& about the same time magpies building in the trees
opposite to my windows.   I am told that a single
martin commonly arrives first, as it were to explore ; &
again withdraws, as it were to fetch a colony.   Mr
Morris, who has lived several years at Rome, related,
that the boys there angle for the *Swallows* with a line
at the end of a reed, & instead of a hook, a noose
baited with a feather, & hung out at the corners of the
streets.   Many are taken by this method, & carried
home to be roasted & eaten ;   .   .   .   or to supply
the markets, where they are commonly sold in the
season.   At Chamberry in Savoy I observed in the
evening a joyous croud, & a great bustle.   My
curiosity led me to see what was the matter.   A net
was spread from one house to an other across a street.
These brutes tied the birds they intercepted (chiefly
swifts) in pairs by two of their legs, & dismissed
them from the windows to flutter down, & become
the sport of the mob below.   I turned away with
horror, & disgust.   The first *quail* that I have heard
this year 1787 was near Rolle, on May 20th in the
evening."

May 28.   A fly-catcher has built a nest in the great
apricot-tree, in which there is one egg.

June 1.   Monks rhubarb seven feet high ; makes a
noble appearance in bloom.

June 5.   Sowed some white cucumber-seeds from
S. Lambeth under an hand-glass.   Moon-shine.

June 6.   *Aphides* begin to appear on the hops : in
some places they are called smother-flies.   Farmer
Spencer's Foredown hops are much injured, & eaten
by the chrysomēlae : while Mr Hale's adjoining are
not much touched.

June 8. The bloom of hawthorns is vast : every bush appears as if covered with snow. Brother Thomas left us, & went to Fyfield.

June 9. Field-crickets shrill on the verge of the forest. Cuckoos abound there. Thinned the apricots, & took off many hundreds.

June 10-*July* 2. [South Lambeth.]

June 10. Rye in ear. Green pease at supper, a large dish. Young Cygnets on the Mole river at Cobham. Hay made, & carying at Wandsworth. Roses, & sweet-briars beginning to blow in my brother's outlet.

June 11. Straw-berries cryed about.

June 12. Bro.^r Benjn cuts his grass, clover & rye, a decent burden, but much infested with wild chamo-mile, vulg : margweed : may weed.

June 13. My brother's barley begins to come into ear. The squirrel is very fond of the cones of various trees. My niece Hannah's squirrel is much delighted with the fruit of the coniferous trees, such as the pine, the fir, the larch, & the birch; & had it an opportunity would probably be pleased with the cones of alders. As to Scotch firs, Squirrels not only devour the cones, but they also bark large boughs, & gnaw off the tops of the leading shoots; so that the pine-groves belonging to Mr Beckford at Basing-park are much injured & defaced by those little mischievous quadrupeds, which are too subtile, & too nimble to be easily taken, or destroyed. The Cypress-trees, & passion-flowers mostly killed by the late hard winter.

June 14. A patent machine, called a *Fire escape* (rather perhaps a *'Scape fire*) was brought along Fleet street. It consisted of a Ladder, perhaps 38 feet in length, which turned on a pivot, so as to be elevated or depressed at will, & was supported on timber-frame-work, drawn on wheels. A groove in each rail of this ladder-like construction admitted a box or hutch to be drawn up or let down by a pulley at the top round & by a windlass at bottom. When the

ladder is set up against a wall, the person in danger is
to escape into the hutch, then drawn to the top.   That
the ladder may not take fire from any flames breaking

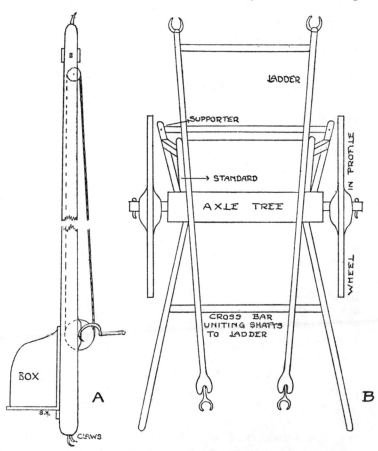

DUFOUR'S FIRE-ESCAPE.

Details from Specification No. 1652, year 1788 (Patent Office Library)
A. Profile of ladder, which could be of any length.   B. Ladder in per-
spective, showing the wheels for transport.

out below, it is defended all the way by a sheathing of
tin.   Several people, it seems, had illiberally refused
the Patentee the privilege of trying his machine against

their houses : but Mr White [Benjamin], on application, immediately consented ; when the ladder was
applyed to a sash on the second story, & a man was
hoisted up, & let down with great expedition, &
safety, & then a couple of boys together. Some
spectators were of opinion that the hutch or box was
too scanty or shallow, & that for security it ought to
be raised on the sides & lower end by a treillis of strong
wire, or iron-work, lest people in terror & confusion
should miss of their aim & fall over to the ground.
This machine was easily drawn by four men only.
The ladder, the owner told us, would reach to a third
story, when properly elevated. The name of the
Inventor is Moun^r Dufour.[8]

June 17.   [London] Cauliflowers.   The Opera-
house[9] in the Hay-market burnt down.

June 21.   Vines begin to blossom : corn-flags blow.
My brother trenched his field, & sowed it with barley :
but the corn seems as if it would be too big, & begins
already to lodge.   [Particulars added in another hand :
" one Acre & a Quarter "   . . .   " The Produce,
when thrashed, was 9 Quarters one Bushel, & an half,
viz. 73½ Bushels.][10]   My brother has set up a may-pole
55 feet in height : it is constructed out of two slender
deal spars, & for support cramped to the corner of a
garden wall.

June 23.   Scarlet strawberries are cryed about at
six pence the pottle : they are not finely flavoured.

June 24.   Mazagan beans come in. The barley much
lodged.   No house-martins appear at S.L.[ambeth], a
very few swallows, & only three pairs of swifts that
seem to belong to the place.   No wonder then that
flies abound so in the autumn as to become a nuisance.

June 25.   Crop-gardeners sell their pease at market
at 20^d the sack, & their cauliflowers at 18^d per dozen :
pease abound, so as hardly to pay for gathering.

June 27.   My brother cuts his first melon, a small
cantaleupe.   Barley in bloom, that which was lodged
rises a little.

June 28.   Daws come on the cherry-trees, for the fruit.   While Mrs J. White, & I were at S. Lambeth, we visited a Mrs Delhust of that place, the wife of a officer, who being at Gibraltar all the time of the siege, underwent all the horrors of that long blockade, & bombardment.   Even at this distance of time, some-what of terror, & uneasiness seem to be imprinted on her features, so as to occasion a lasting impression. Nor is there any room for wonder ;   for fear is a violent passion, which frequently repeated like other strong emotions, must leave traces behind.   Thus, thro' the transports of inebriation, where men habituate themselves to excess in strong liquors, their faces contract an air of intoxication, even when they are cool & sober.   This Lady, with many others, lodged for more than a twelve month in a cave of the rock to avoid the bombs & shot from the gun-boats, which annoyed the Southern part of the Istmus every night, as soon as it began to grow dark.

June 29.   Marrow-fat pease come in.

July 1.   [London] The price of wheat rises on account of the cold, wet, ungenial season.   The wet & wind injures the bloom of the wheat.

July 2.   [S. Lambeth] Cherries sold in the streets, but very bad.   Young fly-catchers come out at Selborne.

July 3.   [Alton] Young swallows on the top of a chimney.   The western sun almost roasted us between Guildford & Farnham [along the Hog's Back], shining directly into our chaise.

July 4.   [Selborne] A cock red-backed butcher-bird, or flusher, was shot in Hartley-gardens, where it had a nest.   My garden is in high beauty, abounding with solstitial flowers, such as roses, corn-flags, late orange-lillies, pinks, scarlet lychnises, &c. &c.   The early honey-suckles were in their day full of blossoms, & so fragrant, that they perfumed the street with their odour : the late yellow honey-suckle is still in high perfection, & is a most lovely shrub ; the only objection is that having a limber [pliant] stem, & branches, it does not make a good standard.

July 5. My scarlet straw-berries are good : what we eat at S. Lambeth were stale, & bad. A peat-cutter brought me lately from Cranmoor [Cranmer] a couple of snipe's eggs which are beautifully marbled. They are rather large, & long for the size of the bird, & not bigger at the one end than the other. The parent birds had not sat on them. [*Later note.*] These eggs, I find since, were the eggs of a Churn-owl : the eggs of Snipes, differ much from the former in size, shape & colour. The peat-cutter was led into the mistake by finding his eggs in a bog, or moor.

July 11. The fly-catchers in the vine bring out their young.

July 12. Wag-tails bring their young to the grass-plots, where they catch insects to feed them.

July 14. Benham skims the horse-fields. Rasps come in : not well flavoured. On this day a woman brought me two eggs of a *fern-owl* or *eve-jarr*, which she found on the verge of the hanger to the left of the hermitage, under a beechen shrub. This person, who lives just at the foot of the hanger, seems well acquainted with these nocturnal swallows, & says she has often found their eggs near that place, & that they lay only two at a time on the bare ground. The eggs were oblong, dusky, & streaked somewhat in the manner of the plumage of the parent-bird, & were equal in size at each end. The dam was sitting on the eggs when found, which contained the rudiments of young, & would have been hatched perhaps in a week. From hence we may see the time of their breeding, which corresponds pretty well with that of the Swift, as does also the period of their arrival. Each species is usually seen about the beginning of May. Each breeds but once in a summer ; each lays only two eggs.

July 15. We have planted-out vast quantities of annuals, but none of them thrive. Grapes do not

blow, nor make any progress. The wet season has
continued just a month this day. Dismal weather!

July 16. Wall-cherries are excellent. Lime-trees
blossom, & smell very sweet. M^r & Mrs Sam. Barker,
& Miss Eliz. Barker came from the county of Rutland.

July 19. When old beech-trees are cleared away,
the naked ground in a year or two becomes covered
with straw-berry plants, the seeds of which must have
lain in the ground for an age at lest. One of the
slidders or trenches down the middle of the hanger,
close covered over with lofty beeches near a century
old, is still called *strawberry slidder*, though no straw-
berries have grown there in the memory of man.
That sort of fruit, no doubt, did once abound there,
& will again when the obstruction is removed.

July 20. Began to cut my hay, a vast burden, but
over-ripe.

July 21. *Anthericum ossifragum*, Lancashire asphodel
a beautiful plant, found by Mr Barker in bloom among
the bogs of Wolmer forest. *Monotropa Hypopithys*
blossoms on the hanger. Thistles begin to blow.
The naked part of the hanger is now covered with
thistles; but mostly with the *carduus lanceolatus*. There
are also the *carduus nutans*, the musk thistle; *carduus
crispus*, the thistle on thistle; *carduus palustris*, the
marsh-thistle. The seeds of these thistles may have
lain probably under the thick shade of the beeches for
many Years; but could not vegetate till the sun & air
were admitted.

July 23. Farmer Knight sold two loads[11] of wheat
for 36 ae! Brisk gale. Hay makes well.

July 25. No garden-beans gathered yet. Threw
the hay in the meadows into large cocks. The lime-
trees with their golden tassels make a most beautiful
show. Hops throw out their side branches, which are
to bear the fruit. Cran-berries at bins pond not ripe.
Hog pease are hacking at Oakhanger.

July 26. By observing two glow-worms, which
were brought from the field to the bank in the garden,

it appeared to us, that those little creatures put-out their lamps between eleven & twelve, & shine no more for the rest of the night.

July 27.   Farmer Spencer & Farmer Knight are beginning to lime their respective farms at Grange & Norton.

July 28.   Lapwings leave the bogs, & moors in large flocks, & frequent the uplands.

July 29.   The land-springs [bournes] have run for some time, & especially in the hollow lane that leads from the village to Rood.

July 30.   John Hale brings home a waggon-load of woollen-rags, which are to be strewed on his hop-grounds in the spring, & dug in as manure.   These rags weighed at ton weight & cost brought home near six pounds.   They came from Gosport.

July 31.   Louring, vast rain, blowing.   This rain was very great at Malpas, in Cheshire.

Aug. 1.   Strong wind in the night which has injured the hops ; & particularly farmer Spencer's in Culver croft.   Trenched out several rows of celeri ; but the plants are of a red ugly colour, & seem not to be a good sort.   The seed came from the gardener at Alton.

Aug. 2.   The goose-berries are bent to the ground with loads of fruit.

Aug. 3.   Wheat reaped at Ropley.   Ripening weather.   Ant-flies begin to come forth on their business of emigration.

Aug. 4.   *Sedum Telephium*, orpine, & *Hypericum Androsaemum*, tutsan, growing in Emshot lane leading to Hawkley mill.

Aug. 5.   Mrs Brown brought to bed of a daughter, who makes the number of my nephews & nieces 54. Forest-fuel brought in.   Beechen fuel brought in. Wood-straw-berries are over.

Aug. 6.   *Rhus Cotinus*, sive Coccygria [smoke-tree] blows; it's blossom is very minute, & stands on the extremities of it's filiform bracteols, which have a sort of feather-like appearance that gives the shrub a singular, & beautiful grace. This tree does not ripen it's berries with us. Is a native of Lombardy, & to be found at the foot of the Apennine, & in Carniola.

Aug. 7. M^r & Mrs Barker, & Miss Eliz. Barker rode to Blackdown [near Haslemere] to see the prospect, & returned by three o'clock : they set out at six in the morning.

Aug. 8. Two poor, half-fledged fern-owls were brought me : they were found out in the forest among the heath. Farmer Hewet of Temple cut 30 acres of wheat this week. This wheat was lodged before it came into ear, & was much blighted. It grew on low grounds [Gault] : the wheat on the high malms [Upper Greensand] at Temple is not ripe.

Aug. 9. The country people have a notion that the *Fern-owl* or *Churn-owl*, or *Eve-jarr*, which they also call a *Puckeridge*, is very injurious to weanling calves by inflicting, as it strikes at them, the fatal distemper known to cow-leeches by the name of *Puckeridge*. Thus does this harmless, ill-fated bird fall under a double imputation, which it by no means deserves ; in Italy, of sucking the teats of goats, whence it is called *Caprimulgus*; & with us, of communicating a deadly disorder to cattle. But the truth of the matter is, the malady above-mentioned is occasioned by the *Oestrus bovis*, a dipterous insect, which lays it's eggs along the backs (chines) of kine,[12] where the maggots, when hatched, eat their way thro' the hide of the beast into the flesh, & grow to a very large size. I have just talked with a man, who says he has, more than once stripped calves who have dyed of the *puckeridge* ; that the ail, or complaint lay along the chine, where the flesh was much swelled, & filled with purulent matter.[13] Once myself I saw a large rough maggot of this sort taken (squeezed) out of the back of a cow. These maggots in Essex are called *wornils*. The least observation & attention would convince men, that these birds neither injure the goatherd, nor the grazier, but are perfectly harmless, & subsist alone, being night birds, on night-insects, such as *scarabaei* & *phalaenae*; thro' the month of July mostly on the *scarabaeus solstitialis*, which in many districts abounds at that season. Those that we have opened, have always had their craws stuffed with large night-moths & their eggs, & pieces of chafers: nor does it any-wise appear how they can, weak & unarmed as they seem, inflict any harm upon kine, unless they possess the powers of animal magnetism, & can affect them by fluttering over them. Mr Churton informs me " that the disease along the chine of calves, or rather the maggots that cause them, are called by the graziers in Cheshire *worry brees*, & a single one *worry-bree*." No doubt they mean a *breese*, or *breeze*, one name for the gad-fly or *Oestrus*, which is the parent of these maggots, & lays it's eggs on the backs of kine. Dogs come into my garden at night, & eat my goose-berries. Levant weather.

Aug. 10. *Monotropa Hypopithys* abounds in the hanger beyond Maiden dance, opposite to coney-croft hanger.

Aug. 11. Got-in forest-fuel in nice order. Farmer Knight begins wheat-harvest. Lovely weather.

Aug. 12. The planters think these foggy mornings, & sunny days, injurious to their hops.

Aug. 17. Cool air. Wheat gleaned.

Aug. 18. Many pease housed. Harvest-scenes are now very beautiful! Turnips thrive since the shower.

Aug. 19. Timothy Turner's brew-house on fire: but much help coming in & pulling off the thatch, the fire was extinguished, without any farther damage than the loss of the roofing. The flames burst thro' the thatch in many places. We are this day annoyed in the brown parlor by multitudes of flying ants, which come forth, as usual, from under the stairs.

Aug. 22. Mrs Ben White came to us from Newton.

Aug. 23. Boy brought me the rudiments of a hornet's nest, with some maggots in it. [Entry about ants (6 Aug. 1781), repeated *verbatim*, but with this addition: " The males that escape being eaten, wander away & die."]

Aug. 24. A fern-owl sits about on my field walks.

Aug. 25. Sweet harvest weather. Wheat ricked & housed. Mʳ & Mʳˢ S. Barker, & Miss E. Barker left us.

Aug. 27. Tho. Holt White comes from Fyfield.

Aug. 28. *Colchicum autumnale*, naked boys, blows. Wheat-harvest goes on finely.

Aug. 30. Michaelmass daiseys begin to blow.

Aug. 31. Gathered a *bushel-basket* of *well-grown* cucumbers, 238 in number. Molly White, & T. H. White left us, & went to London.

Sept. 2. Bees feed on the plums, & the mellow goose-berries. They often devour the peaches, & nectarines.

Sept. 3. Mʳ Charles Etty returns from Canton.

Sept. 4. Mʳ Thomas Mulso[14] comes from London. *Wry-necks*, birds so called, appear on the grass-plots & walks: they walk a little as well as hop, & thrust their bills into the turf, in quest, I conclude, of ants, which are their food. While they hold their bills in the grass, they draw out their prey with their tongues, which are so long as to be coiled round their heads.

Sept. 6. Fog, sun, pleasant showers, moonshine. Rain in the night. Mushrooms begin to come. I see only now & then a wasp.

Sept. 7. Mʳ Thomas Mulso left us & went to Wintõn.

Sept. 8. Broʳ T. W. & Th. H. W. came from London.

Sept. 9. Hops are not large. The *fly-catchers*, which abounded in my outlet, seem to have withdrawn themselves. Some grapes begin to turn colour. Men bind wheat. Sweet harvest, & hop-picking weather. Hirundines congregate on barns, & trees, & on the tower. The hops are smaller than they were last year. There is fine clover in many fields.

Sept. 11. *Ophrys spiralis*, ladies traces, in bloom in the long Lythe, & on the top of the short Lythe. Wasps seize on butter-flies, &, shearing off their wings, carry their bodies home as food for their young : they prey much on flies.

Sept. 12. Some wheat is out. Trimming has a large field not cut. *Gentiana Amarella*, autumnal gentian, or fell-wort, buds for bloom on the hill. Sent 12 plants of *Ophrys spiralis* to Mr Curtis[15] of Lambeth marsh.

Sept. 13. After a bright night, & vast dew, the sky usually becomes clouded by eleven or twelve o'clock in the forenoon ; & clear again towards the decline of the day. The reason seems to be, that the dew, drawn-up by evaporation, occasions the clouds, which towards evening, being no longer rendered buoyant by the warmth of the sun, melt away, & fall down again in dews. If clouds are watched of a still, warm evening, they will be seen to melt away, & disappear. Several nests of gold-finches, with fledged young, were found among the vines of the hops : these nestlings must be second broods.

Sept. 15. The hops at Kimbers grow dingy & lose their colour. T.H.W. left us, & went to Fyfield.

Sept. 16. Timothy the tortoise is very dull, & inactive, & spends his time on the border under the fruit-wall.

Sept. 17. No mushrooms on the down.

Sept. 18. Began to light fires in the parlors. Some young martins in a nest at the end of the brew-house. Small uncrested wrens, chif-chaffs, are seen in the garden.

Sept. 19. No mushrooms in the pastures below Burrant-hangers. Here & there a wasp. The furze-seed which Bro. Tho. sowed last may on the naked part of the hanger, comes up well. Some raspberry-trees in the bushes on the common. Trees keep their verdure well.

Sept. 20. Black-birds feed on the elder berries.

Sept. 21. Myriads of Insects sporting in the sun-beams.

Sept. 23. We find no mushrooms on the down, nor on Nore hill. Women continue to glean, but the corn is grown in the ears. Will. Trimming has wheat still abroad. Gathered-in the white pippins, a large crop.

Sept. 24. M^r & Mrs Ben White came from London.

Sept. 25. Men bag-their hops; & house seed-clover. A fern-owl plays round the Plestor. As we were walking this day, Sept. 22nd: being the King's coronation, on Nore-hill at one o' the clock in the afternoon, we heard great guns on each side of us, viz. from the S. & from the N.E., which undoubtedly were the cannons of Portsmouth & Windsor; the former of which is at least 26 miles distant, & the latter 30. If the guns heard from the N.E. were not from Windsor, they must be those of the Tower of London.

Sept. 26. Multitudes of Hirundines. Sweet Mich: weather.

Sept. 27. A man brought me a *land-rail* or *daker-hen*, a bird so rare in this district, that we seldom see more than one or two in a season, & those only in autumn. This is deemed a bird of passage by all the writers; yet from it's formation seems to be poorly qualifyed for migration; for it's wings are short, & placed so forward, & out of the centre of gravity, that it flies in a very heavy & embarrassed manner, with it's legs hanging down; & can hardly be sprung a second time, as it runs very fast, & seems to depend more on the swiftness of it's feet than on it's flying. When we came to draw it, we found the entrails so soft & tender, that in appearance they might have been dressed like the ropes of an woodcock. The craw or crop was small & lank, containing a mucus; the gizzard thick & strong, & filled with many shell-snails, some whole, & many ground to pieces thro' the attrition which is occasioned by the muscular force

& motion of that intestine. We saw no gravels among the food: perhaps the shell-snails might perform the functions of gravels or pebbles, & might grind one another. *Land-rails* used to abound formerly, I remember, in the low, wet bean-fields of Xtian Malford in North Wilts; & in the meadows near Paradise-Gardens at Oxford, where I have often heard them cry Crex, Crex.[16] The bird mentioned above weighed seven ounces and an half, was fat & tender, & in flavour like the flesh of a woodcock. The liver was very large & delicate.

Sept. 29. Swallows not seen: they withdraw in bad weather, & perhaps sleep most of their time away like dogs & cats, who have a power of accumulating rest, when the season does not permit them to be active.

Oct. 3. Gathered in bu(e)rgamot, & Creson bu(e)rgamot pears. Gathered some grapes, but they are not good. B. Th. White sowed two pounds of furze-seed from Ireland[17] on the naked part of the hanger. The furze-seed sown by him on the same space in May last is come-up well.

Oct. 4. The breed of hares is great: last year there were few. Some have remarked that hares abound most in wet summers.

Oct. 5. Gathered in Chaumontel pears: tied endive. Mr Ben, & Mrs Ben White left us.

Oct. 6. Grapes do not ripen: they are as backward as in the bad summer 1782: the crop is large.

Oct. 7. Many loads of hops set-out for wey hill.

Oct. 9. A bag of hops from Master Hale, weight 36 pounds, & an half.

Oct. 10. Two hop-waggons return with loads of woollen rags, to be spread & dug in as manure for the hop-gardens.

Oct. 11. A trufle-hunter called on us, having in his pocket several large trufles found in this neighbourhood. He says these roots are not to be found in deep woods, but in narrow hedge rows & the skirts of

coppices. Some trufles, he informed us, lie two feet
within the earth; & some quite on the surface : the
latter, he added, have little or no smell, & are not so
easily discovered by the dogs as those that lie deeper.
Half a crown a pound was the price which he asked
for this commodity. Some few bunches of grapes
just eatable. Some of the latter nectarines well-
flavoured. On this day, Dᴿ Chandler saw several
swallows, flying as usual, near Cologne : he had

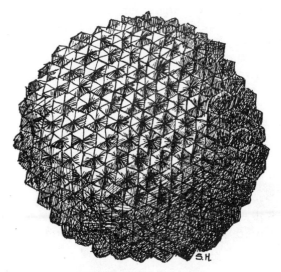

TRUFFLE (*Tuber aestivum*).
Diameter varies from 1-3 inches.

observed none at Rolle since the beginning of
September, nor none on his way to Cologne. On
the 12, in his way to Brussels, he saw more.

    Oct. 15. Mackarel sky. A wheat-ear seen on the
down.

    Oct. 16. Colchicums, a fine double sort, still in
bloom. Ivy blows. Some mushrooms with thick
stems, & pale gills.

    Oct. 19. [Newton] fierce, driving rain !

Oct. 20. Gathered in nonpareils, & some royal russets.

Oct. 21. *Woodcock* seen on the down, among the fern. Finished gathering the apples, many of which are fair fruit. Shoveled the zigzag. Leaves fall. My wall-nut trees, & some ashes are naked.

Oct. 22. Mended the planks of the zigzag. Bro. Tho. White sowed the naked part of the hanger with great quantities of hips, haws, sloes, & holly-berries. In May last he sowed a pound of furze seeds on the same naked space; many of which appear to have grown: & lately he sowed two pounds more. Decembr 1790. As fast as any of these seeds have sprouted, they have constantly been brouzed off, & bitten down by the sheep, which lie very hard on them, & will not suffer them to thrive.

Oct. 23. The quantity of haws is prodigious !

Oct. 26. Bror Th. W. sows laburnum seeds on the hanger, & down. A wood-cock killed in the high wood.

Oct. 27. Planted out many young laurustines, & Portugal laurels from the old stools.

Oct. 28. The young men of this place found a stray fallow deer at the back of the village, which they roused, & hunted with grey hounds, & other dogs. When taken it proved to be a buck of three years old.

Oct. 29. Bror Th. W. left us, & went to London.

Oct. 30. My horses taken into the stable & not to lie out any more a nights. New coped the top of my kitchen-chimney, mended the tiling, & toached[18] the inside of the roofing to keep out the drifting snow.

Nov. 3. Planted 150 cabbages to stand the winter: dunged the ground. Grapes all very bad. Two swallows were seen this morning at Newton vicarage house, hovering & settling on the roofs, & out-buildings. None have been observed at Selborne since Octobr 11. It is very remarkable, that after the hirundines have disappeared for some weeks, a few

are occasionally seen again sometimes, in the firſt week
of Novemʳ, & that only for one day.   Do they withdraw & slumber in some hiding-place during the
interval ? for we cannot suppose they had migrated
to warmer climes, & so returned again for one day.
Is it not more probable that they are awakened from
sleep, & like the bats are come forth to colleɕt a little
food ?   Bats appear at all seasons through the autumn
& spring months, when the Thermomʳ is at 50, because
then phalaenae & moths are ſtirring.   These swallows
looked like young ones.

Nov. 4.   The wind on Saturday laſt [Oɕt. 31ˢᵗ]
occasioned much damage among the shipping in the
river, & on the E. coaſt.

Nov. 5.   Bro. & Siſter Benjⁿ came to us from
Newton.

Nov. 6.   The hermitage capped with snow.

Nov. 11.   The tortoise is going under ground, but
not quite buried : he is in motion, & pushing himself
beneath the turf.

Nov. 12.   Broʳ & Siſter Benjⁿ left us, & went to
Newton.   Tortoise almoſt covered.

Nov. 15.   A flock of red-wings.   Men have not
finished their wheat season : some low grounds too
wet to be sown.

Nov. 16.   Few woodcocks ; & few pheasants left.
Many hares have been found on our hill : the wetness
of the season, it is supposed, induces them to leave the
vales, & to retreat to the uplands.   Reb. & Hannah
White came from Newton.

Nov. 17.   Dᵒ left us.   Flood at Gracious ſtreet.

Nov. 24.   The miller supplies us with cold, damp
flour, & says he can get no other : he adds, that the
beſt wheat is at the bottom of the mows[19], & will not
come forth till the spring.   The latter part of the wheat
harveſt was very wet.

Nov. 28.   Rime on the hanger.

Nov. 29.   Housed 8 cords[20] of beech billet, which
had taken all the rains of the late wet summer, &

autumn; & is therefore of course in but indifferent order.

Nov. 30.   After the servants are gone to bed the kitchen-hearth swarms with minute crickets not so large as fleas, which muſt have been lately hatched. So that these domeſtic inſects, cherished by the influence of a conſtant large fire, regard not the season of the year; but produce their young at a time when their congeners are either dead, or laid up for the winter, to pass away the uncomfortable months in the profoundeſt slumbers, & a ſtate of torpidity.

Dec. 3.   Beautiful picturesque, partial fogs along the vales, representing rivers, islands, & arms of the sea!   These fogs in London & other parts were so deep that much mischief was occasioned by men falling into rivers, & being over-turned into ditches, &c.

Dec. 5.   Mrs Ben White brought to bed of a Son, who makes my nephews & nieces 55 in number.

Dec. 6.   A bushel of American wheat, which Bro. Tho. sent laſt year to one of his tenants in the hundreds of Essex from Nore hill, produced this harveſt 40 bushels of seed :—and is much admired in that diſtrict, because from the ſtiffness of it's ſtraw it does not lodge. Wheat is so apt to lodge in these parts, that they are often obliged to mow it down in the blade about May, leſt it should fall flat to the ground.   This process they call swonging.

Dec. 8.   The Bramshot hounds kill a leash of hares on the hill.

Dec. 9.   The Emshot hounds kill a leash of hares on the hill.

Dec. 13.   One of my neighbours, shot a ring-dove on an evening as it was returning from feed, & going to rooſt.   When his wife had picked & drawn it, she found its craw ſtuffed with the moſt nice & tender tops of turnips.   These she washed & boiled, & so sate down to a choice & delicate plate of greens, culled & provided in this extraordinary manner.   Hence we

may see that granivorous birds, when grain fails, can
subsist on the leaves of vegetables. There is reason to
suppose that they would not long be healthy without ;
for turkies, tho' corn fed, delight in a variety of plants,
such a cabbage, lettuce, endive, &c., & poultry pick
much grass; while geese live for months together on
commons by grazing alone.

"Nought is useless made ; . . .
    . . . On the barren heath
The shepherd tends his flock, that daily crop
Their verdant dinner from the mossy turf
Sufficient : after them the cackling *Goose*
Close-grazer, finds wherewith to ease her want."

Philips's Cyder.

Dec. 19. Walked down to short heath : the sands
were very comfortable, & agreeable to the feet : the
grass grounds, & arable paths very wet, & unpleasant.

Dec. 23. Dark & dismal. Mr Churton came from
Oxford.

Dec. 25. Our rivulets were much flooded ; & the
water at Oakhanger ran over the bridge, which in old
days was called tun-bridge.

Dec. 31. Storm in the night, that blew down my
rain-measurer. The newspapers say that there are
floods on the Thames.

CHAPTER XXIII

" Only that day dawns to which we are awake."
                                        THOREAU : *Walden*.

1790

Jan. 1.   Frost, ice, sun, pleasant, moon-light.   The hounds found a leash of hares on the hill.

Jan. 3.   The spotted Bantam lays a second time.

Jan. 7.   Mr Churton left us, & went to Waverley. Sweet weather : gnats play in the air.   Paths dry.

Jan. 8.   Boys play at taw on the plestor.

Jan. 9.   Water-cresses come in.

Jan. 10.   A ripe wood-straw-berry on a bank, & several blossoms.   Grass grows on the walks.

Jan. 11.   The white spotted Bantam hen lays.

Jan. 12.   Snow-drops blow.   We have in the window of the stair-case a flower-pot with seven sorts of flowers, very sweet & fragrant.

Jan. 13.   driving rain all day.

Jan. 14.   A large *speckled diver* or *loon* was sent to me from the Holt, where it was shot by one of Lord Stawell's servants as it was swimming & diving on a large lake or pond.   These birds are seldom seen so far S. in mild winters.

Jan. 16.   Turnip-greens come in.

Jan. 19.   A trufle-hunter came with his dogs, & tryed my tall hedges, where, as he told us, he found only a few small bulbs, because the season was over : in the autumn, he supposes, many large trufles might be met with.   He says, trufles do not flourish in deep woods, but in hedge-rows, & the skirts of coppices within the influence of the sun & air.

Feb. 1.   A fine young hog salted & tubbed ; weight 7 scores, & 18 pounds.

Feb. 6. The great *titmouse*, or *sit-ye-down*, sings. One crocus is blown-out. Insects abound in the air : bees gather much on the snow-drops, & winter-aconites. Gossamer is seen streaming from the boughs of trees.

Feb. 10. Bull-finches pick the buds of damson-trees.

Feb. 11. Three gallons of best french brandy from London.

Feb. 19. The moon & Venus in the S.W. & Jupiter & Mars in the E. make nightly a charming appearance.

Feb. 20. As the Surveyor of Gosport-turnpike was mending the road in *Rumsdean bottom*, he found several Roman coins,[1] one of which was silver. Hence we may conclude that the remarkable entrenchments in that valley, whatever use may have been made of them since, were originally Roman. There is a tradition that they were frequently occupied during the grand rebellion in the time of Charles the first, a period in which many skirmishes happened in these parts, as at Cherriton, Alton, &c. These trenches must have been a post of consequence, because they are on a great road, & between large sloping woods. At the S.W. end of this valley, towards *Filmer-hill*, in a place called *Feather-bed-lane*, are three large contiguous barrows which seem to indicate that near the spot some considerable battle must have been fought in former times.

Feb. 21. Frost, ice, bright, red even, prodigious white dew.

Feb. 24. Dr Chandler came.

Feb. 25. Cabbage sprouts come in. Both the pullets of last summer lay.

Feb. 27. Daffodils begin to open. Dr Chandler left us.

Feb. 28. Violets abound.

Mar. 2. Sowed the meadow with ashes ; of my own 22 bushels, bought 39 : total 61.

Mar. 3. Sheep turned into the wheat.

Mar. 4. Timothy the tortoise comes forth : he does not usually appear 'till the middle of April.

Mar. 5. The tortoise does not appear. The trufle-man still follows his occupation : when the season is over, I know not.

Mar. 6. A couple & an half of woodcocks, & several pheasants were seen in Hartley-wood. [On a blank leaf is the draft of White's lines : " On the dark, still, dry warm weather occasionally happening during the winter-months."]

Mar. 7. The wheat in the N. field looks well : there has been no good crop since the year 1780.

Mar. 10. About this time Charles Etty sailed for Bengal direct, as second mate to the Earl Fitzwilliam India-man : Dundas captain.

Mar. 11. Several hundreds of fieldfares on the hill : they probably congregate in order to migrate together.

Mar. 13. Planted curran-trees. The garden hoed, & cleaned.

Mar. 14. About this time Ned White [nephew] is to sail for Antegoa in the Lady Jane Halliday : Ross, Captain.

Mar. 15. A vast snake ²appears at the hot-beds.

Mar. 16. Dog's toothed violets blow.

Mar. 17. Timothy the tortoise lies very close in the hedge.

Mar. 20. That noise in the air of some thing passing quick over our heads after it becomes dark, & which we found last year proceeded from the *Stone-curlew*,³ has now been heard for a week or more. Hence it is plain that these birds, which undoubtedly leave us for the winter, return in mild seasons very soon in the spring ; & are the earliest summer birds that we have noticed. They seem always to go down from the uplands towards the brooks, & meads. The next early summer bird that we have remarked is the smallest *Willow-wren*, or *chif-chaf* ; it utters two sharp, piercing notes, so loud in hollow woods as to occasion an echo, & is usually first heard about the 20th of March.

Mar. 21. *Bombylius medius*, a hairy fly, with a long projecting snout, appears : they are seen chiefly in Mar. & April. "Os rostro porrecto, setaceo, longissimo, bivalvi." A dipterous insect, which sucks it's aliment from blossoms. On the 21st of March a single *bank*, or *sand*-martin was seen hovering & playing round the sand-pit at short heath, where in the summer they abound. I have often suspected that *S. martins* are the most early among the hirundines.

Mar. 25. Chaffinches pull-off the finest flowers of the polyanths. Ned White sailed on this day.

Mar. 28. Small birds, Tanner says green finches, pull off my polyanth blossoms by handfulls. A neighbour complained to me that her house was over-run with a kind of *black-beetle*, or, as she expressed herself, with a kind of *black-bob*, which swarmed in her kitchen when they get up in a morning before day-break. Soon after this account, I observed an unusual insect in one of my dark chimney-closets ; & find since that in the night they swarm also in my kitchen. On examination I soon ascertained the Species to be the *Blatta orientalis* of Linnaeus, & the *Blatta molendinaria* of Mouffet. The male is winged, the female is not ; but shows somewhat like the rudiments of wings, as if in the pupa state. These insects belonged originally to the warmer parts of America, & were conveyed from thence by shipping to the East Indies ; & by means of commerce begin to prevail in the more N. parts of Europe, as Russia, Sweden, &c. How long they have abounded in England[4] I cannot say ; but have never observed them in my house 'till lately. They love warmth, & haunt chimney-closets, & the backs of ovens. Poda says that these, & house-crickets will not associate together ; but he is mistaken in that assertion, as Linn. suspected that he was. They are altogether night insects, *lucifugae*, never coming forth till the rooms are dark, & still, & escaping away nimbly at the approach of a candle. Their antennae are remarkably long, slender, & flexile.

Mar. 31. When h. crickets are out, & running about in a room in the night, if surprized by a candle, they give two or three shrill notes, as it were a signal to their fellows, that they may escape to their crannies & lurking-holes to avoid danger.

Apr. 1. Sharp, & biting wind. Some crude oranges were put in a hot cupboard in order that the heat might mellow them, & render them better flavoured : but the crickets got to them, & gnawing holes thro' the rind, sucked out all the juice, & devoured all the pulp.

Apr. 2. *Nightingales* heard in honey-lane.
" The Nightingall, that chaunteth all the springe,
Whose warblinge notes throughout the wooddes are
        harde,
Beinge kepte in cage, she ceaseth for to singe,
And mourns, bicause her libertie is barde."
        Geffrey Whitney's Emblemes : 1586, p. 81.

Apr. 4. Sharp, cutting wind ! Heath-fire in the forest makes a great smoke.

Apr. 5. [Alton] Frost, sun & clouds, sharp wind.

Apr. 6. [Reading] Young goslings on commons.

Apr. 7-11. [Oxford] Thames very full & beautiful, after so much dry weather : wheat looks well ; meadows dry, & scorched ; roads very dusty.

Apr. 11. Deep snow at Selborne: five inches deep! *Red-starts*, *Fly-catchers*, & *Black-caps* arrive. If these little delicate beings are birds of passage (as we have reason to suppose they are, because they are never seen in winter) how could they, feeble as they seem, bear up, against such storms of snow & rain ; & make their way thro' such meteorous turbulencies, as one should suppose would embarrass & retard the most hardy & resolute of the winged nation ? Yet they keep their appointed times & seasons, & in spite of frosts & winds return to their stations periodically, as if they

had met with nothing to obstruct them.   The with-
drawing & appearance of the *short-winged* summer birds
is a very puzzling circumstance in natural History !

Apr. 12.   [Reading.]
Apr. 13.   [Alton.]
Apr. 14.   [Selborne.]
Apr. 18.   A boy has taken three little young
Squirrels in their nest, or *drey*, as it is called in these
parts.   These small creatures he put under the care
of a cat who had lately lost her kittens, & finds that
she nurses & suckles them with the same assiduity &
affection, as if they were her own offspring.   This
circumstance corroborates my suspicion, that the
mention of deserted & exposed children being
nurtured by female beasts of prey who had lost their
young, may not be so improbable an incident as many
have supposed :—& therefore may be a justification
of those authors who have gravely mentioned what
some have deemed to be a wild & improbable story.
So many people went to see the little squirrels suckled
by a cat, that the foster mother became jealous of her
charge, & in pain for their safety ; & therefore hid
them over the ceiling, where one died.   This circumstance shews her affection for these foundlings, & that
she supposes the squirrels to be her own young.   The
hens, when they have hatched ducklings, are equally
attached to them as if they were their own chickens.
For a leveret nursed by a cat see my Nat : History,
p. 214.   I have said " that it is not one whit more
marvellous that Romulus, & Remus, in their infant
& exposed state, should be nursed by a she wolf, than
that a poor little suckling leveret should be fostered &
cherished by a bloody grimalkin."

Apr. 20.   Set the old Bantam speckled Hen with
eleven eggs.   My cook-maid desired there might be
an odd egg for good luck :   . . .   numero Deus
impare gaudet [Virgil : Ecl. 8, l. 75].

Apr. 24.   Planted potatoes & beans in the meadow
garden.   Much thunder & hail at Alton.

Apr. 28. Full moon. Total eclipse.

Apr. 29. Doctor Chandler, & lady came to the parsonage house.

May 6. Mrs Chandler brought to bed of a daughter at the parsonage-house.

May 8. Began to mow the orchard for the horses.

May 9. Master Trimming is taken with the small-pox. Timothy the tortoise eats dandelion leaves & stalks : he swallows his food almost whole.

May 10. The Bantam hen hatches seven chickens. Young red breasts. Made some tarts with the stalks of the leaves of the garden, or Monks rhubarb. Only three swifts ; one was found dead in the church-yard.

May 12. The rhubarb-tart good, & well-flavoured.

May 13. Bro. Tho. came from London.

May 15. Timothy the tortoise weighs 6 ae 12 oz. 14 drs.

May 16. One polyanth-stalk produced 47 pips or blossoms. Mrs Edmund White brought to bed of a boy, who has encreased the number of my nephews & nieces to 56. The bloom of apples is great : the white pippin, as usual, very full. It is a most useful tree, & always bears fruit. The dearling in the meadow is loaded with fruit : last year it produced only one peck of apples, the year before 14 bushels. [Later note.] This year it bore 10 bush. of small fruit. The white pippin produced a good crop again this year : the apples of this tree come in for scalding, & pies in August.

May 22. Monks rhubarb in full bloom.

May 25. Sowed a specimen of some uncommon clover from farmer Street. Sowed a pint of large kidney beans, white : also Savoys, Coss lettuces, & bore-cole [kale, or " curled greens "].

May 27. Thunder : damage done in London.

May 30. John Carpenter brings home from the Plashet at Rotherfield some old chest-nut trees which are very long. In several places the wood-peckers had begun to bore them. The timber & bark of these trees are so very like oak, as might easily deceive an indifferent observer, but the wood is very shakey, & towards the heart *cup-shakey*,[5] so that the inward parts are of no use. They were bought for the purpose of cooperage, but must make but ordinary barrels, buckets, &c. Chestnut sells for half the price of oak ; but has some times been sent into the King's docks, & passed off instead of oak.

May 31. Bottled-out the port-wine which came here in October, but did not get fine.

June 4. *Ophrys nidus-avis,*and *ophrys apifera* blossom.

June 6. [Note about the shearing of ewes and lambs (1 Aug. 1777) repeated, with these additions.] Thus dogs smell *to* persons when they meet, when they want to be informed whether they are strangers or not. After sheep have been washed, there is the same confusion, for the reason given above.

June 7. [London] Went to London by Guildford & Epsom. Spring-corn & grass look well. Hay making near town.

June 17-*July* 1. [S. Lambeth.]

June 12. Cauliflowers abound. Pease sold for ten pence the peck.

June 13. Artichokes, & chardons,[6] come into eating. Cucumbers abound.

June 14. Sweet hay-making weather.

June 16. My Brother finishes a large rick of hay in very nice order.

June 20. Muck laid on a gardener's field poisons my Brother's outlet. A martin at Stockwell chapel[7] has built its nest against the window : it seems to stick firmly to the glass, and has no other support. In former summers I remember similar instances.

June 21. Scarlet-straw-berries good. A small praecox melon. The longest day :

" The longest daye in time resignes to nighte ;
The greatest oke in time to duste doth turne ;
The Raven dies ; the Egle failes of flighte ;
The Phoenix rare in time herselfe doth burne ;
The princelie stagge at lengthe his race doth ronne ;
And all must ende that ever was begonne."

   Geffrey Whitney's Emblemes ; p. 230, 1586.

June 22. Thermometer at Mr Alexander's—87 on
a N. wall ; a S. wall near. Fruit-walls in the sun are
so hot that I cannot bear my hand on them. Bro^r Tho.
therm^r was 89 on an E. wall in the afternoon. [Later
note.] Much damage was done, & some persons
killed by lightening on this sultry day. My Bro.
Tho^{s's} therm^r in Blackfriars road against an eastern
wall in the afternoon was 89. My thermom^r after the
sun was got round upon it, was 100 : Thomas forgot
to look in time.

June 27. Roses make a beautiful show. Orange-
lillies blossom. Sr George Wheeler's tutsan
blows.

July 2. [Alton] Two heavy showers at Guildford
with thunder.

July 3. [Selborne] My hay made into small cocks.
Young swallows come out, & are fed on the wing.
Wood straw-berries ripen.

July 4. The woman, who brought me two fern-
owls eggs last year on July 14, on this day produced me
two more one of which had been laid this morning,
as appears plainly, because there was only one in the
nest the evening before. They were found, as last
July, on the verge of the down above the hermitage,
under a beechen shrub on the naked ground. Last
year those eggs were full of young, & just ready to
be hatched. These circumstances point out the exact
time when these curious nocturnal, migratory birds
lay their eggs & hatch their young. Fern-owls,
like snipes, stone-curlews, & some other birds, make
no nest. Birds that build on the ground do not
make much of nests.

July 7. Grasshopper-lark whispers in my outlet.
Turned the cocks of hay.

July 9. Gathered our first beans, long pods.
Planted-out annuals.

July 10. [Entry on "Instinct," copied in D.B.,
LVI. After the reference to the "hemispheric"
shape of the house-martin's nest, there is the following
addition.] In confirmation of what has been
advanced above, there are now two new martin's nests
at Tim Turner's, which are tunnel-shaped, & nine or
ten inches long, in conformity to the ledge of the wall
of the eaves under which they are built.

July 11. Now the meadow is cleared, the brood-
swallows sweep the face of the ground all day long ; &
from over that smooth surface collect a variety of
insects for the support of their young.

July 14. Tempest, & much thunder to the N.W.
Neither cucumbers, nor kidney beans, nor annuals
thrive on account of the cold blowing season.
Timothy the tortoise is very dull, & spends most of
his time under the shade of the vast, expanded leaves
of the monk's rhubarb.

July 15. Continual gales all thro' this month,
which interrupt the cutting my tall hedges.

July 17. Mr Churton came. A nightingale continues to sing ; but his notes are short & interrupted,
& attended with a *chur*.[8] A fly-catcher has a nest in
my vines. Young swallows settle on the grass-plots
to catch insects.

July 18. Mrs Clement & daughters came.

July 22. A man brought me a cuckoo, found in the
nest of a water-wagtail among the rocks of the hollow
lane leading to Rood. This bird was almost fledge.

July 24. Trenched four rows of celeri, good
streight plants. Lime trees in full bloom. Large
honey-dews on my great oak, that attract the bees,
which swarm upon it. Some wheat is much lodged
by the wind & rain. There is reason to fear from the
coldness & wetness of the season that the crop will

not be good.   Windy, wet, cold solstices are never
favourable to wheat, because they interrupt the bloom,
& shake it off before it has performed it's function.

July 25.   Lime trees are fragrant: the golden
tassels are beautiful.   Dr Chandler tells us that in the
south of France, an infusion of the blossoms of the
lime-tree, tilia, is in much esteem as a remedy for
coughs, hoarsenesses, fevers, &c., & that at Nismes
he saw an avenue of limes that was quite ravaged &
torn to pieces by people greedily gathering the bloom,
which they dryed & kept for their purposes.   Upon
the strength of this information we made some tea of
lime-blossoms, & found it a very soft, well-flavoured,
pleasant, saccharine julep, in taste much resembling
the juice of liquorice.

July 27.   Honey-dews, which make the planters in
pain for their hops.   Hops are infested with aphides;
look badly.

July 28.   Children gather strawberries every morning from the hanger where the tall beeches were felled
in winter 1788.

July 29.   Some mushrooms, & funguses appear on
the down.   Mrs J. White[9] made Rasp, & strawberry
jam, & red curran jelly, & preserved some cherries.

Aug. 1.   [Extract from a letter from Mr R. Marsham,
about tree planting 24 July 1790.   See R.H.W., II,
p. 221.   White gives the substance, rather than an
exact copy.]   The circumference of trees in my outlet
planted by myself, at one foot from the ground.

		f.	inch.
Oak by alcove in	1730	4	5
Ash by Do in	1730	4	6½
Great fir, bakers hill	1751	5	0
Greatest beech	1751	4	0
Elm	1750	5	3
Lime over at Mr Hale's[10]			
planted by me in	1756	5	5
My single great oak in the			
meadow, age unknown		10	6½

[H. G. Osborne, Alresford

The survivors of White's four lime trees planted in 1756 "in the butcher's yard." The shop was built later

[face p. 362

The diameter of it's boughs three ways is 24 yards,
or 72 feet : circumference of it's boughs 72
yards.

Mʳ White's single great oak at Newton measures at
one foot above the ground 12 feet 6 inch : the exact
dimensions of that belonging to, & planted by Mʳ
Marsham. A vast tree must that be at Stratton
[Norfolk] to have been planted by a person now living!

Aug. 3.   Mʳ Churton left us.

Aug. 5.   Piled & housed all the cleft wood of eight
cords of beech : the proportion of blocks was large.

Aug. 6.   The fern-owl churs still ; grass-hopper
lark has been silent some days.

Aug. 7.   Strawberries from the woods are over ;
the crop has been prodigious.   [Entry dated July 22.]
The decanter, into which the wine from a cool cellar
was poured, became clouded over with a thick condensation standing in drops. This appearance, which is
never to be seen but in warm weather, is a curious
phaenomenon, & exhibits matter for speculation to the
modern philosopher. A friend of mine enquires
whether the " *rorantia pocula* " of Tully in his " *de
senectute* " had any reference to such appearances.
But there is great reason to suppose that the ancients
were not accurate philosophers enough to pay much
regard to such occurrences. They knew little of
pneumatics, or the laws whereby air is condensed, &
rarifyed; & much less that water is dissolved in air, &
reducible therefrom by cold. If they saw such dews
on their statues, or metal utensils, they looked on them
as ominous, & were awed with a superstitious horror.
Thus Virgil makes his weeping statues, & sweating
brazen vessels prognostic of the violent death of Julius
Caesar : . . . " maestum illacrymat templis ebur,
aeraq sudant."   Georgic 1ˢᵗ [l. 480].

The phaenomenon in question is finely explained by
the following quotation.—" If a bottle of wine be
fetched out of a cool cellar in the hottest & driest
weather in summer, it's surface will presently be covered

with a thick vapour, which when tasted appears to be pure water. This watery vapour cannot proceed from any exudation of the wine thro' the pores of the bottle ; for glass is impervious to water, & the bottle remains full, & when wiped dry is found to weigh as much as when taken out of the cellar. The same appearance is observable on the outside of a silver, or other vessel in which iced water is put in summer time ; & it is certain, that the water which is condensed on the surface of the vessel does not proceed merely from the moisture exhaled by the breathing of the people in the room where this appearance is most generally noticed, because the same effect will take place, if the vessel be put in the open air." Watson's chemical essays, Vol. 3rd, p. 92.

Aug. 10. A labourer has mown out in the precincts of Hartley-wood, during the course of this summer, as many pheasant's nests as contained 60 eggs ! Bro. Thomas White came.

Aug. 12. Sister Barker, & nieces, Mary, & Eliz : came.

Aug. 14. Young Hirundines cluster on the trees. Harvest-bugs[11] bite the ladies.

Aug. 15. The last gathering of wood-straw-berries. Bull-finches & red-breasts eat the berries of the honeysuckles.

Aug. 16. Cut 43 cucumbers. Wheat is binding. *Blackstonia perfoliata*, yellow centory, blossoms, on the right hand bank up the North field hill. The *Gentiana perfoliata* Linnaei. It is to be found in the marl-dell[12] half way along the N. field lane on the left ; on the dry bank of a narrow field between the N. field hill, & the Fore down ; & on the banks of the Fore down.

Aug. 19. Mrs Barker & her daughters Mary & Elizabeth, & Mrs Chandler, & her infant daughter and nursemaid went all in a cart to see the great oak in the Holt, which is deemed by Mr Marsham of Stratton in Norfolk to be the biggest in this Island.[13] Bro. Thos. & Dr Chandler rode on horse-back. They all dined

THE GRINDSTONE OAK IN HOLT FOREST

under the shade of this tree. At 7 feet from the ground it measures in circumference 34 feet : has in old times lost several boughs, & is tending towards decay. Mr Marsham computes that at 14 feet length this oak contains 1000 feet of timber.

Aug. 20. On this day farmer Spencer built a large wheat-rick[14] near his house the contents of which all came from a field near West-croft barn at the full distance of a mile. Five waggons were going all day.

Aug. 22. There is a covey of partridges in the North-field, seventeen in number.

Aug. 23. John Hale made a large wheat-rick on a staddle.[15]

Aug. 26. Planted out a bed of borecole, & three long rows of curled endive. Bat comes out before the swallows are gone to roost.

Aug. 27. Cold & comfortless weather.

Aug. 30. Cut 152 cucumbers. A fine harvest day : much wheat bound, & much gleaning gathered.

Aug. 31. Farmer Spencer's wheat-rick, when it was near finished, parted, & fell down. Charles, & Bessy White came from Fyfield.

Sept. 3. Some hop-poles blown down. Mr Prowting of Chawton begins to pick hops.

Sept. 5. Boiled a mess of autumnal spinage, sown Aug. 3rd. Nep. J. White left us, & returned to Sarum.[16] There is a fine thriving oak near the path as you go to Combwood, just before you arrive at the pond, round which, at about the distance of the extremities of the boughs, may be seen a sort of circle in the grass, in which the herbage appears dry & withered, as if a fairy-ring was beginning. I remember somewhat of the same appearance at the same place in former years.

Sept. 6. Hardly here & there a wasp to be seen.

Sept. 9. Two stone-curlews in a fallow near Southington [Selborne parish]. A fern-owl flies over my House.

Sept. 10. Cut 140 cucumbers. Hops light, & not very good. Sister Barker & Molly & Betsy left us,

& went to London: Charles White also, & Bessy
returned to Fyfield.

Sept. 13.   But 158 cucumbers.   Nep. Ben White,
& wife, & little Ben & Glyd came from Fyfield.

Sept. 14.   Onions rot.   Barley round the village
very fine.

Sept. 16.   Cut 100 cucumbers.   Sweet autumnal
weather.

Sept. 17.   Martins congregate on the weather-cock,
& vane of the may-pole.   The boys brought me their
first wasps nest from Kimber's ; it was near as big as a
gallon.   When there is no fruit, as is remarkably the
case this year, wasps eat flies, & suck the honey from
flowers, from ivy blossoms, & umbellated plants :
they carry-off also flesh from butchers shambles.

Sept. 18.   My tall beech in Sparrow's hanger,
which measures 50 feet to the first fork, & 24 afterwards, is just 6 feet in girth at 2 feet above the ground.
At the back of Burhant house, in an abrupt field which
inclines towards nightingale-lane, stand four noble
beech trees on the edge of a steep ravin or water gully
the largest of which measures 9 ft. 5 in. at about a yard
from the ground.   This ravin runs with a strong
torrent in winter from nightingale-lane, but is dry in
the summer.   The beeches above are now the finest
remaining in this neighbourhood, & carry fine heads.[17]
There is a romantic, perennial spring in this gully, that
might be rendered very ornamental was it situated in a
gentleman's outlet.

Sept. 19.   On this day Lord Stawell sent me a rare
& curious water-fowl, taken alive a few days before
by a boy at Basing, near Basingstoke, & sent to the
Duke of Bolton at Hackwood park, where it was put
into the bason before the house, in which it soon dyed.
This bird proved to be the *Procellaria Puffinus* of
Linnaeus, the *Manks* [Manx] *puffin,* or *Shear-water* of
Ray.   Shear-waters breed in the *Calf of Man,* & as *Ray*
supposes, in the *Scilly Isles,* & also in the *Orknies* : but
quit our rocks & shores about the latter end of August ;

& from accounts lately given by navigators, are dispersed over the whole Atlantic. By what chance or accident this bird was impelled to visit Hānts is a question that can not easily be answered.[18]

Sept. 21. Mrs Clement, & six of her children, four of which are to be inoculated, & Mrs Chandler, & her two children the youngest of which is also to undergo the same operation, are retired to Harteley great house. Servants & all, some of which are to be inoculated also, they make 14 in family.[19]

Sept. 23. Coss-lettuce finely loaved & bleached! Nep. B. White left us, & went to London.

Sept. 24. Thomas cut 130 cucumbers.

Sept. 25. A vast flock of lapwings, which has forsaken the moors & bogs, now frequents the uplands. Some ring-ouzels were seen round Norehill.

Sept. 27. The inoculated at Harteley sicken.

Sept. 30. Cut 81 cucumbers. On this day Mrs Brown was brought to bed at Stamford of twins, making my nephews & nieces 58 in number. The night following this poor, dear woman dyed, leaving behind her nine young children.

Oct. 2. Bro. Thomas, & his daughter Mrs Ben White left us, & went to London. Lord Stawell sent me from the great Lodge in the Holt a curious bird for my inspection. It was found by the spaniels of one of his keepers in a coppice, & shot on the wing. The shape, & air, & habit of the bird, & the scarlet ring round the eyes, agreed well with the appearance of a cock pheasant; but then the head & neck, & breast & belly, were of a glossy black: & tho' it weighed 3 ae 3½ oun., the weight of a large full-grown cock pheasant,* yet there were no signs of any spurs on the legs, as is usual with all grown cock pheasants, who have long ones. The legs & feet were naked of feathers; & therefore it could be nothing of the Grous kind. In the tail were no long bending feathers, such as cock pheasants usually have, & are characteristic of the sex.　The tail was much shorter than the tail of an hen pheasant, & blunt & square at the end.　The back, wing-feathers, & tail, were all of a pale russet, curiously streaked, somewhat like the upper parts of an hen partridge.　I returned it to the noble sender with my verdict, that it was probably a spurious or hen bird, bred betwen a cock pheasant and some demestic fowl.[20]　When I came to talk with the keeper who brought it, he told me, that some Pea-hens had been known last summer to haunt the coppices & coverts where this mule was found.　*Hen pheasants usually weigh only 2 ae 1 oun.　My advice was that his Lordship would employ Elmer of Farnham, the famous game-painter, to take an exact copy of this curious bird.　[Later.] His Lordship did employ Elmer, & sent me as a present a good painting of that rare bird.

Oct. 3.　The row of ten weeks stocks under the fruit-wall makes a beautiful show.

Oct. 4.　Three martin's nests at Mr Burbey's are now full of young !

Oct. 5.　Cut 3 bunches of grapes : they were just eatable.

Oct. 7.　Timothy the tortoise came out into the walk, & grazed.　Mr Edmd White, while he was at South Lambeth, this summer, kept for a time a regular journal of his Father's barometer, which, when compared with a journal of my own for the same space, proves that the Mercury at S. Lambeth at an average stands full *three* tenths of an inch higher than at Selborne.　Now as we have remarked [D.B., LX] that the barometer at Newton Valence is invariably three tenth lower than my own at Selborne, it plainly appears that the mercury at S. Lambeth exceeds in height at an average the mercury at Newton by six tenths at least.　Hence it follows, according to some calculations, that Newton vicarage house is 600 feet[21] higher than the hamlet of S. Lambeth, which, as may be seen by the tide coming-up the creek[22] before some of the houses, stands but a few feet above high water

mark. It is much to be wished that all persons who attend to barometers would take care to use none but pure distilled Mercury in their tubes : because Mercury adulterated with lead, as it often is, loses much of it's true gravity, & must often stand in tubes above it's proper pitch on account of the diminution of it's specific weight by lead, which is lighter than mercury.[23] The remarks above show the futility of marking the plates of barometers with the words— *fair, changeable,* &c., instead of *inches,* & *tenths* ; since by means of different elevations they are very poor directions, & have but little reference to the weather. After the servants are gone to bed, the kitchen-hearth swarms with young *crickets, Blattae molendinariae,* of all sizes from the most minute growth to their full proportions. They seem to live in a friendly manner together, & not prey the one on the other.
" there the Snake throws her enamal'd skin."
Shakespear, Mids. night's dream [Act. II, Sc. 1].
About the middle of this month [September] we found in a field near a hedge the slough of a large snake, which seemed to have been newly cast. From circumstances it appeared as if turned wrong side outward, & as drawn off backward, like a stocking, or woman's glove. Not only the whole skin, but scales from the very eyes are peeled off, & appear in the head of the slough like a pair of spectacles. The reptile, at the time of changing his coat, had intangled himself intricately in the grass & weeds, so that the friction of the stalks & blades might promote this curious shifting of his exuviae.
" lubrica serpens
Exuit in spinis vestem."
Lucretius [*De Rerum,* IV, 59-60].
It would be a most entertaining sight could a person be an eye-witness to such a feat, & see the snake in the act of changing his garment. As the convexity of the scales of the eyes in the slough are [*sic*] now inward, that circumstance alone is a proof that the skin has

been turned : not to mention that now the present inside is much darker, than the outer. If you look through the scales of the snake's eyes from the concave side, viz : as the reptile used them, they lessen objects much. Thus it appears from what has been said that snakes crawl out of the mouth of their own sloughs, & quit the tail part last ; just as eels are skinned by a cook maid. While the scales of the eyes are growing loose, & a new skin is forming, the creature, in appearance, must be blind, & feel itself in an awkward uneasy situation.

Oct. 11. Gathered the Cardillac pears, a bushel ; the knobbed russets 2 bushels ; the kitchen, ruddy apple at the end of the fruit-wall, near a bushel.

Oct. 12. Gathered in near 4 bushels of *dearling* apples from the meadow tree : the crop is great, but the fruit is small.

Oct. 13. Gathered in a bushel more of dearlings. Mrs Chandler returns home from the Harteley inoculation.

Oct. 14. Gathered in more dearlings : the fruit is small, but the crop on that single tree amounts to nine bushels, & upwards.

Oct. 15. Gathered in the royal russets, & the nonpareils, a few of each. Gathered the berberries.

Oct. 16. *Red wings* return, & are seen on Selborne down. There are no haws this year for the redwings, & field fares.

Oct. 17. Gracious street stream is dry from James Knight's ponds, where it rises, to the foot bridge at the bottom of the church litton closes. Near that bridge, in the corner, the spring is perennial, & runs to Dorton, where it joins the Well-head stream.

Oct. 19. My well is very low, & the water foul.

Oct. 20. Spring-keepers[24] come up in the well-bucket. How they get down there does not appear : they are called by Mr Derham—*squillae aquaticae.*

Oct. 21. I conclude that the Holiburne trufler finds encouragement in our woods, & hangers, as he

frequently passes along the village : he is a surly
fellow, & not communicative. He is attended by
two little cur-dogs, which he leads in a string.

Oct. 24. Dr Chandler buys of the Holiburne trufle-
man one pound of trufles ; price 2s. 6d.

Oct. 25. A flock of 46 ravens over the hanger.
Slipped-out pinks, & fraxinals ; planted out dames
violets from cuttings.

Oct. 26. This morning Rear Admiral Cornish,
with six ships of the line, & two smaller ships of war,
sailed from St. Hellen's.

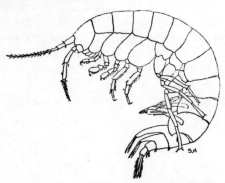

THE " SPRING-KEEPER " (*Gammarus pulex*).
A freshwater shrimp
(after Sars).
Average length: ½ inch.

Oct. 27. Grapes better.

Oct. 28. Wet & uncomfortable.

Oct. 29. Dug & cleansed the border in the orchard,
& planted it with polyanths slipped-out.

Oct. 30. Large *fieldfares*, a great flock, seen on the
hill. Ravens on the down. Wild *wood-pigeons*, or
stock-doves, are seen at my wood at Holtham [in East
Tisted, near Selborne].

Nov. 1. Bro^r Benjn & his wife came to us.

Nov. 4. Green wheat comes up well. Stewed
some trufles : the flavour of their juice very fine, but
the roots hard, & gritty. They were boiled in water,
then sliced, & stewed in gravy.

Nov 6.   Very rough weather at Portsmouth : boats over-set, & people drowned in coming from Spit-head.[25]

Nov. 11.   Two or three wood-cocks seen in the high wood : one was killed. Fyfield improves, & promises to make a good cock-dog [Cocker].

Nov. 13.   Bro^r & Sister Benjn left us.

Nov. 15.   Timothy the tortoise gone under ground in the laurel-hedge. Paths very dry : boys play at taw on the plestor.

Nov. 16.   Paths greazy from the frost. Raked, & swept up the leaves in my outlet. The hanger naked.

Nov. 20.   The parish church of Calstock in Cornwall destroyed by lightening. The tempest was of vast extent, & in many places mischievous, & awful ! This village lies up the Tamer, above Saltash.

Nov. 21.   A vast tempest at Sarum ; & an house beat down. The mast of a man of war was struck at Spit-head by the lightening.

Nov. 23.   The water in my well is risen three or four rounds of the winch, viz. five or six feet : the spring that runs in may be seen, & heard. The water is now clear. Thus will three or four inches of rain replenish my well, deep as it is, after it has been very low, & foul, & almost dry for several months. I have made the same remark in former years. Our stream has been so low for many weeks that the miller at Kingsley could not grind ; but was obliged to send his corn to Headleigh, where the Blackdown stream never fails. At Headleigh park-corner the Blackdown stream joins the Selborne rivulet : & at Tilford bridge they are met by the Farnham river, where together they form so considerable a body of water as within a few miles to become navigable, viz : at the town of Godalming ; & there take the name of Wey. [Cf. T.P., I.]

Dec. 6.   Mr Richardson came.

Dec. 9.   Mr Richardson left us. Water-cresses come in.

Dec. 13.   Blowing, rough day.

Dec. 16.   Thatch torn by the wind.

Dec. 23.   Thunder, lightening, rain, snow! A severe tempest. Much damage done in & about London : damage to some ships at Portsmouth. Vast damage in various parts! Two men were struck dead in a wind-mill near Rooks-hill on the Sussex downs : & on Hind-head one of the bodies on the gibbet was beaten down to the ground.[26] Harry & Ben Woods came.

Dec. 25.   H. & Ben Woods left us.

Dec. 29.   On this day Mrs Clements was delivered of a boy, who makes my nephews & nieces again 57 in number. By the death of Mrs Brown & one twin they were reduced to 56.

Dec. 31.   Total of rain in 1790, 32 inch.   27 h.

CHAPTER XXIV

" What though to-day the ways
    Are chill and bare?
The year has many days,
    And some are fair."

(St. John Adcock : *The Wider View.*)

1791

Jan. 1.   Many horse-beans sprang up in my field-walks in the autumn, & are now grown to a considerable height.   As the Ewel was in beans laſt summer, it is moſt likely that these seeds came from thence; but then the diſtance is too considerable for them to have been conveyed by mice.   It is moſt probable therefore that they were brought by birds, & in particular, by jays, & pies, who seem to have hid them among the grass, & moss, & then to have forgotten where they had ſtowed them.   Some pease are also growing in the same situation, & probably under the same circumſtances.   Mr Derham has recorded that mice hide acorns one by one in paſtures in the autumn; & that he has observed them to be hunted-out by swine, who discovered them by their smell.

Jan. 5.   The great oak in Harteley avenue, juſt as you enter the paſture-field, measures in girth 14 feet. It is a noble tree, & if sound worth many pounds. Why it was left at the general sale does not appear. The girth was taken at four feet above the ground.

Jan. 11.   Ten weeks ſtocks blow: crocus's sprout, & swell.

Jan. 12.   Mr Churton left us.

Jan. 13.   The earth is glutted with water: rills break out at the foot of every little hill: my well is near half full.   The wind in the night blew down the rain-measurer.

WHITE'S
BAROMETER
(after
R. Holt-White).

Jan. 14. Ivy berries swell, & grow : there have been no frosts to check them. —Tubbed, & pickled a good young fat pig :—weight—12 scores, & 4 pounds.

Jan. 16. A disorder prevails among the horses : but has not reached this village yet.

Jan. 17. Sam & Ned White came from Fyfield.

Jan. 19. This morning the Barom^r at Newton was only 28 ! [Selborne 28$\frac{3}{10}$].

Jan. 20. Rain & wind in the night. Thomas [Hoar] says, that when he got-up the Barom^r was down at 27-8 ! !

Jan. 21. Sam & Ned White left us. Late in the evening the planet Jupiter shines in the E.

Jan. 27. One of the Bantam hens begins to lay. Mice devour the crocus's.

Jan. 29. Three gallons of brandy from London.

Jan. 30. [Seleburn] Dark, & harsh.

Feb. 1. My apricot trees were never stripped of their buds before ; & therefore seem to have suffered from a casual flight of gross-beaks [hawfinches], that came into these parts.

Feb. 2. Prodigious high tide at London & in it's environs ! it did much damage in various parts.

Feb. 3. [Newton] Covered the asparagus beds, & the artichokes with muckle [strawy manure] : these were grown out very tall.

Feb. 4. Benham finished mending the hedges.

Feb. 5. [Selborne] Thaw, sun, grey. Hot-bed heats.

Feb. 7. Bull-finches make sad havoc among the buds of my cherry, & apricot trees ; they also destroy the buds of the goose-berries, & honey-suckles ! Green-finches seem also to be concerned in the damage done : many neighbouring gardens have suffered. [*Later note.*] These birds were not observed at the time, nor do they seem to abound. It appeared afterward, that this damage was done by a flight of gross-beaks. [Cf. 20 March.]

Feb. 9. Sowed cucumber-seeds in pots plunged in the hot-bed : bed heats well.

Feb. 10. Brewed strong beer.

Feb. 13. As there has been little frost, the antirrhinum cymb : flourishes, & blossoms thro' the winter.

Feb. 14. Potted cucumbers : bed warm.

Feb. 21. Chaffinches destroy the buds in the honey-suckles.

Feb. 22. Men dig in the hop-gardens.

Feb. 23. The farmers are very much behind in their plowings for a spring crop thro' the wetness of the season.

Feb. 25. Mr Edmd White took down my Barometer, & cleaned tube, & frame. It had not been meddled with for just 18 years, when my Bror John also took it down.

Feb. 26. Deep snow, which damaged & broke my plum-trees, & hedges. This is much the greatest snow that we have seen this year. Some of the deep lanes are hardly passable.

Feb. 27. Snow covers the ground. A large bough broken from the yew-tree, in the church yard, by the snow.

Mar. 2. Seven cart-loads of hot dung carried in for the cucumber-bed : 5 loads from Hale, 1 from Parsons, & 1 of my own.

Mar. 3. Sent me by Lord Stawell a *Sea-mall*, or *Gull*, & a *Coccothraustes*, or *Gross-beak* : the latter is seldom seen in England, & only in the winter.

Mar. 5. Boys play at hop-scotch, & cricket. Some snow under hedges. The *wry-neck* returns, & pipes.

Mar. 7. Coltsfoot blows. Stopped cucumbers. Sowed dwarf lark-spurs. Turned the dung.

Mar. 9. Tapped the new hay-rick : the hay but moderate.

Mar. 11. Sowed radishes, & parsley. Weeded the garden, & dug some ground.

Mar. 12. No frost. Planted four rows of broad beans in the orchard. Some snow still under hedges.

Mar. 13. Crocus's in high glory. Some snow under hedges. Vast halo round the moon.

Mar. 14. Daffodil blows. Timothy the tortoise heaves-up the earth.

Mar. 15. Sweet weather. Mackerel.

Mar. 17. The *Stone-curlew* is returned again : & was heard this evening passing over the village from the uplands down to the meadows & brooks. Planted ½ hundred of cabbages. Timothy comes out.

Mar. 18. Snow lies deep in Newton-lane, & under hedges in the uplands. The hounds find no hare on all Selborne hill.

Mar. 19. Sowed my own ashes on the great meadow. Timothy hides himself again. Men turn their sheep into the green wheat. The hunters killed a female hare, which gave suck : so there are young leverets already. Dr Chandler's labourer,[1] in digging down the bank in the midst of the parsonage garden called the grotto, found human bones among the rocks. As these lay distant from the bounds of the church-yard, it is possible that they might have been deposited there before there was any church, or yard.[2] So again, in 1728, when a saw-pit was sunk on the Plestor under the wall of the court-yard, many human bones were dug-up at a considerable distance from the church-yard.

Mar. 20. Mr Burbey shot a cock *Gross-beak* which he had observed to haunt his garden for more than a fortnight. Dr Chandler had also seen it in his garden.

I began to accuse this bird of making sad havock among the buds of the cherries, goose-berries, & wall-fruit of all the neighbouring orchards. Upon opening its crop & craw, no buds were to be seen ; but a mass of kernels of the stones of fruits. Mr B. observed that this bird frequented the spots where plum-trees grow ; & that he had seen it with some what hard in it's mouth which it broke with difficulty ; these were the stones of damasons. The latin Ornithologists call this bird *Coccothraustes*, i.e., *berry-breaker*, because with it's large horny beak it cracks & breaks the shells of stone-fruits for the sake of the seed or kernel. Birds of this sort are rarely seen in England, & only in winter. About 50 years ago I discovered three of these *gross-beaks* in my outlet, one of which I shot.

Mar. 21. A hen gross-beak was found almost dead in my outlet : it had nothing in it's craw.

Mar. 23. Soft wind. The wood-pecker laughs.

Mar. 25. Sowed onions, radishes, & lettuce : the ground harsh, & cloddy.

Mar. 26. Cucumber-plants show bloom : but the bed is too hot, & draws the plants. We sow our seeds too soon, so that the plants want to be turned out of the pots before the great bed can be got to due temperament.

Mar. 28. Sowed a large plot of parsnips, & radishes in the orchard. Crocus's fade & go off. Sowed also Coss lettuce with the parsnips.

Mar. 30. Some rooks have built several nests in the high wood. The building of rooks in the High wood is an uncommon incident, & never remembred but once before. The Rooks usually carry on the business of breeding in groves, & clumps of trees near houses, & in villages, & towns. Timothy weighs 6 Li. 11 ou.

Mar. 31. Made two hand-glasses for celeri. A gross-beak seen at Newton parsonage-house.

Apr. 1. The bearing cucumber-bed becomes milder & more mellow ; & the plants shoot & blow well. Daffodils make a show. Planted potatoes in the meadow-garden, ten rows.

Apr. 2. Crown imperials begin to blow. Pronged the asparagus beds. Wheat looks well. Mrs B. White & Hannah White come from London.

Apr. 3. The *chif-chaf*, the smallest uncrested wren, is heard in the Hanger, & long Lythe. They are usually heard about the 21 of March. These birds, no bigger than a man's thumb, fetch an echo out of the hanger at every note.

Apr. 4. Mary White came from London.

Apr. 6. The *cuckoo* arrives, & is seen, & heard.[3] The Apricots have no blossoms ; they lost all their buds by birds. *Red start* returns, & appears on the grass-plot.

Apr. 8. Mary White left us.

Apr. 10. The early beech in the long Lythe shows leaves fully expanded.

Apr. 11. Timothy the tortoise marches forth on the grass-plot and grazes.

Apr. 12. Mountain snow-drops blow. Black thorns blossom. Hannah White[4] walks up to the alcove before breakfast.

Apr. 15. A nightingale sings in my outlet. Sowed sweet peas, candy-tuft, sweet alyssum, &c. A man brought me half a dozen good mushrooms[5] from a pasture field ! a great rarity at this season of the year !

Apr. 18. Mr Ben White came from London.

Apr. 19. Mr Chandler & son went away on a visit. Began to use the winter lettuce. Tho' a swallow or two were seen in the village as long ago as the 7th yet have they absconded for some time past ! The house-martin is also withdrawn ; no Swift has yet appeared at Selborne ; what was seen was at Bentley.

Apr. 20. Finished weeding, & dressing all the flower-borders. Several nightingales between the village, & comb-wood pond. Comb-wood coppice was cut last winter.

Apr. 22. The merise, or wild cherries in vast bloom. Grass grows, & clover looks very fine. Mr & Mrs B. White, & Hannah left us & went to Newton.

Apr. 25. [Alton] Mowed some coarse grass in the orchard for the horses.

Apr. 26-May 1. [Oxford.]

Apr. 26. Some of the oaks, planted on the commons between Odiham & Reading about the time that I first knew that road, begin to be felled. Swallows. Goslings. Cherries, apples, & pears in beautiful bloom along the road : grass forward, & corn looks well.

May 1. A prodigious bloom of apple trees along the road.

May 2. [Alton] Swifts, & house-martins over the Thames at Pangbourn.

May 3. [Selborne] Dark & harsh.

May 4. Planted some tricolor violets, & some red cabbages sent from South Lambeth.

May 5. The bloom on my white apple is again very great. Set the middle Bantam hen with eleven eggs : the cook desired that there might be an odd one.

May 7. Vast bloom on my nonpareils. The orchard is mown for the horses. Cut the stalks of garden rhubarb to make tarts : the plants are very strong.

May 8. Mrs Clements & four children, & a nurse-maid came.

May 11. The down of willows floats in the air, conveying, & spreading about their seeds, & affording some birds a soft lining for their nests.

May 13. Ashen shoots injured by the late frosts, & kidney-beans & potatoe-sprouts killed.

May 15. Flesh flies get to be troublesome : hung out the meat-safe. Mrs Clements &c. left us.

May 16. Saw a *flie-catcher* at the vicarage, I think.

May 17. *Fly-catcher* returns.   The *fern-owl*, or *eve-jar* returns, & is heard in the hanger.   These birds are the latest summer-birds of passage : when they appear we hope the summer will soon be established.

May 20.   The weather has been so harsh, that the swallows, & martins are not disposed to build. Found a hen redstart dead in the walks.

May 23.   Brother Thomas White came.

May 24.   *Ophrys nidus avis* blows in Comb-wood. Rain is wanted.   Wheat looks yellow.

May 25.   Mole-cricket jars.   An old hunting mare, which ran on the common, being taken very ill, came down into the village as it were to implore the help of men, & dyed the night following in the street.

May 26.   Finished sowing kidney-beans, having used one quart, which makes five rows, half white & half scarlet.

May 27.   Garden red valerian blows : where it sows itself it soon becomes white.

May 28.   Bantam-hen brings out four chickens.

May 29.   The race of field-crickets, which burrowed in the short Lythe, & used to make such an agreeable, shrilling noise the summer long, seems to be extinct. The boys, I believe, found the method of probing their holes with the stalks of grasses, & so fetched them out, & destroyed them.

May 30.   Cinamon-roses blow.

May 31.   Flowers smell well this evening : some dew.

June 1.   Fern-owl, & chur-worm [mole-cricket] jar.   Men wash their fatting sheep ; & bay the stream to catch trouts.   Trouts come up our shallow streams almost to the spring-heads to lay their spawn.

June 3.   Myriads of tadpoles traverse Comb-wood pond in shoales : when rain comes they will emigrate to land, & will cover the paths & fields.   We draw much water for the garden, so that the well sinks. Flowers are hurried out of bloom by the heat; spring-corn & gardens suffer.

June 4.   Saint foin blows, & the S^t foin fly [burnet moth].   *Sphinx filipendula*, appears.   Rain at Emsworth. Fyfield [cocker spaniel] sprung a fern-owl on the zig-zag which seemed confounded by the glare of the sun, & dropped again immediately.   Mr Bridger sends me a fine present of trouts caught in the ſtream down at Oakhanger.   The diſtant hills look very blue in the evenings.

June 5.   Elder, & corn-flags begin to blow already. Thunder to the S.E., N.E., & N.W.   Gardens, & fields suffer.

June 6.   Wheat begins to come into ear : wheat, which was very yellow from the cold winds, by means of the heat has recovered it's colour without the assiſtance of rain.   Dew, cloudless, sultry.   Red even, dead calm.   The lettuces, which ſtood under the fruit-wall thro' the winter, are juſt over.   They have been of great service at the table now for many weeks.

June 7.   Hops grow prodigiously, yet are infeſted with some aphides.   Early cabbages turn hard, but boil well.   Watered kidney-beans, which come-up well.

June 9.   Summer-cabbages, & lettuce come in. Roses red & white blow.   Began to tack the vines. Thomas finds more rudiments of bloom than he expeᵭted.

June 11.   Male glow-worms, attraᵭted by the light of the candles, come into the parlor.   The diſtant hills look very blue.   There was rain on Sunday on many sides of us, to the S. the S.E. & the N.W. at Alton & Odiham a fine shower, & at Emsworth, & at Newbury : & as near us as Kingsley.   No may chafers this year with us.

June 12.   Clouds, hail, shower, gleams.   Sharp air, & fire in the parlor.   Showers about.   Garden-crops much retarded, & nothing can be planted.   Farmer Bridger sends me three real snipe's eggs : they are in shape, & colour exaᵭtly like those of the lapwing, only one half less.   The colour of the eggs is a dull yellow, spotted with chocolate : they are blunt at the great end, & taper much till they become sharp at the smaller. The eggs, sent me for snipe's eggs last year,[6] seem to have been those of a fern-owl.

June 13.   Farmer Spencer mows his cow-grass.

June 14.   White frost, dark & cold ; covered the kidney beans with straw last night. My annuals, which were left open, much injured by the frost : the balsams, which touched the glass of the light, scorched. Kidney-beans injured, & in some gardens killed. Cucumbers secured by the hand-glasses but they do not grow. The cold weather interrupts the house-martins in their building, & makes them leave their nests unfinished. I have no martins at the end of my brew-house, as usual.

June 15.   The kidney-beans at Newton-house not touched by the late frost. Bro.[r] Thomas left us.

June 16.   Snails come out of hedges after their long confinement from the drought. A swallow in Tanner's chimney has hatched. The fern on the forest killed ; but hardly touched by the frost on Selborne down, which is 400 feet higher than Wolmer.

June 17.   Planted out my annuals from Dan. Wheeler. Pricked out some celeri, good plants. My crop of spinnage is just over : the produce from a pint of seed, sowed the first week in August, was prodigious.

June 18.   Pricked out more celeri in my garden, & M.[r] Burbey's [the village shopkeeper]. Planted some cabbages from D.[r] Chandler's. Timothy hides himself during this wintry weather. The dry weather lasted just 3 weeks & three days ; part of which was very sultry, & part very cold.

June 19.   A flock of ravens about the hanger for many days.

June 20.   [Bramshot place] Went round by Petersfield. *Foxgloves* blow. By going round by Petersfield we make our journey to Bramshot 23 miles.

After we had been driven 20 miles we found ourselves
not a mile from Wever's down, a vast hill in Wolmer
forest, & in the parish of Selborne.   Bramshot in a
direct line is only seven miles from Selborne.
   June 21.   Mr Richardson's straw-berries very dry,
& tasteless.
   June 23.   Went to visit Mr Edmund Woods Senr.
Swifts abound at Godalming.
   June 24.   Meadows not cut.   *Nymphaea lutea*
[yellow water lily] in bloom in a watry ditch.   Went
to see the village of Compton, where my father lived
more than sixty years ago, & where seven of his
children were born.   The people of the village
remember nothing of our family.   Mr Fulham's
conservatory richly furnished ; & the grounds behind
his house engaging, & elegant.   The romantic
grounds, & paddock at the west end of Godalming
town are very bold & striking.   The hanging woods
very solemn, & grand ; & many of the trees of great
age & dimensions.   This place was for many years
inhabited by General Oglethorpe.[7]   The house is now
under a general repair being with it's grounds the
property of Mr Godbold a quack Doctor.   The vale &
hanging woods round Godalming are very beautiful :
the Wey a sweet river, & becomes navigable at this
town.   One branch of the Wey rises at Selburne.[8]
At the entrance to the avenue leading to Bramshot-
place are three great, hollow oaks, the largest of which
measures 21 feet in girth.   We measured this tree
at about 5 feet from the ground, & could not come at
it lower on account of a dry stone-wall in which it
stands.   We measured also the largest Sycamore in
the front of the house, & found the girth to be 13.
They are very tall, & are deemed to be 80 feet in
height : but I should suppose they do not exceed
74 feet.   I hear much of trees 80 or 90 feet high ; but
have never measured any that exceed the supposed
height of the Sycamores above.
   June 25 to *July* 18.   [South Lambeth.]

June 25.   My brother's straw-berries well-flavoured.
The vines here in bloom, & smell very sweet.
June 26.   Fifteen Whites dined this day at my Bro.
B. White's table;   as did also a Mr Wells, a great,
great, great grandson of the Revd John Longworth,
in old times vicar of Selborne, who dyed about the
year 1678.⁹   D^r & Mrs Chandler returned to Selburne.
June 27.   Timothy Turner cuts my grass for him-
self, a small crop.   *Scarabaeus solstitialis* first appears
in my brother's outlet :   they are very punctual in their
coming-out every year.   They are a small species,
about half the size of the *May chafer* [cockchafer], &
are known in some parts by the name of *fern-chafer*.
June 28.   When the Barom^r is 30 at S. Lambeth,
it is 29-7 at Selborne, and 29-4 at Newton.   My
brother cut a good Romagna melon.
June 29.   Some swallows in this district, & only
two pairs of swifts, & no martins.   No wonder then
that they are overrun with flies, which swarm in the
summer months, & destroy their grapes.
June 30.   The Passion-flower buds for bloom :
double flowering Pomegranade has had bloom.
July 1.   Large American straw-berries are hawked
about which the sellers call pine-strawberries.   But
these are oblong, & of a pale red ;   where as the true
pine or Drayton straw-berries are flat, & green :   yet
the flavour is very quick, & truly delicate.   The
American new sorts of strawberries prevail so much,
that the old scarlet, & hautboys are laid aside, & out
of use.
July 3.   My brother's cow, when there is no extraordinary call for cream, produces three pounds of
butter each week.   The footman churns the butter
overnight, & puts it in water ;   in the morning one of
my nieces beats it, & makes it up, & prints it.   M^r M.
black cluster-grapes in his pine-house seem to be well
ripened.
July 5.   [London] Rasps come in.   Many Martins
in the green park.   In a fruit-shop near St. James

were set out to sale black cluster-grapes, pine apples, peaches, nectarines, & Orleans plums.

July 6.   Many martins in Lincolns inn fields.

July 7.   [South Lambeth] Fine, showers, clouds.

July 8.   Cut chardon-heads for boiling : artichokes dry, & not well flavoured.   Roses in high beauty. My nieces make Rasp jam.   Goose-berries not finely flavoured.

July 9.   A cuckoo cries in my Bro^rs garden : some birds of that sort have frequented this place all the summer.   Young swallows at Stockwell.   In Mr Malcolm's garden there is a bed of small silver firs, the tops of which are all killed by the frosts in June. The hothouses of this Gent : afford a most noble appearance ; & his plantations are grand, & splendid. Passion-flower begins to blow in the open air. Cucumbers are scarce, & sell for 2½^d a piece.   Crops of pease go off.   Some celeri trenched out from the seedling bed.

July 10.   Grapes swell.   New potatoes.

July 11.   Chardons are usually blanched, & stewed like celeri : but my Brother boils the heads of his, which are very sweet, & in flavour like artichokes ; the chief objection is, that they are very small, & afford little substance in their bottoms.   The heads of chardons are sold in the markets & are thought to be a delicate morsels.   Chardons are strong, vigorous plants, & grow six & seven feet high, & have strong sharp prickles like thistles.

July 12.   On this day My Bro. Benj. White began to rebuild his house in Fleetstreet which he had entirely pulled to the ground.   His grandson Ben White laid the first brick of the new foundation, & then presented the workmen with five shillings for drink. Ben, who is five years old, may probably remember this circumstance hereafter, & may be able to recite to his grandchildren the occurrences of the day.

July 13.   My brother gathered a sieve of mush-rooms : they come up in the flower-borders, which have been manured with dung from the old hot beds.

July 14. A bat of the largest sort comes forth every evening, & flits about in the front of my brother's house. This is a very rare species, & seldom seen. See my history of Selburne.[10]

July 17. Small shower: heavy rain at Clapham, & Battersea. On this day Mrs Edmd White was brought to bed of a daughter, who encreases my nephews & nieces to the number of 58.

July 19. [Alton] Rye cut & bound at Clapham. Wheat looks well, & turns colour. Hay making at Farnham: pease are hacking near that town; hops distempered.

July 20. [Selborne] Mr Budd's annuals very fine. Ground well moistened: after-grass grows.

July 21. My broad beans are but just come in.

July 22. Children bring wood-strawberries in great plenty. Made straw-berry jam. Gathered currans, & rasps for jam: my rasps are fair & fine. The farmers at Selborne had not half a crop of hay. Hops thrive at this place. Merise, wild cherries, over at the vicarage, ripen.

July 24. The foreign Arum in the vicarage court, called by my Grandmother Dragons, & by Linnaeus *Arum dracunculus*, has lately blown. It is an Italian plant, & yet has subsisted there thro' all the severe frosts of 80 or 90 years; & has escaped all the diggings, & alterations that have befallen the borders of that garden. It thrives best under a N. wall, but how it is propagated does not appear. The spatha, & spadix are very long.

July 26. Mrs Henry White, & Lucy came from Fyfield.

July 29. A basket of mushrooms from Honey-lane. Gathered wall-nuts for pickling.

July 30. Made black curran-jelly. Finished cutting the tall hedges. Gathered some lavender.

July 31. " On the last day of this month my Fathr Mr Ben. Wh. shot in his own garden at S. Lambeth, a

Loxia curviroſtra, or *Cross bill*, as it was feeding on the
cones of his Scotch firs.   There were six, four cocks,
& two hens : what he shot was a cock, which was
beautifully variegated with brown, & green, & a great
deal of red : it answered very accurately to Willughby's
description ; & weighed rather more than 1 ounce
& an half.   In the evening, the five remaining birds
were seen to fly over the garden, making a chearful
note." Thus far Mrs Ben White.   To
which we add that flights of *Cross bills* used to frequent
Mrs Snooke's Scotch firs in the month of July only.
Mr *Ray* says, " per autumnum interdum sed rarius
in Angliam venit, non autem apud nos perennat aut
nidificat."   Synopsis.

Aug. 1.   Gathered our whole crop of apricots,
being one large fine fruit.

Aug. 2.   Sowed white turnip radishes.   Planted-
out savoys, & other winter cabbages.

Aug. 3.   Somewhat of a chilly feel begins to prevail
in the mornings and evenings.   Sowed a pint of
London prickly spinage seed to ſtand the winter.
The same quantity laſt year produced an incredible
crop.   Trod & rolled in the seed.   In Mr Hale's
hop-garden near Dell are several hills containing male
plants, which now shed their farina : the female plants
begin to blow.   Men hoe turnips, & hack pease.
Men house hay as black as old thatch.

Aug. 4.   Farmer Tull begins to reap wheat.   The
hop-garden at Kimber's fails again, & looks black.

Aug. 5.   Mrs H. White, & Lucy left us.   Two
dobchicks [dabchicks or little grebes] in Combwood
pond.   Young martins, & swallows cluſter on the
tower, & on trees, for the firſt time.   A pleasing
circumſtance mixed with some degree of regret for
the decline of summer !

Aug. 6.   Boys bring wasp's neſt.   Codlings, &
ſtewed cucumber come in.   Housed, & piled 8 cords
of beechen billet in fine order.   Watered the cucumbers ; well very low.

Aug. 7.   Received from Farnham, well packed in a box, a picture of a mule pheasant, painted by Mr Elmer, & given me by Lord Stawell.   I have fixed it in a gilt, burnished frame, & hung it in my great parlor, where it makes an elegant piece of furniture. [Cf. 2 Oct. 1790.]   The first broods of swallows, & house-martins, which congregate on roofs, & trees, are very numerous : & yet I have not this year one nest about my buildings.

Aug. 8.   Some young broods of fly-catchers fly about.

Aug. 11.   Half hogshead of portwine from Southampton.   Gleaners come home with corn.

Aug. 12.   Men bind their wheat all day.   The harvesters complain of heat.   The hand-glass cucumbers begin to bear well : red kidney beans begin to pod.

Aug. 13.   Farmer Tull makes a wheat-rick at Wick-hill.

Aug. 14.   Hirundines enjoy the warm season.   Late this evening a storm of thunder arose in the S. which, as usual, divided into two parts, one going to the S.W. & W. & the greater portion to the S.E. and E., & so round to the N.E.   From this latter division proceeded strong, & vivid lightening till late in the night.   At Headleigh there was a very heavy shower, & some hail at E. Tisted.   The lightening, & hail did much damage about the kingdom.   Farmer Spencer's char-coal making in his orchard almost suffocated us : the poisonous smoke penetrated into our parlor, & bed-chambers, & was very offensive in the night.

Aug. 15.   Lightening every moment in the W. & N.W.   Cut 114 cucumbers.   Harvesters complain of violent heat.

Aug. 16.   Colchicums, or naked boys appear.

Aug. 17.   Holt White, & Harry Woods came from Fyfield.

Aug. 18.   Timothy grazes.   John White came from Salisbury.   Cut 133 more cucumbers.   Michaelmas daiseys begin to blow. Farmer Spencer, & Farmer Knight make each a noble wheat-rick : the crop very good, & in fine order.

Aug. 19. The young men left us, & went to Funtington. A second crop of beans, long pods, come in. Sweet day, golden even, red horizon. Some what of an autumnal feel.

Aug. 20. John White called in his way from Funtington to Salisbury. The whole country is one rich prospect of harvest scenery ! ! Fern-owl glances along over my hedges.

Aug. 21. Many creatures are endowed with a ready discernment to see what will turn to their own advantage & emolument ; & often discover more sagacity than could be expected. Thus Benham's poultry watch for waggons loaded with wheat, & running after them pick up a number of grains which are shaken from the sheaves by the agitation of the carriages. Thus when my brother used to take down his gun to shoot sparrows, his cats would run out before him to be ready to catch up the birds as they fell.

Aug. 24. Gathered kidney-beans, scarlet. Cut 80 cucumbers.

Aug. 25. Holt White came back from Shopwick [near Chichester].

Aug. 26. My potatoes come in, & are good.

Aug. 27. Cut 179 cucumbers : in all this week —349. A large sea-gull went over my house.

Aug. 29. Hop-picking begins in Hartley gardens. Cut 96 cucumbers.

Aug. 30. Mr Hale begins his hops near the Pound field. Farmer Hoar says that during this blowing weather his well was raised some rounds of the rope.[11]

Aug. 31. Cut 31 cucumbers. Fly-catcher still appears.

Sept. 2. Cut 62 cucumbers. Holt White left us, & went to Newton.

Sept. 3. Bad weather for the hops, & pickers. When the boys bring me wasps nests, my Bantam fowls fare deliciously; & when the combs are pulled to pieces, devour the young wasps in their maggot-state with the highest glee, and delight. Any insect-eating bird would do the same : & therefore I have often wondered that the accurate Mr Ray should call one species of buzzard *Buteo apivorus, sive vespivorus,* or the *Honey-buzzard,* because some combs of wasps happened to be found in one of their nests. The combs were conveyed thither doubtless for the sake of the maggots or nymphs, & not for their honey; since none is to be found in the combs of wasps.[12] Birds of prey occasionally feed on insects : thus have I seen a tame kite picking up the female ants, full of eggs, with much satisfaction.

Sept. 5. Cut 107 cucumbers. Nectarines are finely flavoured, but eaten by bees, & wasps. Churn-owl is seen over the village : fly-catchers seem to be gone.

Sept. 6. Tyed up about 30 endives. A swift still hovers about the brew-house at Fyfield. About a week ago, one young swift, not half-fledged, was found, under the eaves of that building ! The dam no doubt is detained to this very late period by her attendance on this late-hatched, callow young ! The roof of my nephew's brew-house abounds with swifts all the summer.

Sept. 7. Cut 125 cucumbers. Young martins, several hundreds, congregate on the tower, church, & yew-tree. Hence I conclude that most of the second broods are flown. Such an assemblage is very beautiful, & amusing, did it not bring with it an association of ideas tending to make us reflect that winter is approaching; & that these little birds are consulting how they may avoid it.

Sept. 9. Gathered in the white apples, a very fine crop of large fine fruit, consisting of many bushels.

Sept. 10. Young broods of swallows come out. Cut 171 cucumbers; in all 424 this week. Sweet moon light !

Sept. 11. *Grey crow* returns, & is seen near Andover. Red even, sweet moon. [*Note next day.*] Some nightly thief stole a dozen of my finest nectarines.

Sept. 13. My well is very low, & the water foul! Timothy eats voraciously. Winged female ants migrate from their nests, & fill the air. These afford a dainty feast for the hirundines, all save the swifts; they being gone before these emigrations, which never take place till sultry weather in August, & September.

Sept. 14. Hop-picking goes on without the least interruption. Stone-curlews cry late in the evenings. The congregating flocks of *hirundines* on the church & tower are very beautiful, & amusing! When they fly-off altogether from the Roof, on any alarm, they quite swarm in the air. But they soon settle in heaps, & preening their feathers, & lifting up their wings to admit the sun, seem highly to enjoy the warm situation. Thus they spend the heat of the day, preparing for their emigration, &, as it were consulting when & where they are to go. The flight about the church seems to consist chiefly of house-martins, about 400 in number: but there are other places of rendezvous about the village frequented at the same time. The swallows seem to delight more in holding their assemblies on trees.

" When Autumn scatters his departing gleams,
   Warn'd of approaching winter gathered play
   The *Swallow people*; & toss'd wide around
   O'er the calm sky in convolution swift,
   The feather'd eddy floats: rejoicing once
   Ere to their wintry slumbers they retire,
   In clusters clung beneath the mouldring bank,
   And where, unpierced by frost, the cavern sweats.
   Or rather into warmer climes convey'd,
   With other kindred birds of season, there
   They twitter chearful, till the vernal months
   Invite them welcome back :—for thronging now
   Innumerous wings are in commotion all."[13]

Sept. 15.   The springs are very low : the water fails at Webb's bridge.

Sept. 20.   Some neighbours finish their hops. The whole air of the village of an evening is perfumed by effluvia from the hops drying in the kilns.   Began to light a fire in the parlor.

Sept. 24.   Young martins, & swallows come-out, & are fed flying.   Endive well blanched comes in. Bottled-off half hogsh : of port wine.   The port ran eleven doz. & 7 bottles.   Nep. Ben White & wife, & little Ben came.

Sept. 25.   Several wells in the village are dry : my well is very low ; Burbey's, Turner's, Dan Loe's hold out very well.

Sept. 26.   Gathered in the pear-mains, golden rennets, & golden pippins.

Sept. 27.   Strong cold gale.

Sept. 28.   Linnets congregate in great flocks. This sweet autumnal weather has lasted three weeks, from Septr 8th.

Sept. 29.   A gale rises every morning at ten o' the clock & falls at sunset.

Oct. 1.   Nep. B. White left us, & went to London. It was with difficulty that we procured water enough for a brewing from my well.

Oct. 2.   Gathered one fine nectarine, the last. My double-bearing raspberries produce a good crop. Grapes very fine, endive good.

Oct. 5.   Arrived off the isle of Wight the Earl Fitzwilliams Captn Dundas from Madras.   Charles Etty sailed in this India man as second mate about the 10th of March, 1790.   [*Later.*] Poor Charles Etty did not come home in the Earl Fitzwilliams, having unfortunately broke his leg at Madras the evening before the ship sailed for Europe.

Oct. 6.   Received a bag of hops from Mr. Hale, weight 61 pounds.

Oct. 7.   Gathered in Chaumontel, swans-egg, & Virgoleuse pears : the latter rot before they ripen.

Gathered also the kitchen apples at the end of the fruit-wall, & the knobbed russetings : of both there is a great crop.   Gathered the Cadillac pears, a small crop.

Oct. 8.   Earthed up the celeri, which is very gross, & large.

Oct. 9.   It has been observed that divers flies, besides their sharp, hooked nails, have also skinny palms or flaps to their feet, whereby they are enabled to stick on glass & other smooth bodies, & to walk on ceilings with their backs downward, by means of the pressure of the atmosphere on those flaps. [Cf. 9 Feb. 1777] ; the weight of which they easily overcome in warm weather when they are brisk and alert.   But in the decline of the year, this resistance becomes too mighty for their diminished strength ; & we see flies labouring along, & lugging their feet in windows as if they stuck fast to the glass, & it is with the utmost difficulty they can draw one foot after another, & disengage their hollow caps from the slippery surface.   Upon the same principle that flies stick,[14] & support themselves, do boys, by way of play, carry heavy weights by only a piece of wet leather at the end of a string clapped close on the surface of a stone.   Tho' the Virgoleuse pears always rot before they ripen, & are eatable ; yet when baked dry on a tin, they become an excellent sweet-meat.

Oct. 12.   Gathered cucumbers for picklers.   One of my Apricot-trees withers, & looks as if it would die.   Hunter's moon rises early.   Mrs Ben White left us, & took Tom with her, leaving Ben behind.

Oct. 13.   My beeches in the field shed ripe mast. Some of the Bantams sicken.

Oct. 15.   Bro. Ben, & wife, Hannah came.   Wood-cock, & red wings return, & are seen.

Oct. 17.   Saw a wood-cock on the down among the fern : Fyfield [cocker spaniel] flushed it.

Oct. 22.   One young martin in one of Burbey's nests, which the dams continue to feed.   Gracious stream now runs a little.

Oct. 24. The dams continue to feed the poor little martin in the nest at Burbey's with great assiduity !

Oct. 25. There are two young martins in the nest.

Oct. 26. No young martins to be seen in the nest, nor old ones round it.

Oct. 27. Young martins, & their dams again. Wood-cock on the down. Bro. Ben, & wife, & Hannah left us, & went to Newton.

Oct. 28. There are now apparently three young martins in the nest nearly fledged.

Oct. 29. The young martins remain.

Oct. 30. The young martins still in their nest; at least some of them. Dr Chandler saw four hawking round the plestor.

Oct. 31. The young martins not seen in their nest : dams about.

Nov. 1. The young martins are out : one was found dead this morning in the parsonage garden.

Nov. 2. [Newton] The late rains have not had any influence yet on my well-water, which is very low, & foul. Snow on the Sussex downs.

Nov. 4. [Selborne] Grey, gleams. Snow gone.

Nov. 8. Planted one doz. of red hairy gooseberries, & one doz. of smooth amber, from Armstrong, in the quarters of the garden. Gathered-in the grapes : decaying. Two rills run now into my well, the water of which begins to get clear.

Nov. 9. Planted a row of Hyacinths on the verge of the fruit-border ; & tulips along the broad walk. Planted winter-cabbages. Potatoes dug up.

Nov. 13. Thunder in the night. Thomas heard the Portsmouth evening gun.

Nov. 25. Well rises very fast.

Nov. 26. 3 gallons of brandy from London.

Nov. 28. Mr & Mrs Edmd White came.

Nov. 29. Put a large cross on the hermitage.[15] A trufle-hunter tryed my tall hedges, & found some bulbs of those peculiar plants, which have neither roots, nor branches, nor stems.

Dec. 1. M^r & M^rs Ed. White left us. The Hermitage, new capped with a coat of thatch, & embellished with a large cross, makes a very picturesq object on the hanger, & takes the eye agreeably.

Dec. 3. Snow covers the ground, snow shoe deep.

Dec. 5. Cut down, & covered the artichokes: covered the rhubarb plants; & the lettuces under the fruit-wall, & the spinage lightly with straw.

Dec. 7. Ground very wet. Farmer Tull plants Butts-close with hops.

Dec. 8. Timothy has laid himself up under the hedge against Benham's yard in a very comfortable, snug manner: a thick tuft of grass shelters his back, & he will have the warmth of the winter sun.

Dec. 16. Swept-up the leaves in the walks.

Dec. 17. Hard frost, very white, boys slide. Snipes come up from the forest along the meads by the sides of the stream. Hardly here & there a wood-cock to be seen.

Dec. 20. Saw lately a white, & a yellow[16] wagtail about the Well-head rivulet. No farther north than Rutland wagtails, withdraw, & are never seen in the winter.

Dec. 21. [Newton] Dark & cold, frost.

Dec. 23. [Selborne] Mr Churton came from Oxford.

CHAPTER XXV

" Blest, who can unconcern'dly find
Hours, days, and years steal soft away
In health of body, peace of mind,
Quiet by day."

(POPE : *Ode on Solitude.*)

1792

Jan. 6.   Snow-drops, & crocus's shoot.

Jan. 8.   Mᴿ Churton left us, & returned to Oxford.

Jan. 13.   Vast frost-work on the windows.

Jan. 14.   Lord Stawell sends me a cock & an hen brambling.

Jan. 17.   The *Antirrhinum Cymb.* which flourished, & blossomed thro' all last winter, & the summer & autumn following, now killed by the frost.   Hence it is probable that in milder regions it is at least a biennial,[1] if not a perennial.   Before, it has always dyed every winter as soon as the hard frosts began to prevail.

Jan. 19.   The wood-men begin to fell beeches on the hanger.

Jan. 23.   Water-cresses come in.

Jan. 27.   The Swallow, Lord Cornwallis's advice sloop, arriv'd at Bristol from Madras, which it left on the 21ˢᵗ of Septemᴿ.   The weather was so rough, that it could not get up the Bristol channel.

Feb. 1.   Turner's heifers feed down the dead grass in my great mead.

Feb. 2.   Grass-walks are very verdurous.

Feb. 4.   Spring like : crocus blows : gossamer floats : musca tenax comes forth : blackbird whistles.

Feb. 6.   Fairey-rings encrease on my grass-plot.

Feb. 8.   The hasels in my hedges are illuminated by numbers of catkins.   Bantam lays.

Feb. 9.   Tubbed, & pickled a fat porker : weight nine scores, & eleven pounds : price 8ˢ & 4ᵈ, from farmer Hoar.

Feb. 10.   Wood-cock killed in the shrubs above the Hermitage.

Feb. 11.   The *meadow* measures 2 *acres & 19 rods,* besides the dug ground.

Feb. 13.   Sowed the ashes of my own making in the great mead, where the grass is finest.   Finished tacking the fruit wall-trees.   Gossamer streams from the boughs of trees.   Brimstone butterfly, Papilio rhamni.

Feb. 15.   Crown imperials sprout.

Feb. 19.   Frost comes within doors.

Feb. 20.   Snow about four inches deep.   3 Bantam hens lay.

Feb. 21.   Yellow wagtail appears.

Feb. 23.   Began to drink tea by day light.

Feb. 26.   Rain in the night.   Humble bee.   Worms come out on grass plots : a great snail.

Feb. 27.   Mʳ Littleton Etty called.   Long tailed titmouse.   Crocus's blowing very much.   Winter aconites fade.

Mar. 1.   The laurustines, & the young shoots of the honey-suckles are not hurt by the late frosts.

Mar. 9.   Much sharp March weather.   Flights of snow, freezing all day.

Mar. 10.   Broʳ Benjamin, & wife, & Rebecca dined with us.   White water-wagtail.

Mar. 12.   Carted in 6 loads of hot dung for the cucumber bed ; 1 of my own, & 5 from Kimbers.

Mar. 15.   Snow-drops are out of bloom.   Rainbow.

Mar. 16.   Daffodil blows.   .   .   .
　　　" it takes the winds of March
　　Before the Swallow dares."
　　　　　[Winter's Tale, Act IV., Sc. 3.]

Mar. 17.   Dog's toothed violets bud.   Lord Stawell made me a visit on this day, & brought me a white wood-cock; it's head, neck, belly, sides, were milk-white, as were the under sides of the wings. On the back, & upper parts of the wings were a few spots of the natural colour. From the shortness of the bill I should suppose it to have been a male bird. It was plump, & in good condition.

Mar. 23. *Timothy* the *Tortoise* comes out. Crown imperials bud for bloom, & stink much.

Mar. 25. Mrs Clement came with her three daughters.

Mar. 26. Crocus's go off. The Kingsley miller assures me that he saw a *Swallow* skimming over the meadow near the mill. Hirundines are often seen early near mill-ponds, & other waters.

Mar. 27. The ground in a sad wet condition, so that men cannot plow, nor sow their spring-corn. A wet March is very unkind for this district.

Mar. 31. Mrs Chandler was brought to bed of a daughter.

Apr. 1. Stormy, wet night. Mrs Clement, & daughters left us. Berriman's field[2] measured contains 1 acre 3 qu. 25 rds.

Apr. 3. Some players came hither from Alton. A hand-glass of early celeri entirely eaten-up by the *Chrysomela oleracea saltatoria*, vulgarly called the *turnip fly*. Sowed more.

Apr. 5. Wind damages the hedges. Some thatch torn by the wind. Mr White's tank at Newton runs over, & Capt. Dumaresque's is near full.

Apr. 6. Players left us.

Apr. 7. The cucumbers shoot out fibres down their hills: earthed them a little. Thomas mowed the dark green grass growing on the Fairy circles, & segments of circles in my grass plot, which encrease in number every year.

Apr. 9. *Nightingale* sings. *Cuckoo* is heard. Timothy the tortoise weighs 6 ae 11½ oz.

Apr. 10. [Wallingford] Hot sun. Goslins on commons. Black thorn blossoms

Apr. 11.   [Oxford]³ Men hoe their wheat, which is very forward, & fine.   Thomas in my absence planted beans, & sowed carrots, parsnips, cabbage-seed, onions, lettuce, & radishes.

Apr. 12.   Thermometer at Fyfield 72 ! in the shade.

Apr. 13.   A great thunder-storm at Woodstock, & Islip : the Charwel much flooded, & discoloured. No rain at Oxford.   Prodigious was the damage done about the Kingdom on this day by storms of thunder, lightening, & vast torrents, & floods, & hail.   The town of Bromsgrove in Worcestershire was quite deluged, & the shops & sitting rooms filled with water.   A house was burnt at some place ; & in others many people hurt, & some killed.

Apr. 16.   [Alton] Great bloom of cherries, pears, & plums.

Apr. 17.   [Selborne] Saw a pair of swallows at Alton.

Apr. 19.   *Redstart* appears.   Daffodils are gone : mountain-snow-drops, & hyacinths in bloom ; the latter very fine : fritillaries going.   Vast flood at Whitney in Oxfordshire, on the Windrush.

Apr. 21.   Planted 4 rows of my own potatoes in the garden.   Mowed the terrace walk.

Apr. 23.   A nest of young blackbirds destroyed by a cat in my garden.

Apr. 26.   Two nightingales within hearing : cuckoos come round the village.

Apr. 27.   The middle Bantam hen sits in the barn. Planted four rows of potatoes in the home garden.

Apr. 28.   Planted in the mead-garden eleven rows of potatoes ; four of which were *potatoes* from *Liverpool*, sent to Dᴿ Chandler by Mᴿ Clarke.   Planted in the mead four rows of beans.

Apr. 30.   Men tye their hops.   Dressing some of the borders.   Heavy thundrous clouds.   Tulips blow.   On this beautiful evening came all at once *seven Swifts*, which began to dash & play round the church. *Chur-worm* jars down at Dorton in swampy ground. Mrs Ben White, & her son Tom came from London.

May 1. Cut a good mess of aspargus.

May 2. Cut the leaves of Rhubarb for tarts : the tarts are very good. Sent some of the leaves of the crocus's to Edm^d White ; they make good tyings for hops, being both tough, & pliant.

May 4. Began to use the lettuces under the fruit wall.

May 6. During the severe winds it is not easy to say how the Hirundines subsist ; for they withdraw themselves, & are hardly ever seen, nor do any insects appear for their support. That they can retire to rest, & sleep away these uncomfortable periods, as the bats do, is a matter to be suspected rather than proved : or do they not rather spend their time in deep & shelt'red vales near waters, where insects are more likely to be found ? Certain it is, that hardly any individuals of this Genus have been seen for several days together.[4]

May 8. On this day 26 houses, besides a number of barns, stables, granaries, &c. were burnt down at Barton-Stacey near Winchester. Only ten or twelve houses were preserved, among which is the parsonage, a large farm house, & some others out of the line of the street. The people of Selborne subscribed 6 ae. 1s. 0d. on this occasion : the county collection was very large & ample.

May 9. Still for the first time since May 1st. Chalk cart.

May 10. Peat cart begins.

May 12. An army of caterpillars infest my young goose-berry trees, which were planted this spring : & the case is the same at Dr Chandler's. Thomas picked the trees carefully, & gave them a good watering.

May 13. Mrs Ben White came.

May 17.   Sowed some Nasturtion seeds on the bank.   Mr Charles Etty returns from Madras well in health, & not lame from the accident of breaking his leg ; but thinner than he was.   He went first to Bengal, & so home in a Danish India man.

May 18.   The *fern-owl*, or *eve-jarr* is heard to chatter in the hanger.   So punctual are they !

May 19.   The middle Bantam brought forth nine chickens.

May 20.   The missel-thrush has a nest on the orchard pear-tree.   The thunder of this evening burnt the barns, & out houses of a farm between Gosport & Titchfield, & destroyed eight fine horses.

May 21.   The cock missel-thrush sings on the tops of the tall firs.

May 22.   The Fly-catcher comes to my vines, where probably it was bred, or had a nest last year.   It is the latest summer bird, & appears almost to a day ! " Amusive bird,⁵ say where your snug retreat ? " !
The *white apples* are out of bloom, being forward : the *Dearling*, a late keeping apple, but just in bloom. So the earlier the fruit ripens the sooner the tree blossoms.   The Dearling bears only once in two years, but then an enormous burthen.   It has pro-duced 10, & 13 bushels of fruit at a crop.   The bloom this year is prodigious !   [*Later*] the crop moderate, & the fruit small.

May 24.   The old speckled Bantam sits on eight eggs.   *Sorbus aucuparia*, the Quicken-tree, or mountain ash full of bloom.   The bunches of red berries would make a fine appearance in winter : but they are devoured by thrushes, as soon as they turn colour. Tanner shot a hen Sparrow-hawk as she was sitting on her eggs in an old crow's nest on one of the beeches in the High wood.   The bird fell to the ground, &, what was very strange, brought down with her one of the eggs unbroken.   The eggs of Sparrow-hawks, like those of other birds of prey, are round, & blunt-ended, & marked at one end with a bloody blotch. The hen bird of this species is a fine large hawk; the male is much smaller, & more slender. Hawks seldom build any nest. This Hawk had in her craw the limbs of an unfledged lark.

May 27. The missel-thrush has got young.

May 30. My table abounds with lettuces, that have stood the winter; radishes; spinage; cucumbers; with a moderate crop of asparagus.

May 31. Grass grows very fast. Honey-suckles very fragrant, & most beautiful objects! Columbines make a figure. My white thorn, which hangs over the earth-house, is now one sheet of bloom, & has pendulous boughs down to the ground. One of my low balm of Gilead firs begins to throw out a profusion of cones; a token this that it will be a short-lived, stunted tree. One that I planted in my shrubbery began to decay at 20 years of age. Miller in his gardener's Dictionary mentions the short continuance of this species of fir, & cautions people against depending on them as a permanent tree for ornamental plantations.

June 1. Mr & Mrs Ben White left us, & went to Newton.

June 2. Mushrooms are brought to the door.

June 3. No may-chafers this year. The intermediate flowers, which now figure between the spring, & solstitial, are the early orange, & fiery-lily, the columbine, the early honey-suckle, the peony, the garden red valeriam, the double rocket or dames violet, the broad blue flag-iris, the thrift, the double lychnis, spider-wort, monk's hood, &c.

June 4. Hay making about London.

June 5. One Fly-catcher builds in the Virginia Creeper, over the garden-door: & one in the vine over the parlor-window. Between Newton & us we heard three Fern-owls chattering on the hill; one at the side of the *High-wood*, one at the top of the *Bostal*, & one near the *Hermitage*. That at the top of the

Boſtal is heard diſtinctly in my orchard. Fern-owls
haunt year by year nearly the same spots.

June 6. The mare lies out. Sᵗ foin begins to blow.

June 7. Heavy thundrous clouds, copious dew.
Opened, & slipped-out the superfluous shoots of the
artichokes.

June 8. Cut-off the cones of the *balm* of *Gilead fir*
in such numbers that they measured one gallon & an
half. So much fruit would have exhauſted a young
tree. The cones grow sursùm, upright; those of the
Spruce, deorsùm, downward.

June 10. Began to use green goose-berries.

June 11. [Mareland, Alton] Went, & dined with
my Brother Benjamin White at Mareland,⁶ to which he
& his wife were come down for two or three days.
We found the house roomy, & good, & abounding
with conveniences: the out-door accommodations
are also in great abundance, such as a larder, pantry,
dairy, laundry, pigeon-house, & good ſtables. The
view from the back front is elegant, commanding
sloping meadows thro' which runs the Wey (the ſtream
from Alton to Farnham) meandering in beautiful
curves, & shewing a rippling fall occasioned by a
tumbling bay formed by Mʳ Sainesbury, who also
widened the current. The murmur of this water-fall
is heard from the windows. Behind the house next
the turnpike are three good ponds, & round the
extensive outlet a variety of pleasant gravel walks.
Across the meadows the view is bounded by the Holt:
but up & down the valley the prospect is diversifyed,
& engaging. In short Mareland is a very fine
situation, & a very pleasing Gentleman's seat. I was
much amused with the number of Hirundines to be
seen from the windows: for besides the several
martins & swallows belonging to the house, many
Swifts from Farnham range up & down the vale;
& what ſtruck me moſt were forty or fifty bank-
martins, from the heaths, & sand-hills below, which
follow the ſtream up the meadows, & were the whole

day long busied in catching the several sorts of Ephemerae which at this season swarm in the neighbourhood of waters. The stream below the house abounds with trouts. Nine fine coach-horses were burnt in a stable at Alresford.

June 12. [Selborne] Mr Burbey has got eleven martins nests under the eaves of his old shop.

June 15. Beat the banks ; & planted cabbages in the meadow-garden.

June 16. Planted some hand-glass plants in the frames of the fruiting cucumber-bed : cut down the lining, & worked it up with some grass-mowings. Some young fly-catchers are out, & fed by their dams.

June 17. When the servants are gone to bed, the kitchen-hearth swarms with minute crickets not so big as fleas. The Blattae are almost subdued by the persevering assiduity of Mrs J. W.[7] who waged war with them for many months, & destroyed thousands : at first she killed some hundreds every night. The thermometer at George's fields Surrey 82 : on the 21,—51. Saint foin fly, sphynx filipendulae, appears.

June 18. The spotted Bantam hen brings out seven chickens. Took a black birds nest the third time : the young were fledged, & flew out of the nest at a signal given by the old ones.

June 19. Pinks, scarlet-lychnis, & fraxinellas blow. The narrow-leaved blue Iris, called Xiphium, begins to blow.

June 21. Put sticks to some of the kidney-beans. Longest day : a cold, harsh solstice ! The rats have carried away six out of seven of my biggest Bantam chickens ; some from the stable, & some from the brew-house.

June 24. Thunder, & hail. A sad midsum^r day. When the Blattae seem to be subdued, & got under ; all at once several large ones appear : no doubt they migrate from the houses of neighbours, which swarm with them.

June 25. Timothy Turner sowed 40 bushels of ashes on Baker's hill: an unusual season for such manure! Tryed for rats over the stable, & brewhouse with a ferret, but did not succeed.

June 27. The late pliant sort of Honeysuckles, that do not make good standards, begin to show their yellow bloom: the more early are on the decline. Hung the net over the cherry-trees at the end of the house to keep off the magpies, which come to our very windows at three & four in the morning. The daws also from the church have invaded my neighbour's cherries. Pies, & daws are very impudent!

June 28. Glow-worms abound in Baker's hill.

June 29. Straw-berries from the woods are brought; but they are crude, & pale, as might be expected. Cut-off the large leaves of the *Colchicum*, or meadow-saffron, now decaying: towards the end of August the blossoms, called by some *naked boys*, will shoot out, & make a pleasing appearance.

June 30. The Saint foin about the neighbourhood lies in a bad way.

July 1. There is a natural occurrence to be met with upon the highest part of our down in hot summer days, which always amuses me much, without giving me any satisfaction with respect to the cause of it; & that is a loud audible humming of bees in the air, tho' not one insect is to be seen. This sound is to be heard distinctly the whole common through, from the Money-dells, to Mr White's avenue-gate [at Newton]. Any person would suppose that a large swarm of bees was in motion, & playing about over his head. This noise was heard last week on June 28th.[8]
" Resounds the living surface of the ground,
Nor undelightful is the ceaseless *hum*
To him who muses . . . at noon."
" Thick in yon stream of light a thousand ways,
Upward, and downward, thwarting, & convolv'd,
The quivering nations sport."
Thomson's Seasons. [*Summer*: ll. 281-3; 342-4].

July 5.   The Provost of Oriel,[9] & Lady came.

July 6.   Mr Eveleigh says, that the churring of the fern-owl is like the noise of a razor-grinder's wheel.

July 7.   Farmer Hoare's son shot at a hen *Wood-chat* [*Lanius s. senator*], or small *Butcher-bird* as it was washing at Well-head, attended by the cock.   It is a rare bird in these parts.   In it's craw were insects.

July 8.   The Poet of Nature lets few rural incidents escape him.   In his *Summer* he mentions the *whetting* of a *scythe* as a pleasing circumstance, not from the real sound, which is harsh, grating, & unmusical ; but from the train of summer ideas which it raises in the imagination.   No one who loves his garden & lawn but rejoices to hear the sound of the mower on an early, dewy morning.—

" Echo no more returns the *chearful* sound
    Of sharpening scythe."

Milton also, as a pleasing summer-morning occurrence, says,

    . . .   " the mower whets his scythe."

                                        L'Allegro.

July 9.   The Provost & Lady left us.   Thunder in the night, & most part of the day to the S. & S.E. Yellow evening.

July 10.   Guns fire at Portsmouth.

July 13.   Whortle-berries are offered at the door. Cherries have little flavour.

July 14.   The double roses rot in the bud without blowing out : an instance this of the coldness, & wetness of the summer.   Potatoes blossom.

July 16.   Farmer Corps brought me two eggs of a *fern-owl*, which he found under a bush in shrub-wood.   The dam was sitting on the nest ; & the eggs, by their weight, seemed to be just near hatching. These eggs were darker, & more mottled than what I have procured before.

July 18.   Men cut their meadows.   Mr Churton came.

July 19.   My meadow is begun to be mowed.

July 20.   Simeon Etty brought me two eggs of a
Razor-bill from the cliffs of the Isle of Wight : they
are large, & long, & very blunt at the big end, &
very sharp & peaked at the small.   The eggs of these
birds are, as Ray justly remarks, " in omnibus hujus
generis majora quàm pro corporis mole."   One
of these eggs is of a pale green, the other more white ;
both are marked & dotted irregularly with chocolate-
coloured spots.   *Razor-bills* lay but one egg, except
the first is taken away, & then a second, & on to a
third.   By their weight these eggs seem to have been
sat on, & to contain young ones.

July 21.   Made rasp, & curran jam, & jelly.

July 22.   Took the black bird's nest the fourth
time : it contained squab young.   [Here occurs a
long entry, in another hand, giving extracts concerning Perambulations of Alice Holt and Wolmer, and
the Plestor.]

July 24.   Preserved some cherries.   My meadow-
hay was carried, in decent order.   As we were coming
from Newton this evening, on this side of the Money-
dells, a cock Fern-owl came round us, & showed
himself in a very amusing manner, whistling, or piping
as he flew.   Whenever he settled on the turf, as was
often the case, Mr Churton went, & sprung him, &
brought him round again.   He did not clash his
wings over his back, so as to make them snap.
At the top of the Bostal we found a bat hawking
for moths.   Fern-owls & Bats are rivals in their
food, commanding each great powers of wing, &
contending who shall catch the phalaenae of the
evening.

July 26.   This cool, shady summer is not good for
mens fallows, which are heavy, & weedy. Lettuces
have not loaved, or bleached well this summer.

July 29.   Heavy showers.   Apples fall much.
The well at Temple is 77 feet deep : 60 to the water,
& 17 afterward.   My well measures only 63 feet.

July 30.   Mr Churton left us, & went to Waverley.

July 31. The young Hirundines *begin* to congregate on the tower. How punctual are these birds in all their proceedings !

Aug. 1. Floods out in several parts of the kingdom, & much hay & corn destroyed. Young buzzards follow their dams with a piping, wailing noise.

Aug. 5. The guns at the camp on Bagshot-heath were heard distinctly this evening.

Aug. 7. Several of my neighbours went up the Hill[10] (this being the day of the great review at Bagshot heath) whence they heard distinctly the discharges from the ordnance, & small arms, & saw the clouds of smoke from the guns. The wind being N.E. they smelled, or seemed to smell, the scent of the gunpowder. Wickham bushes, the scene of action, is more than 20 miles from hence. The crouds of people assembled upon this occasion were great beyond anything seen at such meetings !

Aug. 8. My lower wall nut-tree casts it's leaves in a very unusual manner. No wall-nuts ; the crop dropped off early in the summer.

Aug. 12. The thermometer for three or four days past has stood in the shade at Newton at 79, & 80.

Aug. 13. Goose-berries wither on the trees.

Aug. 14. Housed two loads of peat.

Aug. 18. Blackcaps eat the berries of the honey-suckles. Mrs J. White, after long & severe campaign carried on against the *Blattae molendinariae*, which have of late invaded my house, & of which she has destroyed many thousands, finds that at intervals a fresh detachment of old ones arrives ; & particularly during the hot season : for the windows being left open in the evenings, the males come flying in at the casements from the neighbouring houses, wch swarm with them. How the females, that seem to have no perfect wings that they can use, can contrive to get from house to house, does not so readily appear. These, like many insects, when they find their present abodes over-stocked, have powers of migrating to fresh quarters. Since the *Blattae* have been so much kept under, the Crickets have greatly encreased in number.

Aug. 19. My shrub, *Rhus cotinus*, [" smoke tree "] known to the nursery-men by the title of *Coccygria*, makes this summer a peculiar shew, being covered all over with it's " bracteae paniculae filiformes," which give it a feathery plume-like appearance, very amusing to those that have not seen it before. On the extremities of these panicles appear about midsumer a minute white bloom, which with us brings no seeds to perfection. Towards the end of August the panicles turn red & decay.

Aug. 20. Thomas, in mowing the walks, finds that the grass begins to grow weak, & to yield before the scythe. This is an indication of the decline of heat. Yucca filamentosa, silk grass, blows with a fine large white flower. It thrives abroad in a warm aspect. Habitat in Virginia.

Aug. 21. My large American Juniper, probably *Juniperus Virginiana*, has produced this summer a few small blossoms of a strong flavour like that of the juniper-berries : but I could not distinguish whether the flowers were male, or female ; so consequently could not determine the sex of the tree, which is dioecious. The order is *dioecia monadelphia*.

Aug. 22. The seeds of the lime begin to fall. Some wheat under hedges begins to grow.

Aug. 23. Some wheat bound ; & some gleaning. I have not seen one wasp.

Aug. 24. John Berriman's hops at the end of the Foredown very fine.

Aug. 26. A fly-catcher brings out a brood of young : & yet they will all withdraw & leave us by the 10th of next month.

Aug. 27. A fern-owl this evening showed-off in a very unusual, & entertaining manner, by hawking round, & round the circumference of my great spreading oak for twenty times following, keeping mostly close to the grass but occasionally glancing up amidst the boughs of the tree. This amusing bird was then in pursuit of a brood of some particular phalaena belonging to the oak, of which there are several sorts; & exhibited on the occasion a command of wing superior, I think, to that of the swallow itself. Fern-owls have attachment to oaks, no doubt on account of food: for the next evening we saw one again several times among the

COCK CHAFER, OR MAY BUG.
White's *Scarabaeus melolontha*
(⅔ natural size).

SUMMER CHAFER.
White's *Scarabaeus solstitialis*
(⅔ natural size).

boughs of the same tree; but it did not skim round it's stem over the grass, as on the evening before. In May these birds find the *Scarabaeus melolontha* on the oak; & the Scarabaeus solstitialis at Midsummer. These peculiar birds can only be watched & observed for two hours in the twenty-four; & then in a dubious twilight, an hour after sun-set & an hour before sun-rise.

Aug. 28. Men make wheat-ricks. Mr Hale's rick fell. Vivid rain-bow.

Aug. 29. Mr Clement begins to pick hops at Alton. *Clavaria's* [club fungi] appear on the hanger.

Aug. 31. Many moor-hens on Comb-wood pond.

Sept. 1. Grass grows on the walks very fast. Garden beans at an end.

Sept. 2. The well at Temple is 77 feet deep : 60 to the water, & 17 afterwards. My well measures only 63 feet to the bottom.

	yards	feet
Goleigh well[11] to the water is	55½	166
„ „ to the bottom	57½	172½
Heards well to the water is	70⅔	212
„ „ to the bottom	83⅓	250

A stone was 4½ seconds falling to the bottom of Heards well ; & 4 seconds to the water of Goleigh. The wells were measured accurately by the Revd Edmund White on the 25th of August 1792, in the midst of a very wet summer. Deep, & tremendous as is the well at Heards, John Gillman, an Ideot, fell to the bottom of it twice in one morning ; & was taken out alive, & survived the strange accident many years. Only Goleigh & Heards wells were measured by Mr E. White.

Sept. 4. Hop-picking becomes general ; & the women leave their gleaning in the wheat-stubbles. Wheat grows as it stands in the shocks.

Sept. 6. Gil. White left us. The flying ants of the small black sort are in great agitation on the zigzag, & are leaving their nests. This business used to be carryed on in August in a warm summer. While these emigrations take place, the Hirundines fare deliciously on the female ants full of eggs. Hop-picking becomes general ; & all the kilns, or as they are called in some counties, *oasts*, are in use. Hops dry brown, & are pretty much subject this year to vinny, or mould.

Sept. 8. Sowed thirteen rods, on the twelfth part of an acre of grass ground in my own upper Ewel close with 50 pounds weight of Gypsum ; also thirteen rods in D^o with 50 pounds weight of lime : thirteen rods more in D^o with 50 pounds weight of wood & peat-ashes : and four rods more on D^o with peat-dust. All these sorts of manures were sown by Bro^r T. W. on very indifferent grass in the way of experiment.

Sept. 9.   As most of the second brood of Hirundines are now out, the young on fine days congregate in considerable numbers on the church & tower :. & it is remarkable that tho' the generality sit on the battlements & roof, yet many *hang* or cling for some time by their claws against the surface of the walls in a manner not practised at any other time of their remaining with us.   By far the greater number of these amusing birds are house-martins, not swallows, which congregate more on trees.   A writer in the Gent. Mag. supposes that the chilly mornings & evenings, at this decline of the year, begin to influence the feelings of the young broods ; & that they cluster thus in the hot sunshine to prevent their blood from being benumbed, & themselves from being reduced to a state of untimely torpidity.

Sept. 11.   On this day my niece Anne Woods was married to Mr John Hounsom, who encreases my nephews, & nieces to the number of 59.   Mr John White came from Salisbury.

Sept. 12.   Began to light fires in the parlour. J.W. left us.

Sept. 13.   The stream at Gracious street, which fails every dry summer, has run briskly all this year ; & seems now to be equal to the current from Wellhead.   The rocky channel up the hollow-lane towards Rood has also run with water for months : nor has my great water-tub been dry the summer through.

Sept. 14.   From London three gallons of French brandy, & two gallons of Jamaica rum.

Sept. 15.   Hop women complain of the cold.

Sept. 16.   Dr Chandler's Bantam sow brought him this last summer a large litter of pigs, several of which were not cloven-footed, but had their toes joined together.   For tho' on the upper part of the foot there was somewhat of a suture, or division ; yet below in the soles the toes were perfectly united ; & on some of the hind legs there was a solid hoof like that of a colt.   The feet of the sow are completely cloven.

Mr Ray in his *Synopsis animalium quadrupedum* takes no notice of this singular variety; but *Linnaeus* in his *Syſtema Naturae* says, " Varietas frequens Upsaliae Suis domeſtici semper *monunguli* : in ceteris eadem species."

Sept. 17. Gathered-in the white pippins, about a bushel : many were blown down laſt week. Oats housed.

Sept. 19. Rain. Hops become very brown, & damaged. The hop-pickers are wet through every day.

Sept. 21. On this day Monarchy was abolished at Paris by the National Convention ; & France became a Republic !

Sept. 22. As I have queſtioned men that frequent coppices respeċting Fern-owls, which they have not seen or heard of late ; there is reason to suspeċt that they have withdrawn themselves, as well as the fly-catchers, & black-caps, about the beginning of this month. Where timber lies felled among the bushes, & covert, wood-men tell me, that fern-owls love to sit upon the logs of an evening : but what their motive is does not appear.

Sept. 23. My Bantam chickens, which have been kept in the scullery every night till now for fear of the rats, that carried away the firſt brood from the brew-house, went up laſt week to the beam over the ſtable. The earneſt & early propensity of the *Gallinae* to rooſt on high is very observable ; & discovers a ſtrong dread impressed on their spirits respeċting vermin that may annoy them on the ground during the hours of darkness. Hence poultry, if left to themselves & not housed, will perch, the winter through on yew-trees & fir-trees ; & turkies & Guinea-fowls, heavy as they are, get up into apple trees ; pheasants also in woods sleep on trees to avoid foxes :—while pea-fowls climb to the tops of the higheſt trees round their owner's house for security, let the weather be ever so cold or blowing. Partridges, it is true, rooſt on the ground, not having the faculty of perching ; but then the same fear prevails in their minds ; for through apprehensions from pole-cats, weasels, & ſtoats, they never truſt themselves to coverts ; but neſtle together in the midſt of large fields, far removed from hedges & coppices, which they love to $\frac{\text{haunt}}{\text{frequent}}$ in the day ; & where ar that season they can skulk more secure from the ravages of rapacious birds.   As to ducks, & geese, their aukward splay web-feet forbid them to settle on trees : they therefore, in the hours of darkness & danger, betake themselves to their own element the water, where amidſt large lakes & pools, like ships riding at anchor, they float the whole night long in peace, & security.

Sept. 25.   Men begin to bag hops.   Celeri comes in.   Vine-leaves turn purple.

Sept. 30.   There is a remarkable hill on the downs near *Lewes* in *Sussex*, known by the name of *Mount Carburn* [Caburn], which over-looks that town, & affords a moſt engaging prospeƈt of all the country round, besides several views of the sea.   On the very summit of this exalted promontory, & amidſt the trenches of its Danish [British] camp, there haunts a species of wild Bee,[12] making its neſt in the chalky soil.   When people approach the place, these inseƈts begin to be alarmed, & with a sharp & hoſtile sound dash, & ſtrike round the heads & faces of intruders. I have often been interrupted myself while contemplating the grandeur of the scenery around me, & have thought myself in danger of being ſtung :— and have heard my Brother *Benjamin* say, that he & his daughter *Rebecca* were driven from the spot by the fierce menaces of these angry inseƈts.   In old days Mr *Hay*[13] of *Glynd Bourn*, the Author of *Deformity*, & other works, wrote a loco-descriptive poem on the beauties of *Mount Carburn*.

Oƈt. 1.   Wheat out at Buriton, Froxfield, Ropley, & other places.

Oct. 2. Flying ants, male & female, usually swarm, & migrate on hot sunny days in August & Septem^r; but this day a vast emigration took place in my garden, & myriads came forth in appearance, from the drain which goes under the fruit-wall; filling the air & the adjoining trees & shrubs with their numbers. The females were full of eggs. This late swarming is probably owing to the backward, wet season. The day following, not one flying ant was to be seen. The males, it is supposed all perish: the females wander away; & such as escape from Hirundines get into the grass, & under stones, & tiles, & lay the foundation of future colonies.

Oct. 3. Hirundines swarm around the Plestor, & up & down the street.

Oct. 6. Many Hirundines: several very young swallows on the thatch of the cottage near the pound. The evening is uncommonly dark.

Oct. 7. The crop of stoneless berberries is prodigious! Among the many sorts of people that are injured by this very wet summer, the peat-cutters are great sufferers: for they have not disposed of half the peat & turf which they have prepared; & the poor have lost their season for laying in their forest-fuel. The brick-burner can get no dry heath to burn his lime, & bricks: nor can I house my cleft wood, which lies drenched in wet. The brick-burner could never get his last makings of tiles & bricks dry enough for burning the autumn thro'; so they must be destroyed, & worked up again. He had paid duty for them; but is, as I understand, to be reimbursed.

Oct. 9. Master Hale houses barley that looks like old thatch. Much barley about the country, & some wheat. Some pheasants found in the manour. The sound of great guns was heard distinctly this day to the S.E. probably from Goodwood, where the Duke of Richmond has a detachment from the train of artillery encamped in his park, that he may try experiments with some of the ordnance.

Oct. 11. Dr Chandler mows the church-litton closes for hay. Farmer Parsons houses pease, which have been hacked for weeks. Barley abroad.

Oct. 12. Gathered in the dearling apples: fruit small, & stunted.

Oct. 19. Made presents of berberries to several neighbours. Ring-ouzel seen in the Kings field.

Oct. 23. Dr Bingham & family left Selborne.

Oct. 26. Hired two old labourers to house my cleft billet wood, which is still in a damp, cold condition, & should have been under cover some months ago, had the weather permitted.

Oct. 27. Some few grapes just eatable: a large crop. Housed all the billet wood. Leaves fall in showers. A curlew is heard loudly whistling on the hill towards the Wadden. On this day Mrs S. Barker was brought to bed of a boy, who advances my nepotes to the round & compleat number of 60.

Oct. 28. Thomas saw a polecat run across the garden.

Oct. 29. Finished piling my wood: housed the bavins: fallows very wet.

Oct. 30. Planted 100 of cabbages, in ground well dunged, to stand the winter.

Nov. 3. Men sow wheat: but the land-springs break out in some of the Hartley malm-fields. [Upper Greensand rock].

Nov. 5. Gossamer abounds. Vast dew lies on the grass all day, even in the sun.

Nov. 8. Planted 3 quarters of an hundred more of cabbages to stand the winter: dug-up potatoes; those in the garden large, & fine, those in the meadow small, & rotting.

Nov. 10. On this day Brother Benjamin quitted South Lambeth, & came to reside at His House at Mareland.

Nov. 12. Planted in the garden 2 codling-trees, 2 damson-trees, & 22 goose-berry trees, sent me by Bro^r T.W.

Nov. 13. Mʳ Ed. White & man brought a good fine young white poplar from his out-let at Newton, & planted it at yᵉ top of Parsons's, slip behind the bench; where it will be ornamental.

Nov. 15. Timothy comes out.

Nov. 17. Baker's hill is planted all over with horse-beans, which are grown four or five inches high. They were probably sown by jays; & spring up thro' the grass, or moss. Many were planted there laſt year, but not in such abundance as now.

Nov. 19. Water-cresses come in.

Nov. 21. Sent 3 bantam fowls to Miss Reb. White at Mareland, a cock & two pullets.

Nov. 22. Timothy comes forth.

Nov. 24. Saw a squirrel in Baker's hill: it was very tame. This was probably what Thomas called a pole-cat. [See 28 Oct. *supra.*]

Nov. 26. Timothy hides.

Nov. 29. This dry weather enables men to bring in loads of turf, not much damaged: while scores of loads of peat lie rotting in the Foreſt.

Dec. 1. Thomas ſtarted a hare, which lay in her form under a cabbage, in the midſt of my garden. It has begun to eat the tops of my pinks in many places. The land-springs, which began to appear, are much abated.

Dec. 2. This dry fit has proved of vaſt advantage to the kingdom; & by drying & draining the fallows, will occasion the growing of wheat on many hundred of acres of wet, & flooded land, that were deemed to be in a desperate ſtate, & incapable of being seeded this season.

Dec. 4. Timothy is gone under a tuft of long grass, but is not yet buried in the ground.

Dec. 5. Timothy appears, & flies come-out.

Dec. 7. Took down the urns, & shut up the alcove.

Dec. 8. Dʳ Chandler brought a vaſt pear from the garden of his niece at Hampton, which weighed 20 ounces, & ¾, & measured in length 6 inches, & ¾,

& in girth eleven inches. It is the sort known by
the name of Dʳ Uvedale's great Saint Germain.

Dec. 9.   Damage by the wind in some places.

Dec. 10.   Mʳ Taylor brought me a pine-apple,
which was, for the season, large, & well-flavoured.

Dec. 14.   [Newton] Grey, & mild, gleams.

Dec. 15.   [Selborne] Grey, sun, pleasant, yellow
even.

Dec. 16.   The season has been so mild that the
Antirrhinum Cymb. still flourishes, & continues in
bloom. [Cf. 17 Jan. *supra*.]

Dec. 20.   Dark & wet.   Shower, a short, but
violent gust.   Lightening.

Dec. 24.   Covered the artichokes, & rhubarb with
litter, & the spinage, & the Yucca filimentosa with
straw ; & the few brown lettuces with straw.   Mʳ
Churton came.

Dec. 26.   Bramblings are seen : they are winter-
birds of passage, & come with the hen-chaffinches.
Nep. Ben. White & wife came.

Dec. 29.   B. White, & wife left us.

CHAPTER XXVI

" When sinks the sun behind the hill,
When all the weary wheels ſtand ſtill."
W. MORLEY PUNSHON.

1793

Jan. 4. Rain, rain, gleams. Venus is very resplendent.

Jan. 6. N. papers mention snow to the northward. On this day Mrs Clement was brought to bed of a boy, her ninth child. My nephews & nieces are now 61.

Jan. 7. Nephew Holt White came.

Jan. 10. Mr Churton left us.

Jan. 11. On this day came my Nep. John White of Sarūm with his bride late Miss Louisa Neave, who encreased my Nep. & nieces to the number of 62.

Jan. 12. Vaſt rain in the night, lightening. Great ſtream in the cart-way.

Jan. 14. [Newton] Snow-drops bud, & winter-aconites blossom. John White, & wife, & Holt White left us.

Jan. 16. [Selborne] dark, & sharp froſt.

Jan. 17. Turnip-greens come in.

Jan. 20. Rime on the hanger. Mr *Marsham*, who lives near Norwich, writes me word, that a servant of his shot a bird laſt autumn near his house that was quite new to him. Upon examination it appeared to him, & me to answer the description of the *Certhia muraria*, the *Wall-creeper*, a bird little known, but some times seen in England. *Ray*, & *Willughby* never met with it, nor did I ever find it wild, or among the vaſt collections exhibited in London; but *Scopoli* had a specimen in his Museum, & says it is to be found in

Carniola. It haunts towers, & castles, & ruins, some times frequents towns, running up the walls of tall houses, & searching the crannies, & chinks for spiders, & other insects. Some of the internal wing-feathers are beautifully marked on the inner web with two white, or pale yellow spots; & the middle of the outer web edged with red. Two of these quills, drawn in water-colours, by a young Lady, & charmingly executed, were sent me by Mr *Marsham* in a frank: the pencilling of these specimens is truly delicate, soft, & feathery. It is much to be regretted that she did not draw the whole bird.[1] The claws of this bird are strong & large, say *Linnaeus*, & Mr *Marsham*; & especially the hind claw.

Jan. 21. Thrush sings, the song-thrush: the missle-thrush has not been heard. On this day Louis 16th late king of France, was beheaded at Paris, & his body flung into a deep grave without any coffin, or funeral service performed.

Jan. 28. Bees come out, & gather on the snow-drops.

Feb. 1. The Republic of France declares war against England & Holland.

Feb. 3. A strong gust in the night blew down the rain-gage, which, by the appearance in the tubs, must have contained a considerable quantity of water.

Feb. 4. Venus is very bright, & shadows.

Feb. 5. [Newton] Mrs. J. White set out for Kingston on Thames.

Feb. 8. [Selborne] War declared & letters of Marque granted against the french Republic.

Feb. 10. Grey, sun, severe wind, with flights of snow, sleet, & hail.

Feb. 11. Paths get dry. Sowed a bed of radishes, & carrots under the fruit-wall.

Feb. 12. Mrs J. White returns.

Feb. 15. Rain & hail in the night. Made a seedling-cucumber bed: mended the frame, & put it on.

Feb. 16.   Sent some winter-aconites in bloom to
Dr Chandler ; & received back some roots of *Arum
dracunculus*.   [Cf. 14 July 1791.]   Tubbed, & salted-up
a fine young hog, bought of Timothy Turner.

Feb. 19.   Sowed half a barrel of American Gypsum,
which was sent for in the autumn by Bro. Tho. on
the *fourth* ridge of Tim Turner's wheat, as you reckon
from the walk in that field.   The powder strewed
about two thirds of the ridge from the Ewel S.E. ward.

Feb. 20.   Wheeled much dung into the garden.

Feb. 21.   Dug the garden-plot in the orchard, &
in the meadow : but the ground is very wet, & heavy.

Feb. 24.   Mr White of Newton sprung a pheasant
in a wheat-stubble, & shot at it; when, notwithstanding
the report of the gun, it was immediately pursued by
the blue hawk, known by the name of the *Hen-harrier*,
but escaped into some covert.   He then sprung a
second, & a third in the same field, that got away in
the same manner ; the hawk hovering round him all
the while that he was beating the field, conscious no
doubt of the game that lurked in the stubble.   Hence
we may conclude that this bird of prey was rendered
very daring, & bold by hunger ; & that Hawks cannot
always seize their game when they please.   We may
further observe that they cannot pounce their quarry
on the ground, where it might be able to make a stout
resistance ; since so large a fowl as a pheasant could
not but be visible to the piercing eye of an hawk,
when hovering over a field.   Hence that propensity
of cowring & squatting till they are almost trod on,
which no doubt was intended as a mode of security ;
tho' long rendered destructive to the whole race of
Gallinae, by the invention of nets, & guns.

Feb. 28.   Planted   50   good   cabbage-plants :
mended the bed planted in the autumn, & eaten in
part by the hares.

Mar. 3.   The wind last night blowed-off some tiles
from my roof.   This storm did much mischief about
the kingdom.

Mar. 4-15. [Mareland.]

Mar. 4.   We are much amused every morning
by a string of Lord Stawell's Hunters that are aired,
exercised, & watered in a meadow opposite to the
windows of this house.   There seem to be two sets,
which appear alternatly on the days that they are not
hunted.   He has in all sixteen.

Mar. 5.   Herons haunt the stream below the house,
where the Wey meanders along the meads.   Lord
Stawell sent me a curious water-fowl, shot on Frinsham
pond, which proved to be the *Shoveler* [*Spatula
clypeata*], remarkable for the largeness of it's bill.   It
is a species of duck, & most exactly described by Mr
Ray.   Large wood-pecker laughs very loud.   My
Brother's lambs frolick before the windows, & run to
a certain hillock, which is their goal, from whence
they hurry back ; & put us in mind of the following
passage in the Poet of nature :
         " Now the sprightly race
Invites them forth ; then swift, the signal given,
They start away, & sweep the mossy mound
That runs around the hill."
            [*Seasons : Spring*, ll. 835-8.]
Mar. 6.   Dogs-tooth violets blow.   Wag-tails on
the grass-plots : they were here all this mild winter.
Goldfinches are not paired.

Mar. 7.   Trouts begin to rise: some angling takes
place in this month.   My Brother's cucumbers are
strong, & healthy.   Lady Stawell tells Mrs White that
they have seen more woodcocks & snipes at their
table this winter than usual.

Mar. 8.   Many redwings feeding in the meadow.

Mar. 10.   The sweet bells at Farnham, heard up
the vale of a still evening, is a pleasant circumstance
belonging to this situation, not only as occasioning
agreeable associations in the mind, & remembrances
of the days of my youth, when I once resided in that
town² :—but also by bringing to one's recollection
many beautiful passages from the poets respecting

this tuneable & manly amusement, for which this island is so remarkable. Of these none are more distinguished, & masterly than the following :—

"   Let the village bells as often wont,
Come swelling on the breeze, & to the sun
Half set, ring merrily their evening round.
.   .   .   .
It is enough for me to hear the sound
Of the remote, exhilerating peal,
Now dying all away, now faintly heard.
And now with loud, & musical relapse
In mellow changes pouring on the ear."[3]

The Village Curate.

Mar. 11.   There is a glade cut thro' the covert of the Holt opposite these windows, up to the great Lodge.   To this opening a herd of deer often resorts, & contributes to enliven & diversify the prospect, in itself beautiful & engaging.

Mar. 12.   Apricot begins to blow.   Red-wings, & starlings abound in the meadow, where they feed in the moist, & watered spots.

Mar. 13.   During my absence Thomas parted-out my polyanths, & planted them in rows along the orchard walk, & up the border of Baker's Hill by the hot beds.   My Brother has a pigeon-house stocked with perhaps 50 pairs of birds, which have not yet begun to breed.   He has in the yard Turkeys, a large breed of ducks, & fine fowls.   On the ponds are geese, which begin to sit.

Mar. 14.   Papilio rhamni, the brimstone butterfly, appears in the Holt.   Trouts rise, & catch at insects. A dob-chick comes down the Wey in sight of the windows, some times diving, & some times running on the banks.   Timothy the tortoise comes forth, & weighs 6 ae 5½ z.   Took a walk in the Holt up to the lodge : no bushes, & of course no young oaks : some Hollies, & here & there a few aged yews : no oaks of any great size.   The soil wet & boggy.

Mar. 17.   On friday last my Brother & I walked up to Bentley church, which is more than a mile from his house & on a considerable elevation of ground. From thence the prospect is good, & you see at a distance Cruxbury [Crooksbury] hill, Guild down [Hogs Back], part of Lethe hill, Hind-head, & beyond it to the top of one of the Sussex downs.   There is an avenue of aged yew-trees up to the church : & the yard, which is large, abounds with brick-tombs covered with slabs of stone : of these there are ten in a row, belonging to the family of the Lutmans.   The church consists of three ailes, & has a squat tower containing six bells.   From the inscriptions it appears that the inhabitants live to considerable ages.   There are hop-grounds along on the north side of the turn-pike road, but none on the south towards the stream. The whole district abounds with streams.   The largest spring on my brother's farm issues out of the bank in the meadow, just below the terrace.   Some body formerly was pleased with this fountain, & has, at no small expence bestowed a facing of Portland stone with an arch, & a pipe, thro' which the water falls into a stone bason, in a perennial stream.   By means of a wooden trough this spring waters some part of the circumjacent slopes.   It is not so copious as Well head [at Selborne]."

Mar. 20.   Planted 30 cauliflowers brought from Mareland ; & a row of red cabbages.   The ground is so glutted with rain that men can neither plow, nor sow, nor dig.

Mar. 21.   Parted the bunches of Hepatica's, that were got weak, & planted them again round the borders.

Mar. 24.   This evening Admiral Gardner's fleet sailed from St Helens with a fair wind.

Mar. 26.   Snow, & rain, harsh.   A sad wintry day !

Mar. 28.   Snow does not lie, ice, frost & icicles all day.

Mar. 29. White sharp frost: thick ice: icicles. Apricots blow: peaches & nectarines begin to open their buds. Some thing again eats off the young celeri.

Mar. 30. Made a new hand-glass bed for celeri in the garden. The crocus's still look very gay when the sun shines.

Apr. 1. In the mid counties there was a prodigious snow; some people were lost in it, & perished.

Apr. 3. The small *willow-wren*, or *chif-chaf*, is heard in the short Lythe. This is the earliest summer bird, & is heard usually about the 20th of March. Tho' one of the smallest of our birds, yet it's two notes are very loud, & piercing, so as to occasion an echo in hanging woods. It loves to frequent tall beeches.

Apr. 4. Timothy Turner ashed a great part of Baker's hill, & dunged one part. Wag-tail on grass-plots.

Apr. 5. The air smells very sweet, & salubrious. Men dig their hop-gardens, & sow spring-corn. Cucumber plants show rudiments of fruit. Planted cuttings of currans, & goose-berries. Dug some of the quarters in the garden, & sowed onions, parsnips, radishes, & lettuces. Planted more beans in the meadow. Many flies are out basking in the sun.

Apr. 6. On the 6th of last October I saw many swallows hawking for flies around the Plestor, & a row of young ones, with square tails, sitting on a spar of the old ragged thatch of the empty house. This morning Dr Chandler & I caused the roof to be examined, hoping to have found some of these birds in their winter retreat: but we did not meet with any success, tho' Benham searched every hole & every breach in the decayed roof.

Apr. 7. The chaffinches destroy the blossoms of the polyanths in a sad manner. Sowed a bed of carrots: the ground hard, & rough, & does not rake fine.

Apr. 9. Thomas Knight, a sober hind, assures us, that this day on Wish-hanger Common between

Hedleigh & Frinsham he saw several *Bank martins* playing in & out, & hanging before some nest-holes in a sand-hill, where these birds usually nestle. This incident confirms my suspicions, that this species of Hirundo is to be seen first of any; & gives great reason to suppose that they do not leave their wild haunts at all, but are secreted amidst the clefts, & caverns of these abrupt cliffs where they usually spend their summers. The late severe weather considered, it is not very probable that these birds should have migrated so early from a tropical region thro' all these cutting winds and pinching frosts : but it is easy to suppose that they may, like bats & flies, have been awakened by the influence of the Sun, amidst their secret latebrae, where they have spent the uncomfortable foodless months in a torpid state, & the profoundest of slumbers. There is a large pond at Wish-hanger which induces these sand-martins to frequent that district. For I have ever remarked that they haunt near great waters, either rivers or lakes. Planted in one of the quarters of the garden, in ground well dunged, 8 long rows of potatoes. Carted in hot dung for the cucumber-bed.

Apr. 10. Dug the asparagus bed, & cleared away the straw laid on. Farmers wish for a gentle rain.

Apr. 11. Hoed & cleaned the alleys.

Apr. 12. The *Nightingale* was heard this harsh evening near James Knight's ponds. This bird of passage, I observe, comes as early in cold cutting springs, as mild ones !

Apr. 13. Bat out. This is the twelfth dry day.

Apr. 15. Sowed fringed bore-cole, & Savoys, & leeks.

Apr. 16. Made a hot bed for the two-light frame with lapped glass.

Apr. 19. Showers of hail, sleet. Gleams. Timothy, who has withdrawn himself for several days, appears.

Apr. 20. The *Cuckoo* is heard on Greatham common.

Apr. 23. Mowed the terrace. Cut the first cucumber. Pulled the first radishes. A swallow over my meadow.

Apr. 24. When Thomas got-up to brew at four o' the clock, he heard some stone-curlews pass by over the house in their way to the uplands. In the evening they flie over the village downwards, towards the brook, & meadows, where they seem to spend the night.

Apr. 27. Men begin to pole their hops. Mountain snow-drop blows.

Apr. 28. Wall-flowers full of bloom, & very fine. Nightingale in my fields.

Apr. 29. I have seen no hirundo yet myself. Sowed Columbines, two sorts; Scabius; Scarlet lychnis; Nigella; 10 weeks stocks; Mountain lychnis.

Apr. 30. Saw two swallows at Gracious street.

May 2. Sad, blowing, wintry weather. I think I saw an house martin. There is a bird of the black-bird kind, with white on the breast, that haunts my outlet as if it had a nest there. Is this a ring-ouzel? If it is, it must be a great curiosity; because they have not been known to breed in these parts.[4]

May 3. Timothy eats. A pair of Missel-thrushes have made a nest in the apple-tree near the fruit-wall. One young half-fledged was found in the garden.

May 4. Some beeches begin to show leaves. Sowed some fine Savoy seed from Newton. Hen *red-start* appears.

May 5. Damson, sloe-trees, & wild Merise blow. *Cock Red start.* There has been so little frost, that the *Antirrhinum Cymb.* flourished & blossomed the whole winter thro', & is now very thriving, tho' its usually dies about Xmass. So that, in mild times, it is at least a biennial with us, & may be perhaps of longer duration in milder regions. [Cf. 17 Jan. 1792.]

James Knight has observed two large fieldfares in the high wood lately, haunting the same part, as if they intended to breed there. They are not wild. A nest of this sort of bird would be a great curiosity.[5]

May 9. The mag-pies, which probably have young, are now very ravenous, & destroy the broods of Missel-thrushes, tho' the dams are fierce birds, & fight boldly in defence of their nests. It is probably to avoid such insults, that this species of thrush, tho' wild at other times, delights to build near houses, & in frequented walks, & gardens.

May 10. M[issel] thrushes do not destroy the fruit in gardens like the other species of turdi, but feed on the berries of missel toe; & in the spring on ivy berries which then begin to ripen. In the summer, when their young become fledge, they leave neighbourhoods, & retire to sheep walks, & wild commons.

May 12. The merise, or wild cherry in beautiful bloom.

May 13. Two nightingales sing in my outlet. Foliage of trees expands very fast. Peat begins to be brought in: it is in good condition. H. martins build. The old Bantam hen began to sit in the barn on eleven eggs. The *fern-owl*, or *churn-owl* returns, & chatters in the hanger.

May 14. Timothy travels about the garden.

May 16. Sowed-in the three-light annual frame African & French marrigolds, China asters, pendulous Amaranths, Orange-gourds. Took the blackbird's nest the second time; it had squab young.

May 17. Set the second Bantam hen over the saddle cup-board in the stable with eleven dark eggs.

May 18. A man brought me a large trout weighing three pounds, which he found in the waste current at the tail of Bins pond, in water so shallow that it could not get back again to the Selborne stream. Made rhubarb tarts, & a rhubarb pudding, which was very good.

May 19. The white apple-tree shows again, as usual, much bloom.

May 20. Cut *two brace* of fine cucumbers; & left one for seed. The 10 weeks stocks, which stood the winter, make a fine show, & are very fragrant. Tulips blow.

May 21. Timothy eats much.

May 22. Nep. Ben. White, & wife came.

May 25. Cut down the greens of the crocus's; they make good tyings for hops; better than rushes, more pliant, & tough.

May 26. The white pippin is covered with bloom. Farmer Spencer's apple-trees blow well. Nep. Ben White, & wife left us.

May 28. The season is so cold, that no species of Hirundines make any advances towards building, & breeding. Brother Benjn & Mrs. White, & Mary White, & Miss Mary Barker came.

May 28. My weeding-woman swept-up on the grass-plot a bushel-basket of blossoms from the white apple-tree : & yet that tree seems still covered with bloom.

May 30. Fyfield sprung a brace of pheasants in Sparrow's hanger. Hail-like clouds about.

May 31. My great oak abounds in bloom, which is of a yellowish cast : the young shoots usually look red. The house-martins at Mareland, in the few hot days, began to build, but when the winds became cold again immediately desisted.

June 1. Timothy is very voracious : when he can get no other food he eats grass in the walks.

June 2. Bror Benjn & I measured my tall beech in Sparrow's hanger, which, at five feet from the ground, girths six feet one inch, and three quarters.

June 3. The ground sadly burnt up. Royal russets show much bloom. Summer cabbage comes in.

June 4. Cinnamon rose blows.

June 5. Men's St foin burns, & dies away. The farmers on the sands complain that they have no grass.

June 6.   Sowed two rows of large white kidney-beans : but the ground is so hard, that it required much labour to render it fit to receive the seed.   The old Bantam brought out only three chickens.

June 7.   Watered well the white poplar at the foot of the bostal.   Cut the slope hedge in Bakers hill. Mrs Clement, & children came.

June 8.   The young Bantam hen brought out only three chickens.   Showers that wetted the blades of corn, & grass, but did not descend to the root. Ground very hard.

June 9.   Early orange-lilies blow.   Few chafers. The water at Kingsley mill begins to fail.   The land-spring in the stoney-lane, as you go to Rood, stops. We draw much water for the garden : the well sinks very fast.

June 10.   Cut *five* cucumbers.

June 11.   A man brought me a large plate of straw-berries, which were crude, & not near ripe.   The ground all as hard as iron : we can sow nothing nor plant out.

June 12.   Bright, sun, golden even.   Cut *eight* cucumbers.   Mrs Clement & children left us.   Many swifts.

June 13.   Cut *ten* cucumbers.   Provence roses blow against a wall.   Dames violets very fine.   ten weeks stocks still in full beauty.

June 14.   Cut four cucumbers.   Mr John Mulso came.

June 15.   Men wash their sheep.   Mr J. Mulso left us.

NOTES

1768

[1] The effects of this frost are described in the letter to D.B., LXI.

[2] *Ananas sativus*, the pineapple, introduced into England about 1690.

[3] A village about 2m. N. of Selborne.

[4] Gilbert White and his brother Thomas were fond of experimenting with fodder crops.

[5] *Bombylius medius*, the humble-bee fly, which has a hairy body and a long proboscis. White's citation probably shows that he has just identified the insect.

[6] Under the name titlark White seems to confuse two species, —the tree pipit (*Anthus t. trivialis*), a summer migrant, which sings " from tree to tree," and the resident meadow pipit (*Anthus pratensis*) which sings as it glides to the ground.

[7] The exclamation marks refer to the early date (Cf. *Luscinia*, the nightingale, on April 5).

[8] The grasshopper warbler (*Locustella n. naevia*).

[9] Apparently the common sandpiper (*Tringa hypoleucos*), a summer migrant (See T.P., XX).

[10] Identified with the old Priory fishponds at Coneycroft.

[11] White told Pennant that the viper, though oviparous, is viviparous also (T.P., XVII). His quotation refers to the curious mode of birth, now called ovoviviparous, in which the living foetis is more or less freed from the egg-coverings.

[12] White's reed-sparrow is apparently the sedge warbler (*Acrocephalus schoenobaenus*), a summer migrant ; his added note distinguishes it from another " reed sparrow," the resident reed-bunting (*Emberiza s. schoeniclus*). Cf. D.B., VI.

[13] *Phallus impudicus*, the stinkhorn, or wood-witch, an erect, pillar-like fungus, with a reticulate cap (pileus). Its odour is very offensive.

[14] *Listera ovata*, one of the Orchidaceae.

[15] Henry White was rector of Fyfield, near Andover.

[16] Thomas Hoar, White's general utility man.

[17] This stood on the Plestor, or village green. It was mended and re-painted in 1779, but disappeared at some unknown date.

[18] The gad-fly, or horse-bot (*Gastrophilus equi*). Confused with *Oestrus bovis* (Sept. 23). Rev. W. Derham (1657-1735) theologian and naturalist.

[19] The spotted flycatcher (*Muscicapa striata*).

²⁰ This observation is disputed by modern naturalists. Some species of caterpillar is suggested as the destroyer.

²¹ Ring-ousel (*Turdus t. torquatus*), a bird of passage in Hampshire.

²² A hill in Priors Dean parish, 4 m. S. of Selborne.

²³ Faringdon, a neighbouring village to Selborne; here White was curate from 1761-84. Lavants and bournes, or intermittent streams.

²⁴ Virgil, *Georg.* I, 400. Apparently White's humorous reference to unrung pigs.

²⁵ Read "brother-in-law." Thomas Barker (1722-1809) married White's sister, Anne.

1769

¹ The common, or corn bunting (*Emberiza c. calandra*).

² Also called the hooded crow (*Corvus cornix cornix*).

³ *Par. Lost*, Bk. VII, l. 438.

⁴ The larvae of a midge (*Chironomus plumosus*).

⁵ Nore Hill, a spur of the chalk downs, about 1¾ miles S. of Selborne.

⁶ Further particulars omitted, since they occur in T.P. XXIV.

⁷ A wryneck noted in London!

⁸ The observation of this transit was the object of Cook's first voyage in the *Endeavour* to Otaheite.

⁹ Sedge Warbler.

¹⁰ White wants the reader to distinguish wheatears from "ears of wheat."

¹¹ Marsh harrier, belonging to the same Order as the sparrow-hawk and kestrel.

¹² Dr W. Sheffield, afterwards Provost of Worcester College.

¹³ First observed by Messier; it was extremely brilliant, and its tail formed a huge luminous arch.

¹⁴ Mrs John White (White's sister-in-law), who was returning to Gibraltar.

¹⁵ White is seeking evidence of the hibernation of swallows.

¹⁶ The pied wagtail (*Motacilla alba Yarrellii*) must be intended.

¹⁷ A large genus of Lepidoptera, according to the scheme of Linnaeus.

1770

¹ *Fauna Calpensis*, a work completed but never published.

² A brown cup-shaped fungus, of the Order *Ascomycetes*.

³ The steel-blue dor, or dung beetle (*Geotrupes stercorarius*).

⁴ White frequently mentions the aberdevine, or siskin, but after many observations, rightly decides that his bird was the

reed-sparrow [=reed bunting], a very different species. (Cf. Note 12, for 1768.)

⁵ Small tortoise-shell butterfly; evidently a hibernated specimen.

⁶ The hazel is monoecious; the separate male and female flowers occur on the same tree.

⁷ A valuable observation. Bustards (*Otis t. tarda*) become extinct in England about the year 1835 or 1836.

⁸ Hooded or Royston crow. (Cf. Note 2, 1769.)

⁹ Cf. account in T.P., XXXV.

¹⁰ *Par. Lost*, VII, 444-6.

¹¹ Probably the birds mentioned in T.P. ii.

¹² See note 4, *supra*.

¹³ The flea of sand-martins is a different species from the bed-flea (*Pulex irritans*).

¹⁴ *Olet.* See Note 13, year 1768.

¹⁵ The maggot of the ox warble fly, *Hypoderma* (White's *Oestrus*), of which two species are known.

¹⁶ *Eucera longicornis*, a black hairy, burrowing bee.

¹⁷ The horse-bot (*Gastrophilus equi*). See T.P., XXXIV. White's name is quoted from Derham; why one specific name is in Greek characters is uncertain.

¹⁸ White first discovered this bat as a British species. (Cf. T.P., XXVI, XXXVI.)

¹⁹ Mrs Snooke, of Ringmer, was White's aunt; she was a sister of the elder John White, Gilbert's father.

²⁰ White was riding from Ringmer to Selborne.

²¹ See map for these places.

²² *Columba oenas*, really a resident bird. (Cf. D.B., X.)

²³ Cf. D.B., X. One of the few birds whose notes can be expressed in our musical scale.

²⁴ Local name for bournes, or intermittent streams.

²⁵ This must refer to the alarm cry, when prolonged. The " whistle " in inapplicable to the song, or to the ordinary harsh alarm cry.

1771

¹ The so-called " silver-fish,"insects belonging to the family *Thysanura*.

² This sentence may refer to the previous year's crop; it was evidently written on March 2.

³ Some confusion of dates (see March 13), but the identification is sound.

⁴ John Woods, a connection of White by marriage.

⁵ Adapted from *Georg.* III, ll. 538-9. *Illum* in original.

⁶ Apparent reference to illness. (Cf. Apr. 18.)

7 Also called the forest-fly (*Hippobosca equina*).   Cf. D.B., XV.

8 Cf. D.B., XLVIII.

9 The six-spot burnet moth (*Zygaena filipendulae*).   The cocoon is attached to grasses or plants like the dropwort.   The larva feeds on the trefoil group of plants.

10 Cart-loads, dialect word in Midlands and Southern England. (See N.O.D.)

11 " Curious," in obsolete meaning, fine or excellent in quality. (See N.O.D., S.V. II., 14.)

12 Hack, dialect word for a reaping-hook with a long handle.

13 14 At this period of his life White had come to believe in the migration of the majority of the Hirundines.

15 Timothy, the tortoise, frequently mentioned in the *Journal* and the " *Natural History*."

16 White was on his way home.   The kite seems to be now restricted to a few pairs in Wales.

17 The so-called " Ripon flood."

18 Cf. entry March 13th, 1771, there properly called *Larus ridibundus*.

1772

1 There are no entries for this period ; the *Journal* seems to have been left at home.

2 The great grey shrike, a casual visitor to England.

3 The locality is given as [East] Tisted Park in T.P., XXXIX. Rotherfield Park is in East Tisted parish.

4 Cf. T.P., XXXIX.   The question is discussed in the editor's " *Gilbert White : Pioneer, Poet, and Stylist*," 1928, pp. 113-14.

5 The death-watch of superstition is the beetle *Anobium tessellatum*.   White probably means this species.

6 West of Rowland's Castle, in South Hants.

7 White's abbreviation of Brighthelmstone, the spelling of his day.

8 John White was chaplain at Gibraltar.   (Cf. July 27th 1772.)

9 Dial. word for drizzling, or spray-like rain ; here applied to clouds.

10 Qy. " shuttering " :   dial. word for continuous or fast-falling rain.

11 The carder, or hoop-shaver bee (*Anthidium manicatum*).

12 See D.B., XXIX.   The question is discussed in W.J., pp. 217-23.

13 Cf. T.P., XXXIV.   White recognized the " fly " as a beetle—the " black dolphin."   It is now called *Haltica* (=*Phyllotreta nemorum*).

¹⁴ The exceptionally late date helped to confirm White's mistake about the hibernation of the Hirundines. The " Northern naturalist " is Olaus Magnus.

¹⁵ The garden nasturtium (*Tropaeolum majus*) is a native of Peru and Chile.

¹⁶ Cf. D.B., IX. Queest, quist, or queist, is a name recorded from the Midland and Southern counties generally.

¹⁷ Apparently *Helvella crispa,* an edible fungus.

1773

¹ From 24 Dec. 1772 to 6 Jan. 1773, weather records only are entered.

² This separation of the sexes is often recorded by White. (Cf. T.P., XIII, XV, XXXIX.)

³ *Tuber aestivum,* a globose, warty, underground fungus, is the best-known of our edible truffles. *T. brumale* is a winter species.

⁴ The tree pipit is probably intended : White seems to confuse the tree with the meadow pipit.

⁵ A fine description : the noun was borrowed from John White. (R.H.W., I, p. 219.)

⁶ *Pyrus Aria,* the white beam tree.

⁷ Also called *turgid,* or Rivett wheat, of which there are many varieties, recognized by the prominent keels of the hairy lower glumes.

⁸ Cf. T.P., XXXVII. Question discussed in W.J., pp. 111-12.

⁹ Surely a superstition !

¹⁰ A member of the weevil family (*Curculionidae*). Possibly the plum weevil (*Otiorhynchus*), but this beetle feeds chiefly on shoots and leaves.

¹¹ Rev. Richard Yalden, vicar of Newton Valence, near Selborne.

¹² Dial. word for a brood of pheasants, recorded from Cheshire and Lincolnshire, and Southwards. (*Eng. Dial. Dict.*)

¹³ John White spent the winter of this year at Selborne. (R.H.W., I, p. 213.)

¹⁴ An error : the yellow wagtail is a summer visitor only. White doubtless meant the grey wagtail, which has bright yellow underparts.

¹⁵ A simple bridge across a stream, or a raised side-walk formed of planks or stone slabs.

¹⁶ The white wagtail is a scarce summer visitor to Southern England : the bird noted was apparently an albino " water wagtail," as indicated.

1774

1 In front of Grange Farm.

2 Monkshood, or winter aconite (*Aconitum Napellus*).

3 White wrote " yews," and then erased the y. On Feb. 27 he uses the modern spelling.

4 The songs of the two birds seem to be reversed : the great tit is the saw-sharpener.

5 William Lucombe was a nurseryman at Exeter. His oak is sub-evergreen.

6 There is a full description in D.B., XLV.

7 Thomas Hoar noted the weather and temperature, but not the barometer readings, and White appears to have copied these records from a separate sheet.

8 This derivation, which is sound, is repeated in Chap. X of the " *Antiquities.*"

9 Old gardening term for " forming a head," applied to cabbages and lettuces.

10 Midland and Southern dialect word for shrivelled, wrinkled. (*Eng. Dial. Dict.*)

11 Apparently caused by the leaf-curl fungus (*Exoascus deformans*).

12 Circular shafts sunk vertically to obtain chalk ; known variously as draw-pits, chalk wells, and dene-holes.

13 Siberian or " naked " barley (*Hordeum coeleste*), which has the grains loose in the husk. It is said to have been introduced about 1768 ; but did not obtain much popularity.

14 Some of the notes on swifts made about this time are incorporated in D.B., XXI.

15 " *Summer,*" ll. 1656-8. A few species of thistles flower in June.

16 Reproduced, in part, in D.B., XXXV, but some facts there omitted.

17 From Oct. 10 to Oct. 22 the journal chiefly notes on the weather, seems to have been kept by Thomas White (R.H.W., I, p. 267), but there is apparent collaboration.

18 Or Bin's Pond, described in T.P., VIII.

19 There is a gap from Nov. 22 to Dec. 9; footnotes were evidently added on White's return to Selborne.

20 This is an error; the mature sprat greatly resembles a herring, and belongs to the same genus, *Clupea*, but the species differs.

21 The bird was evidently the female goosander (*Mergus merganser merganser*), known in the North as the sparling (i.e., smelt) fowl.

1775

[1] *The Squire's Tale*, ll. 250-2.

[2] The spoonbill formerly bred in England; it is now but a rare straggler.

[3] The subject is developed in D.B., LVI.

[4] Cf. June 24th 1774.

[5] The tree-creeper (*Certhia familiaris britannica*).

[6] Red admiral butterfly. In the 18th century " fly " was still sometimes used for any winged insect (N.O.D., s.v. " Fly ").

[7] *Dictamnus*, or dittany, belonging to the Order *Rutaceae*.

[8] The cry of the female cuckoo is the so-called " water-bubble " sound.

[9] From this point onwards, notes about the Hirundines, if embodied in D.B., XV, et seqq. are more strictly excluded.

[10] White assumes that birds prefer sweet fruits to acid.

[11] The ringed plover (*Charadrius apricarius*), or, less probably, the rock pipit.

[12] The traps are described in A. Beckett's " *The Spirit of the Downs*," 3rd edn., 1923, pp. 253-4.

[13] Cf. D.B., XXXI. White's views about vipers swallowing their young are given in T.P., XVIII.

[14] Dial. word for flooded lands (*Eng. Dial. Dict.*).

[15] A native of S. Europe and Asia Minor, belongs to a genus of the *Graminaceae*.

[16] Cf. remarks in Antiq., II, n.

[17] An under-estimate: the British list now comprises 500 birds, but more than a quarter are racial forms, or mere waifs. (H. Saunders, *Man. Brit. Birds*, 1927, p. vii.)

[18] Archaic sense, " to be deranged " (N.O.D.).

[19] White notes the short quantity of the first u, while popular usage makes it long.

1776

[1] Several of the entries which follow re-appear, in substance, in D.B., LXII.

[2] An unexplained gap of six weeks, without entries; White was still in London.

[3] Apparently *Asilus crabroniformis*, one of the beaked " robber flies "; it is nearly an inch in length.

[4] *Calopteryx Virgo*, the " demoiselle." The male only is blue; the female is greenish or brownish.

[5] Probably a " swallet " caused by the solution of calcareous rock, but the circumstances suggest human agency, in part.

[6] *Abies balsamea*, a N. American species, which yield Canada balsam.

7 Either the eyed hawk moth or the humming-bird hawk-moth.

8 Cook's third and fatal voyage. Omai, a native of Ulietea, was brought to England by Cook on his second voyage.

9 S. H. Grimm (1734-1794), a Swiss, who illustrated the "Natural History."

10 *Georgics*, IV, 245. Read "*immiscuit*," which is correctly quoted on 19 Sept. 1780.

11 Dare=to stupefy, paralyze with fear. Dial. word in Midlands, S. and S.W. (Examples in *Eng. Dial. Dict.* & N.O.D.)

12 Cf. Nov. 10. The "Great fieldfare" is the common species. White sometimes calls the redwing, the redwing-fieldfare.

13 The hawfinch (*Coccothraustes*). White uses "cross-beak" and "gross-bill" for the crossbill (*Loxia*), but he clearly marks off the hawfinch by his descriptions.

1777

1 Gad-fly or horse-bot (cf. 8 Aug. 1770).

2 Caused by the attack of a fungus (*Empusa muscae*).

3 This theory is now abandoned : a sticky fluid is exuded from the pads. (C. G. Hewitt, *The House Fly*, 1914, p. 26.)

4 The next regular entry is dated from Guildford, on April 10th. White had had an attack of illness in London. (R.H.W., II, p. 8.)

5 The entries for March 26 and 27 are evidently based on correspondence.

6 For details of this " new parlour " see R.H.W., II, chap. I.

7 " Witches' butter," a unicellular green alga.

8 *Georgics*, IV, ll. 557-8.

9 The common snipe (*Capella g. gallinago*) is both a resident and a winter visitor ; the jack snipe (*Lymnocryptes minimus*), a winter migrant only.

10 For ten days the notes refer to the weather only.

11 White's favourite tree ; see T.P., I.

12 Reversing the Norfolk, or four-course rotation, in which turnips follow wheat.

13 A valuable note, because the Norway rat reached England only about 1728. (M. A. C. Hinton, "*Rats and Mice*," 1918, p. 6.)

14 This explanation is doubtful, since the bird's own velocity must be added to that of the wind. (Cf. F. W. Headley, *Structure and Life of Birds*, 1895, p. 358.)

1778

[1] After Feb. 6, the notes for the rest of the month, chiefly relating to the weather, are on leaves specially ruled by White; some leaves had evidently been torn out.

[2] *Anecdote* here has the primary meaning, "something not yet published" (ἀνέκδοτα).

[3] The barnacle goose (*Branta leucopsis*) visits England from September to May.

[4] Read *L. Squamaria*, the toothwort; parasitic on the roots of hazel (Cf. 21 April, 1779).

[5] Sir Ashton Lever's private museum at Leicester House, London.

[6] White may be referring to a skull of the urus, or primitive ox (*Bos primigenius*) extinct in England before the Roman invasion.

[7] The black-fruited gean, or wild cherry (*Prunus avium*) is locally known as the merry in S. England.

[8] Autumnal lady's tresses, an orchid (*Spiranthes spiralis*).

[9] The fruit of the common barberry (*Berberis vulgaris*) was used for making jam.

[10] The orange-red lily (*Lilium bulbiferum*), which bears bulbils in its leaf-axils.

[11] White discusses yew trees fully in *Antiq.*, V.

[12] The click beetle (*Elater*), the larva of which is the wireworm.

[13] The rise usually occurs a few months after heavy rains.

1779

[1] The next thermometer record is Feb. 1, but regular records do not appear until March.

[2] Presumably a reference to "quartering," to avoid the deep ruts of winter (cf. R.H.W., II, p. 139).

[3] Grasshoppers moult several times before maturity, and in this sense only may be said to "grow." (Cf. Note on crickets, April 24th, *infra*.)

[4] The brown tail moth, or one of the small ermines, seems to be indicated.

[5] White discusses this bird in D.B., XLIX. His mathematical error is shown in W.J., pp. 262-3.

[6] Partial migration is still admitted in this note.

[7] *Lilium croceum*. White's "Martagon" would be *L. Martagon*, and his "Turk's cap," *L. chalcedonicum*.

[8] Bramshott stands on the Lower Greensand (Hythe Beds).

[9] Believed to be caused by minute hymenopterous insects.

10 This is White's first record of measured rainfall. His brother John had presented him with a " rain measurer " in the preceding April.

11 The " New " Hermitage, an arbour or summer-house, on Selborne Hanger.

12 The rainfall entries, which have been few, now cease for some time.

1780

1 White left London to attend his aunt's funeral at Ringmer.

2 A fine anticipation of the modern theory of local variations in bird-song.

3 Tables for the years 1782 to 1793 inclusive are published in many editions of Selborne.

4 Sim Etty, son of the Rev. Andrew Etty, then vicar of Selborne.

5 The Gordon, or " No-Popery " Riots.

6 Notes of this kind were either the result of passing correspondence, or were entered after a return to Selborne.

7 William Curtis (1746-99) had a botanic garden in St. George's Fields, Lambeth Marsh. Thomas White had assisted Curtis to transfer his enterprise from Bermondsey.

8 Gilbert White's niece, and daughter of Benjamin White.

9 The carapace of Timothy, now in the British Museum, is identified as *Testudo ibera*. (Cf. W.J., p. 83.)

10 White's friend, Dr Richard Chandler, the antiquary, who assisted in the " Antiquities."

11 Cf. 9 Feb. 1777, and note.

12 The swallow-tail butterfly, now found only in the Fens and a small area in Suffolk.

13 The introduction of this spelling, here and later, is probably due to White's antiquarian researches.

14 The Zigzag was constructed in 1753 (R.H.W., I, p. 71). *Bostal*, or *borstal*=a steep path or bridle-way on a hillside (S.E. England).

15 These balls would be composed of marcasite, from the Lower Chalk. Marcasite has a radiating structure, and though a disulphide of iron, like pyrites, is more easily decomposed.

16 White is approaching a notable discovery : the fungoid origin of fairy-rings.

17 Sir George Baker (1722-1809) who was nine times president of the College of Physicians, made notable discoveries concerning lead-poisoning (*Dict. Nat. Biog.*).

1781

[1] A heliotrope was a contrivance to mark the variation-points of the sun's rising and setting.

[2] The site is now mostly occupied by Battersea Park.

[3] In Berkshire, so-called from the famous White Horse on the downs above Uffington.

[4] Newton " dew-pond," on the down; distinct from the " Great Pond " in the village.

[5] The grey, or grey-brown larvae of the turnip moth (*Agrotis segetum*).

[6] White's brother Thomas, with his daughter, Mary (Molly).

[7] The main source of the Selborne, or Oakhanger stream.

[8] Notes on Sept. 20 and 22 are in the handwriting of Thomas White.

[9] Near the village of Privett, 6 miles N.W. of Petersfield.

[10] Iron-stained sands of Lower Greensand (Folkestone Beds) rounded by wind, waves and currents.

[11] Both the name (darlings) and the variety of apple, seem to have disappeared.

[12] Lord Stawell of Marelands, in the parish of Bentley, near Selborne.

[13] Short Heath, near Oakhanger hamlet.   Varying degrees of coarseness and angularity are met with in these Folkestone Beds.

[14] Female of black cock, or black grouse.   Cf. T.P., VI.

[15] Cottage near the junction of the bourne and the permanent Selborne stream.

[16] Described in D.B., XXIX.   (Wood Pond.)

[17] Medsted, 4 m. S.W., Bentworth, 4 m. N.W. of Alton.

[18] Rev. R. Yalden, vicar of Newton Valence, about 2 miles distant.

[19] Question discussed in W.J., pp. 226-7.

[20] Cf. T.P., XV, and remarks in W.J., p. 49.

1782

[1] Cf. the poetical rendering in White's " *Selborne Hanger*," ll. 15, 16.

[2] A plain proof that White did not deny the migration of swifts entirely.

[3] Influenza epidemic of 1782. Kempenfelt's squadron, among others, was affected, and had to return home. (Cf. entry, 18 June, 1782.)

[4] Gilbert's brother, Thomas White (1724-1797), a retired London ironmonger and a well-known naturalist.

[5] White must at this period be gleaning information from correspondents.

⁶ Wheat was beginning to be cut much nearer the ground than formerly, and tying the sheaves became necessary. A 15th century illustration shows very large sheaves with two bands— a probable explanation of White's expression.

⁷ White's "weeding woman"; called "Hampton" in a letter, 9 April, 1781.

⁸ Probably calcined flint, a common object in barrows. (W.J., p. 227.)

⁹ There is the least twilight at the equinoxes (H. Spencer Jones, "General Astronomy," 1923, p. 42).

¹⁰ Corn was then cut closer to the ear, and stored in barns; not stacked.

1783

¹ An obvious slip of the pen : the woodlark or the nuthatch was probably intended.

² The pond on Ockham Common, midway between Cobham and Ripley.

³ See 26 April 1779, and note 4 thereon.

⁴ A mistaken theory ; see W.J., p. 129. (Cf. D.B., LXIV.)

⁵ Dr Henry Becke (1751-1837), fellow of Oriel, professor of modern history, and a botanist.

⁶ A new sign is used in the "wind column" on this date. It is a spiral , either right-handed or left-handed. Its occurrence is only occasional, and it evidently denotes a veering wind.

⁷ An apt quotation from *Georgics*, I, l. 467.

⁸ Selborne continued to suffer from the drought.

⁹ Such honey would be fit only for the bees themselves, during a period of scarcity.

¹⁰ These red skies were associated with distant volcanic eruptions. (D.B., LXV.)

¹¹ Son of the Rev. Andrew Etty, vicar of Selborne from 1758 to 1784.

¹² Church-litton=churchyard : O.E. *líc-tún*, corpse-enclosure. (N.O.D.)

¹³ A thatched summer-house, "the new Hermitage." (R.H.W., II, p. 235.)

¹⁴ Most likely a species of *Paxillus*, or *Clitocybe*.

¹⁵ Rev. John Mulso (1721-91), an Oriel contemporary and life-long friend.

¹⁶ Rev. Ralph Churton (1754-1831), a friend and correspondent on miscellaneous topics.

1784

[1] The next day was Sunday, and White was still curate of Faringdon. For " Hirn," see 21 Dec. 1775.

[2] An error : the grasshopper warbler is a summer migrant. White's earliest date elsewhere is 16 April.

[3] Large ladles, attached to long handles, used to empty ponds, or to fill water carts. (*Eng. Dial. Dict.*)

[4] *Rip* = to put on fresh laths ; *fur* = to repair rafters and bring them to a level by affixing strips of wood.

[5] Or " hot codlings " (baked apples). See discussion *N. and Q.*, 7th Ser., IX (1890), pp. 108, 153. " Scalded " has dialect meaning of scorched, baked. (Cf. 2 July 1785.)

[6] Plan now recommended in Leaflet on Hops (Board of Agric. and Fisheries), 1925, p. 21.

[7] The brothers Montgolfier had experimented with fire balloons in 1782 and 1783.

[8] The Rev. C. Taylor succeeded the Rev. A. Etty, who had died early in the year.

[9] Relinquishing the curacy of Faringdon. White remained curate of Selborne until his death in 1793.

[10] The letter is a reproduction, with variations in spelling and punctuation, of the main part of a letter to White's sister, Mrs Barker. (R.H.W., II, pp. 134-6.)

[11] The injury must have been trivial : White does not mention it again, nor is it alluded to in Mulso's letters.

1785

[1] A vane, representing a fox, over the parlour chimney.

[2] Benjamin White also now lived at South Lambeth.

[3] The tortoise seems to have been weighed during White's absence.

[4] A near approach to the modern theory of bournes,—the rise and fall of the water-table.

[5] The pea green moth (*Tortrix viridana*) must be meant. See discussion, W.J., pp. 131-2.

[6] About 5½ miles by the nearest road.

[7] John Mulso was now rector of Meonstoke, a village about 16 miles S.W. of Selborne.

[8] The custom of stacking corn was not yet general. (Cf. Note 10, for 1782.)

[9] Through clearance of crops.

[10] Usually, after a drought, it takes two or three months for rains to affect the springs materially.

¹¹ They were probably going to Ireland, but were driven back by the W. wind.

¹² An exceptionally late date; can it be a slip of the pen for " ivy " ? Or was it a second flowering ?

1786

¹ The date was Jan. 6th; 386 lives were lost.

² Information supplied by Dr Chandler. (Cf. Feb. 27.)

³ Rev. Christopher Taylor, the non-resident vicar, inducted 1784.

⁴ The water-ouzel, or dipper (*Cinclus c. gularis*) occasionally visits S.E. England.

⁵ This name appears as " bantam " in all editions of " Selborne," but the word is not known in this connection, even in Hampshire (*Eng. Dial. Dict.*). Qy. Bantham, near Kingsbridge, Devon. But cf. 3 Dec. 1787.

⁶ Quoted from White's own poem, " The Naturalist's Summer-Evening Walk."

⁷ Apparently " cumulus " clouds. Luke Howard did not introduce this and similar terms until 1803.

⁸ Quoted from his own fine stanzas on the Crocus.

⁹ Doubtless a quotation from a letter from Dr Chandler in Switzerland.

¹⁰ An experiment on wrong principles; White's own ideas were sounder. (Cf. 15 Oct. 1780.) See W.J., pp. 163-6.

¹¹ See *Antiq.* I. ; and cf. *Vict. Hist. of Hampshire and the Isle of Wight*, 1900, I, pp. 340-1.

¹² White's cousin, Rev. Basil Cane, curate of W. Deane, Wilts.

¹³ A rare entry: evidently a settling of accounts with Mr Randolph, rector of Faringdon, on resigning the curacy. (Cf. 11 Sept. 1784.)

1787

¹ The diary for 1787 is hand-ruled throughout, and the earlier entries are brief.

² Communicated evidently by Dr Chandler. (Cf. 23 Feb., 27 March, &c.)

³ Samson White, son of Gilbert White's youngest brother, Henry. (See R.H.W., II, pp. 148, 172.)

⁴ Cf. particulars given in *Camb. Nat. Hist.*, Mammalia, 1902, p. 392.

⁵ Practice now recommended ; see pamphlet on " *Cultivation . . . of the Hop Crop* " (Bd. of Agric. and Fisheries), 1925, p. 23.

⁶ Potatoes were not much cultivated at Selborne before c. 1758. (D.B., XXXVII.)

⁷ A clear proof that White still believed in a partial migration.

⁸ A registered, or " testimonial" man who bore a letter from his native place, testifying to his character, and promising to take him back if he became chargeable. (S. and B. Webb, *Eng. Poor Law Hist.*, 1927, Pt. I, pp. 336-7.)

⁹ The players were the children of Rev. Henry White and their friends. (R.H.W., II, p. 171.)

¹⁰ The question of the Norway rat is discussed in W.J., pp. 80-1.

¹¹ Barked at the sight, or scent, of the quarry.

1788

¹ *Slidders* = trenches on a hillside, mostly bare of vegetation.

² This note must have been inserted late in the year.

³ Letter to relative: " Mrs C. . . . and her Nursemaid *Zebra White sic.*" (R.H.W., II, p. 180.)

⁴ Cf. " Solomon's Song," II, 12.

⁵ A small cultivated form of wild bean [*Vicia Faba*] found at Magazan, in Morocco.

⁶ The " purse-gall " (*Pemphigus bursarius*).

⁷ Evidently Mrs John White, now a widow, who resided with her brother-in-law, Gilbert. (R.H.W., II, p. 67.)

⁸ The best authorities here read *nux* as the almond.

⁹ To " pipe " pinks is to propagate them by slips cut off at a joint of the stem. (N.O.D.)

¹⁰ " *Joanettings,*" or *Jennetons* (Durham), *jennetens* (Suffolk), a very early apple. Orig. *pomme de Saint-Jean.* (N.O.D.)

¹¹ John Philips (1676-1708), wrote " *The Splendid Shilling* " and " *Cyder.*" White was fond of his poems. " Mass " is the correct reading.

¹² This field, at that period, would be one of the " open " fields of the village.

¹³ Reference to " Old Style." " New Style " was adopted in 1752, when eleven days were left out of the calendar : 3 Sept. became 14 Sept.

¹⁴ A plant of the Order *Graminaceae*, cultivated by Sir Joseph Banks under the name of Tuscanora.

¹⁵ Jonathan Carver (1732-80), an American traveller, explored part of Minnesota and the district around L. Superior. (D.N.B.)

¹⁶ The lime is a doubtful native, but may often be considered as half-wild.

¹⁷ The Chilmark quarry, near Salisbury, yields a siliceous (Portland) limestone which was used in the construction of Salisbury Cathedral.

¹⁸ Chalk marl, at the base of the Lower Chalk.   [Cf. T.P., I.]

¹⁹ Or Chloritic Marl, which would be thereby improved in texture and fertility.   The dressing was very liberal in quantity.

²⁰ Chalk Marl: from this pit White obtained casts of ammonites [T.P., III.]

²¹ An early reference : Paul was manufacturing aërated waters at Geneva in 1790 ; but in England Thomson's apparatus was not patented until 1807.

1789

¹ The " blue rag " was a calcareous layer in the Upper Greensand, fully described in T.P., IV.

² One of White's rare appeals to teleology.

³ Near the junction of the Upper Greensand and the Gault clay.

⁴ This depth was exceeded in London during the frost of 1895.

⁵ " Burhunt Farm " of the Ordnance map.   It is on the Malmstone, or Upper Greensand.

⁶ The floating manna grass (*Glyceria fluitans*), very attractive to cattle.

⁷ See Note 8, referring to 21 Oct. 1787.

⁸ Wm Dufour describes himself in his specification as of " Cantshill, Buckinghamshire, Gentleman."

⁹ The Italian, or Queen's Opera House, opened 1705 ; re-opened after the fire, 1791.

¹⁰ 59 bushels per acre : the average yield of barley in England for 1929 was about 34½ bushels (56 lbs. to the bushel).

¹¹ 72s. per quarter.   The average for the year was 61s. 6d. (Mulhall, *Dict. of Statistics*).   Rogers, *Agric. and Prices*, gives 67s. as the highest price.

¹² Curved brackets here indicate that White wrote alternative words, one above the other.

¹³ Most of this information was repeated in White's first letter to Robert Marsham, and again to Rev. R. Churton.   (R.H.W., II., pp. 205-6.)

¹⁴ Son of White's friend, John Mulso.

¹⁵ See Note 7, 1780.   Curtis occasionally corresponded with White.

¹⁶ Hence its alternative name, corn-crake.

¹⁷ Possibly the very rare Irish furze, *Ulex strictus*.

¹⁸ *Toach (torch)*=to point the inside joints with plaster, lime and hair, or, in former times, with cow-dung.

¹⁹ Mow=a heap of unthreshed corn stored in a barn.

²⁰ The statutory cord measured 8 ft. by 4 ft. by 4 ft.   Billet was wood sawn ready for burning.

1790

¹ A Roman camp, with relics, is recorded at Ridgehanger, to the North of this valley, and entrenchments, probably pre-Roman from the adjacent villages of Froxfield and West Meon. (Williams-Freeman, *Field Archaeol . . . Hampshire*, 1915, pp. 285-95.

² The ring, or grass snake, which lays its eggs in such places.

³ A summer visitor; a few birds remain for the winter in S. England.

⁴ Introduced from the East, probably South Russia, late in the 17th century, the cockroach at first spread slowly.

⁵ Breaking up into cup-like portions. (This information has been inserted in the *Journal*, but by another hand.)

⁶ Cardoon (*Cynara*), a kind of cultivated thistle, of which the fleshy receptacles and scale-bases are edible. The Jerusalem artichoke (*Helianthus*) has tuberous roots.

⁷ Near Stockwell Green; now St. Andrew's Church; built 1767.

⁸ White seems to have been the first to notice that the nightingale introduces a harsh note late in the season.

⁹ Widow of John White; she lived after her husband's death in 1781.

¹⁰ One of the four limes planted by White " in the Butcher's Yard, to hide the sight of Blood, & filth from ye windows." (*Garden Kal.*, 31 March, 1756). Two limes, and the stump of a third, remain.

¹¹ The harvest mite (*Tetranychus*), in an immature stage. (Cf. T.P., XXXIV.)

¹² Circular holes, tree-studded, whence marl was formerly dug as a top-dressing. (Cf. Note 12, 1774.)

¹³ Letter from Marsham 24 July 1790. This oak was called the Grindstone Oak because a grindstone once stood there, when much timber was felled. (See R.H.W., II, pp. 221, 241, 269.)

¹⁴ Stacking wheat was evidently newly-introduced. (Cf. 21 Aug. 1790.)

¹⁵ To keep away vermin, and provide aeration.

¹⁶ John White (" Jack "), son of Rev. John White, and then a surgeon at Salisbury.

¹⁷ A portion of this matter was included in a letter to Marsham, 18 Jan. 1791.

¹⁸ Apparently a storm-tossed bird; the prevailing winds from Sept. 9th to Sept. 16th had been W., S.W., and N.W.

¹⁹ Inoculation by injection of the small-pox virus; not vaccination. (See W.J., p. 243.)

²⁰ Bell's opinion (edition of "Selborne," 1877, I, p. 431) that the bird was a hybrid between a black-cock and a pheasant is now generally accepted.

²¹ Newton Vicarage is close to the 600 feet O.D. contour-line.

²² The mouth of the Effra, near Vauxhall. The stream is said to have once been navigable for barges up to Kennington, or even Brixton.

²³ Specific gravity of lead, 11.4 approx.; of mercury 13.6.

²⁴ The fresh-water shrimp (*Gammarus pulex*), which is found in springs and streams, and keeps near the bottom.

²⁵ "From Spit-head." (Cf. 21 Nov., *infra*.) The name was originally applied, not to the channel, but to the sandbank lying S.E. from Gillkicker Point, on the Hants coast.

²⁶ In a letter quoted in Buckland's edition, 1875, p. xxii, White says "two murderers" were beaten down. There were three bodies in all.

1791

¹ Dr Chandler was now living at Selborne vicarage.

² The bones seem to indicate an early cemetery, either Christian or pagan, probably the latter.

³ The same note had been entered opposite March 29, but had been carefully crossed out, letter by letter. Either there was an error in placing the entry, or White had been misinformed.

⁴ Daughter of Gilbert's brother Benjamin. The alcove was an arbour built on the Hanger.

⁵ Doubtless the St. George's mushroom (*Tricholoma gambosum*).

⁶ Two years previously, 5 June 1789, White added a correction under that date.

⁷ Gen. J. E. Oglethorpe, founder of Georgia, was M.P. for Haslemere in the middle of the century.

⁸ The Selborne, or Oakhanger Stream. (Cf. T.P., I.)

⁹ Vicar, 1632-78. See *Antiq.*, VI. Dispossessed during the Commonwealth.

¹⁰ The noctule bat, which White added to the British list. (Cf. T.P., XXII, XXVI, XXXVI.)

¹¹ Due, not to the winds, but to the heavy rainfall in July: more than 5½ inches.

¹² "Honey buzzard" is wrongly attributed to Ray. See also T.P., XLIII.

¹³ Thomson: "*Autumn*," ll. 834-46. Concerning the migration problem, Thomson, like White, keeps a foot in either camp.

¹⁴ Cf. Note 3, 1777. The "labouring" flies had doubtless been attacked by a fungus. (Cf. Note 2, 1777.)

¹⁵ A new structure, close to the Bostal, and opposite the Wakes. (R.H.W., II, p. 235.)

¹⁶ For " white " read " pied." White's yellow wagtail was evidently the grey wagtail ; the former is a summer migrant only. (Cf. Note 14, 1773.)

1792

¹ The plant is a perennial. It buries its seeds in crannies, hence White would have renewals each year.

² Apparently rented from White by Berriman.

³ This proved to be White's last visit to Oxford.

⁴ Two phrases, which elucidate the sense, are inserted in all editions of the " Miscell. Obser.," are omitted, since they are by another hand, and not original.

⁵ White here quotes from his own poem, " *The Naturalist's Summer Evening Walk.*"

⁶ Benj. White left S. Lambeth, to live at Mareland, near Farnham, in Nov. 1792.

⁷ Mrs John White. (Cf. 18 Aug. 1792.)

⁸ Caused by minute diptera and hymenoptera.

⁹ Dr John Eveleigh, Provost of Oriel since 1781.

¹⁰ White repeats the substance of this note in his 6th letter to Robert Marsham.

¹¹ Goleigh (Priors Dean) and Heard's (E. Tisted) are on the Chalk ; White's Selborne well was sunk in the Upper Greensand, hence the differences in depth.

¹² The bee appears to have been *Bombus lapidarius*, which resents the presence of intruders.

¹³ William Hay (1695-1755), traveller, essayist, and minor poet, was himself deformed. (*D.N.B.*)

1793

¹ The matter is fully discussed in the 8th and 9th letters to Marsham.

² The statement seems to confirm the supposition that White went to school at Farnham. (But see R.H.W., II, p. 262, where " the " is erroneously copied for " that.")

³ James Hurdis (1763-1801) here quoted, is believed to have been an acquaintance of White. For " round " read " song," and for " pouring " read " huddling." White quoted from memory.

⁴ Was this a pied blackbird ?

⁵ Very improbable ; a missel-thrush seems to be indicated. Compare " not wild " with entry on 9 May, *infra.*

INDEX

Medlar, 122, 133
Melons, 308, 336
Meonstoke, 55, 69, 90, 110, 127, 155, 265
Mercury (metal), 370
Mercury (planet), 33, 135, 218
Merris-tree, 154
Meteors, 187, 226
Meymot, Mr, 105
Mice, 331, 375
Midhurst, 51, 85
Migration, 73, 82, 102, 114, 121, 304
Miller, P., 404
Milton, quoted, 252, 408
Missel thrush, 13, 152, 403
Mist, 6, 24
Mistletoe, 181, 287
Mock-suns, 102
Mole cricket, 40
Moles, 162
Monk's rhubarb, 289, 333, 358
Moor buzzard, 17, 90
Mortar, 148
Moths, 125
Mount Caburn, 75, 416
Mulso, Rev. John, 231, 246, 252
Mulso, T., 432
Musca, 110, 136
Mushrooms, 106, 131, 161, 212, 280, 380
Music, 216
Mussels, 314

Naples, 311, 332
Nasturtiums, 60, 232
Nectarines, 60, 84, 138, 228
Nephews and nieces, White's, 211, 233, 247, 257, 276, 279, 283, 289, 303, 306, 321, 368
Newcastle, 45
Newton, 16, 51, 146, 184, 212, 253, 327, 369, 396; pond, 191, 196, 198, 210, 264

Nightingale, 151, 162, 361, 428
Nore hill, 204, 345
Norway rat, 300
Nose-fly, 108, 135-6, 175
Nostoc, 138, 171, 208, 228
Nuthatch, 97, 99, 131, 133, 183, 293, 328

Oakhanger Ponds, 66, 101, 149, 326
Oaks, 112, 375; felling, 82, 137, 381; ravaged, 261; stripped by chafers, 33, 87
Oasts, 413
Oestrus, 29, 92, 107-8, 341
Oglethorpe, Gen., 385
Opera House, burning of, 336
Ophrys, 84, 139, 177, 344
Oranges, 356
Orchis, 28
Orion, 198
Osprey, 64
Ostrich, eggs, 243
Otter, 249
Owls, 33, 152, 204, 288, 298, 360
Oxford, 14, 17, 26, 58, 72, 93, 101, 113, 121, 144, 150, 160, 185, 200, 219, 290, 305

Pantry, White's, 176, 206
Parhelia, 102
Parker, Adm. Hyde, 190
Parlour, the new, 138, 141, 142, 145, 146, 302
Parnassia, 267
Partridges, 72, 110, 125, 128, 208, 244, 317, 415
Pasque flower, 14
Peaches, 65; blight, 70, 83
Peacock, 24, 127
Pear, vast, 419
Peas, 293, 336, 337
Peat, 197, 206; ashes, 237, 272, 413; cutters, 417